Glossary of Plant-Pathological Terms

Malcolm C. Shurtleff
Professor Emeritus, University of Illinois
Urbana-Champaign

and

Charles W. Averre III
Professor Emeritus, North Carolina State University
Raleigh

APS PRESS
The American Phytopathological Society
St. Paul, Minnesota

Reference in this publication to a trademark, proprietary product, or company name by personnel of the U.S. Department of Agriculture or anyone else is intended for explicit description only and does not imply approval or recommendation to the exclusion of others that may be suitable.

Library of Congress Catalog Card Number: 97-70274
International Standard Book Number: 0-89054-176-0

©1997 by The American Phytopathological Society

All rights reserved.
No portion of this book may be reproduced in any form, including photocopy, microfilm, information storage and retrieval system, computer database, or software, or by any means, including electronic or mechanical, without written permission from the publisher.

Copyright is not claimed in any portion of this work written by U.S. government employees as a part of their official duties.

Printed in the United States of America on acid free paper

The American Phytopathological Society
3340 Pilot Knob Road
St. Paul, Minnesota 55121-2097, USA

Preface

This glossary is written for those interested in plant pathology as a science or for managing plant diseases. This includes professional agriculturists such as county agents, plant science specialists, plant diagnosticians, agribusiness personnel, agricultural inspectors, and private practitioners or consultants. Students in plant pathology courses, as well as those involved with plant health problems, disorders, and diseases in the field, plant clinic, or laboratory will also benefit. The glossary should also be useful to the teacher, researcher, or extension (advisory) worker. To this end, we have tried to make it readable to those interested in all the subdisciplines within plant pathology. Some entries are presented as a brief discussion of a subject rather than as a short, cryptic synonym.

A fundamental requirement of a language, especially where scientific terms are concerned, is that it accurately convey concepts not only among close colleagues, but among all interested, informed, and intelligent persons in diverse occupations, locations, backgrounds, and time periods. This glossary also attempts to serve this purpose. English is increasingly used as an international language, and the glossary should help facilitate communication in the world community of plant pathologists and other plant scientists.

In compiling this glossary, a major problem was to decide what criteria to use for including or excluding a term. In most cases, the choice was to include the word if it pertained to plant pathology or was a term in a related plant science discipline that a plant pathologist might need to know. Modern biology is a continuum, and any attempt to subdivide it is bound to fail. However, plant pathologists, like other plant scientists, have a propensity to make up new words, give special meaning to words, and to use terms from other languages. In general, we chose not to include such words unless they are used extensively by plant pathologists in a specific area. In general, words in common English usage are not included unless they have a peculiar, second meaning in plant pathology.

This glossary was assembled over many years, from over 100 sources. It has been used and modified in teaching plant pathology courses at both our institutions and elsewhere. An attempt was made to gather together in one volume as many as possible of the terms used by plant pathologists and other plant scientists, whatever their origin or application, whether modern or obsolete, popular or trivial, traditional terms or scientific phrases, and to set down their meanings as accurately and precisely as possible. However, our personal biases of a combined 85 years of experience, over much of the United States and some of Central America and Europe, cannot be entirely eliminated. Also, in many cases, it was difficult to find precisely what was meant by some words. The many coiners of new terms, in the various disciples within the plant sciences, have often been content to allow readers to "puzzle out" the idea that a new word was intended to convey or the limits to be set for the meaning. In addition, old terms once widely used have come to have new meanings, and many terms now are used quite differently than originally intended. It is inevitable that usages of words will change, that new terms will become commonplace, and that any glossary or dictionary will soon have omissions.

We are grateful to many individuals in the departments of plant pathology at the University of Illinois and North Carolina State University, as well as staff members in sister plant science departments, for their assistance in the attempt to produce order and clarity. We encourage readers who have suggestions for future editions, corrections, and comments to send them to us. They will always be appreciated.

Malcolm C. Shurtleff
Charles W. Averre, III

Glossary of
Plant-Pathological Terms

A

Å (Ångstrom)—*See* Angstrom.

"a"—Nematode body length/greatest body width usually; often given as a range, e.g., "a = 50–60"; a deMan's value.

A horizon—The surface layer of varying thickness of a mineral soil having maximum organic matter accumulation, maximum biological activity, and/or eluviation by water of materials such as iron and aluminum oxides and silicate clays.

a-, an- (prefix)—Not having; not; without; lacking; denotes the opposite.

Ab- (prefix)—Position away from.

Ab 1—The first antibody produced by an animal's B lymphocytes in response to the stimulus provided by an antigen.

Ab 2 alpha—An antibody that recognizes an idiotope on Ab 1 that is extraneous to the paratope of Ab 1. Ab 1/Ab 2 alpha interactions are not generally inhibitable by antigens.

Ab 2 beta—An antibody (also known as a homobody, an internal-image antibody, or an antigen-mimicking antibody) that interacts with the paratope of Ab 1. Ab 1/Ab 2 beta interactions are not inhibitable by antigens.

Ab 2 epsilon—An antibody (also known as an epibody) that interacts with an idiotope on Ab 1 and the antigen against which Ab 1 was derived.

Ab 2 gamma—An antibody that interacts with an idiotope on Ab 1 that is closely associated with the paratope of Ab 1. Ab 1/Ab 2 gamma interactions are inhibitable by antigens.

Abaxial—Away from the axis or central line; turned toward the base; dorsal. Often used to designate the lower leaf surface; (of a basidiospore) the side away from the long axis of the basidium.

abbr.—Abbreviation.

Aberrant (n. aberration)—Atypical; deviate; straying from the normal type.

Abhymenial—Opposite the spore-producing surface.

Abietine—Of, or growing on firs (*Abies*).

Abiotic—Of or pertaining to the nonliving. Synonym inanimate.

Abjection—Separation of a spore from a sporophore, sterigma, conidiophore, etc. by an act of the fungus.

Abjunction—The cutting off of a spore from a hypha by a septum.

Aboospore—A parthenogenetically developed oospore.

Abortive infection—Viral infection of a cell in which the virus does not fully replicate or produces a defective progeny. Its effect on the cell can still be cytopathogenic.

Abortive transduction—Transduction in which the donor DNA is not integrated with the recipient chromosome (as in complete transduction) but persists as a nonreplicating fragment.

Abortive transformation—Temporary transformation of a cell by a virus that fails to integrate into the host DNA.

Abrupt—Appearing as if cut off transversely; truncate; without a thin, sterile margin.

Abscise (n. **abscission**)—To fall off, as with leaves, flowers, fruits, spores, or other plant parts. Premature abscission can occur in response to infection by pathogens; but abscission is a natural process involving formation of a separation layer or wall, causing the plant part(s) to fall or shed.

Abscisic acid—Plant hormone involved in abscission, dormancy, stomatal closure, growth inhibition or promotion, and other plant responses.

Abscission layer or zone—A layer of thin-walled cells extending across the base of the petiole or peduncle, the breakdown of which separates the leaf or fruit from the stem, causing the leaf or fruit to fall; separation layer.

Absorbance—Amount of light absorbed by a substance at a particular wavelength.

Absorption spectrum—Graphical representation of absorbance of a substance at different wavelengths. Valuable in obtaining an approximate estimate of the percentage of nucleic acid in a virus from the ratio of absorbance values at 254 and 280 nm.

Absorption—The taking up of water by assimilation or imbibition. The taking up of substances by capillary, osmotic, chemical, or solvent action such as taking up of water from air, taking up of gases by water, or taking up of mineral nutrients by plant roots; (of viruses) use of an antibody or antigen to remove corresponding antigen or antibody from a mixture; (in serology) nonspecific binding of molecules to solid surfaces typically in very thin or monomolecular layers, e.g., binding of antibodies or antigens to glass, plastic, or nitrocellulose surfaces; (of pesticides) entrance into the body through the skin. Also, entrance into a plant, animal, or microorganism.

Abstriction—First abjunction and then abscission, especially by constriction.

Acantha—A sharp pointed process; a spine.

Acanthocyte—A spiny cell produced on a short branch from the vegetative mycelium of species of *Stropharia*.

Acanthohyphidium—*See* Hyphidium.

Acanthophysis (pl. **acanthophyses**)—Bottle-brush paraphysis.

Acapsulate—Without a capsule.

Acaricide—A chemical or physical agent that kills or inhibits the growth of mites and ticks.

Acaryallagic (of reproduction)—*See* Caryallagic.

Acaudate—Without a tail.

Accession number—The number given to a culture or specimen in a culture collection of microorganisms.

Accessory fruit—A structure consisting of a true fruit (ripened ovary) plus other parts, such as calyx and receptacle.

Acclimatization—The adaptation of an individual plant to a changed climate, or the adjustment of a species or a population to a changed environment, often over several generations.

Accumbent—Resting or lying against something.

-aceae (suffix)—Attached to the stem of a name to form the family name.

Acentric—Describes pieces of chromosomes that lack a centromere.

Acephalous—Headless.

Acerose, acerous—Needle-shaped; stiff like a pine needle.

Acervate—Heaped or massed up; growing in heaps or groups.

Acervulus (pl. **acervuli**, adj. **acervular**)—A subepidermal but later erumpent, open, colorless to dark, more or less saucer-shaped, asexual fruitbody (conidioma) of a fungus bearing closely packed, short conidiophores, conidia, and sometimes setae, typically found on leaves or stems; characteristic of the Melanconiales (Deuteromycotina).

Acetabuliform—Saucer- or cruet-shaped.

Acetylation—The addition of acetyl groups either chemically or enzymatically.

Achene—Simple, dry, one-celled, one-seeded, indehiscent fruit with the seed attached to the ovary wall at only one point.

Achroic, achromatic, achromous, achrous—Without color or pigment.

Achromogenic—Failing to produce pigment.

Acicular—Needle- or bristle-shaped; slender and sharp-pointed.

Acid curd—Coagulation (precipitation) of milk protein due to acid production. *See* Rennet curd.

Acid delinting—Process used to remove short fibers (lint) from seed cotton.

Acid dye—Dye consisting of an acidic organic grouping of atoms (anions), which is the actively staining part, combined with a metal; has affinity for cytoplasm.

Acid hydrolases—Hydrolytic enzymes with a low pH optimum; term usually refers to phosphatases, glycosidases, nucleases, and lipases found in lysosomal compartments. They are considered to operate as intracellular digestive enzymes.

Acid phosphatases—Enzymes with an acidic pH optimum. They catalyze cleavage of inorganic phosphate from a variety of sources; found particularly in lysosomes and secretory vesicles.

Acid proteases—Proteolytic enzymes with an acidic pH optimum; generally found in lysosomes. *See* Proteases.

Acid soil—A soil with an acid reaction (pH below 7.0); more technically, a soil having a preponderance of hydrogen ions over hydroxy ions in solution. It is commonly measured from a soil-water extract with a pH meter. *See* pH.

Acid-fast—Describes bacterial cells that are relatively impermeable to simple stains but that keep carbol fuchsin stain (basic fuchsin in aqueous 5% phenol applied with heat) after the addition of 25% sulfuric acid.

Acidophilous, acidiphilous—Growing on acidic substrates.

Aciduric—Acid tolerant; capable of resisting acid.

Acquired immunity—Immunity developed during a lifetime from having a disease naturally or experimentally (may not be present in plants). In modern terms, the clonal expression of a population of immune cells in response to a specific antigenic response and the persistence of this clone.

Acquired resistance (syn. **induced resistance** or **acquired immunity**)—A noninherited resistance response developed by a normally susceptible host following a predisposing treatment, such as inoculation with a virus, fungus, bacterium, or treatment with certain chemicals. *See* Cross-protection.

Acquisition access period—The period of time that a vector has feeding access to a source of inoculum.

Acquisition feeding period—The time during which a vector feeds on an infected plant to acquire a virus for subsequent transmission (e.g., to become viruliferous).

Acquisition feeding—The feeding of a vector on a source of inoculum in transmission tests.

Acquisition threshold period—The minimum feeding time for a vector to acquire an infective dose of a pathogen.

Acrasiales, Acrasidae—The cellular slime molds.

Acrasin—A chemotactically active substance (cyclic AMP) that controls the aggregation of the myxamoebae of *Dictyostelium discoideum* and other Acrasiales.

Acre—Unit of land measure = 160 sq rods = 43,560 ft^2 = 0.40469 ha = 4,046.87 m^2.

Acrid—Biting to the tongue; peppery; irritating to more than just the sense of taste.

Acridine orange—A fluorescent vital dye used for determining the nature of the nucleic acid in virus particles or cells. Nuclei of stained cells fluoresce green, and cytoplasmic RNA fluoresces orange. It also stains acid mucopolysaccharides.

Acriflavine—A dye that activates viruses in the presence of light (photodynamic inactivation) by binding to nucleic acid.

Acro- (prefix)—At the end or tip; apical; terminal.

Acroaquatic fungi—Fungi that grow under water but form their spores in the air above.

Acroauxic—Growing and elongating at the apex.

Acrochroic—Especially colored at the hyphal tips.

Acrogenous—Growing or formed at the tip, apical.

Acronym—A word made up of the first letters of the principal words in a compound term.

Acropetal—Describing the development of structures (such as conidia) in succession from the base towards the apex, with the youngest at the tip; basifugal.

Acropleurogenous—Forming at the tip and along the sides.

Acrospore—Spore formed at the apex.

Acrosporogenous (of conidial maturation)—Having delimited cells that mature in sequence from the base to the apex as the tip of the conidium expands.

Acroton—A spinule in lichens bearing side branches.

Acrylamide—A chemical polymerized using a cross-linking agent to give polyacrylamide, a commonly used support for gel electrophoresis.

Acrylamide-gel disk electrophoresis—Process for physically separating proteins based on the charge or size of macromolecules.

Actidione—Trade name for cycloheximide (which see).

Actinogyrose, actinogyr (of apothecia)—Disc gyrose and lacking a proper margin.

Actinolichen—A lichen-like association between a green alga and an actinomycete, e.g., *Chlorella xanthella* and *Streptomyces* sp.

Actinomycetes—Organisms characterized by fine branching filaments, usually less than 1 µm in diameter, that readily break into spore fragments resembling bacterial cells. They are classified as Gram-positive bacteria in the order Actinomycetales, except the Mycobacteriaceae. Certain streptomycetes have been used to develop a host-vector system of cloning.

Actinomycin D—An antibiotic from *Streptomyces* spp. that inhibits transcription by interacting with the guanine residues of helical DNA, blocking the movement of RNA polymerase and preventing RNA synthesis in both prokaryotes and eukaryotes.

Actinophage—Virus that infects members of the order Actinomycetales (filamentous branching bacteria), including *Streptomyces* and *Corynebacterium*.

Action spectrum—The relationship between the frequency (wavelength) of a form of radiation and its effectiveness in inducing a specific chemical or biological effect.

Activated sludge—Material containing a very large and active microbial population produced in one method of sewage disposal by vigorous aeration of sewage.

Activation energy—The energy required to bring a system from the ground state to a level where a reaction will proceed.

Activator—(In molecular biology) a protein that binds to DNA upstream of a gene and activates transcription from that gene; (in enzymology) a small molecule that binds to an enzyme and increases its catalytic activity; (of pesticides) a material added to increase its toxicity.

Active immunity—Protection due to development of an immune response by an individual following stimulation with antigen, e.g., in a vaccine or during infection.

Active immunization—Stimulation of an individual's immune responses so as to confer protection against disease; effected by exposure to protective antigens either during the course of infection (which may be subclinical) or by vaccination. The protection takes a week or more to develop but then is long-lasting and rapidly revived by a booster shot.

Active ingredient—The active component in a formulated product.

Active site—The region of a protein that binds to substrate molecule(s) and facilitates a specific chemical conversion.

Active transport—Absorption of ions as a result of expenditure of energy by living protoplasm and usually against a concentration gradient; ion accumulation; unidirectional or vectorial transport produced within a membrane-bound protein complex by coupling an energy-yielding process to a transport process.

Active water absorption—Osmotic uptake of water by living cells whose osmotic concentration has been raised by ion accumulation; process responsible for guttation and root pressure.

Aculeate—Having narrow spines; slender and sharp-pointed.

Aculeolate—Somewhat spiny; having spinelike processes.

Acuminate—Gradually tapering to a slender point.

Acute (of disease)—As opposed to chronic; symptoms of disease, more severe than in the chronic condition, that develop suddenly; having a sudden onset and a sharp rise within a short period of time; sharp or tapered to a fine point; less than a right angle; (of pileus margins) tapering to a thin edge. *See* Chronic.

Acute toxicity (poisoning)—A single or limited exposure to a pesticide or poison that may result in injury or death.

Acyclovir—Antiviral agent, an analogue of guanosine, that inhibits DNA replication of viruses.

Adanal (of nematodes)—Located near the anus; pertains to the bursa, which does not envelop the entire tail.

Adanate—(Of basidiocarps) attached tightly to the substratum; (of lamellae) attached squarely, over the entire gill width, to the stem.

Adaptation (adj. **adaptive**)—From an evolutionary standpoint, a characteristic of a living organism that improves its chances for survival in the environment of its habitat; change brought about in a population of an organism as a result of exposure to a particular set of environmental conditions; the change in structural or functional development of an individual or population enabling the organism to adjust to the environmental conditions in question.

Adapted race—Physiologic race.

Adaptive enzyme—An enzyme produced by an organism in response to the presence of its substrate or a related substance. Also called induced enzyme.

Adaxial—(Of a basidiospore) describing the side next to the long axis of the basidium, usually that with the apiculus; (of plants) toward the axis or central line; turned toward the apex; ventral.

Adelphogamy—Pseudomictic copulation of mother and daughter cells, as in some yeasts.

Adenine—One of the purine bases found in DNA, RNA, nucleosides, and nucleotides. In DNA it pairs with thymine.

Adenophorea—One of the two classes of the phylum Nematoda.

Adenose, adenous—Having glands; glandular.

Adenosine 5′-triphosphate (ATP)—A compound of one molecule each of adenine and D-ribose with three molecules of phosphoric acid; is the phosphorylated condition of ADP. It conveys energy needed for metabolic reactions, then loses one phosphate group to become ADP.

Adenosine diphosphate (ADP)—A nucleotide composed of adenine and D-ribose with two phosphate groups attached; ADP and ATP participate in metabolic reactions (both anabolic and catabolic). These molecules, through the process of being phosphorylated or dephosphorylated, transfer energy within cells to drive metabolic processes.

Adenosine monophosphate (AMP) or adenylate—Unless otherwise specified, 5′-AMP, the nucleotide having a phosphate in ribose-O-phosphate ester linkage at position 5 of the ribose moiety. *See* Cyclic AMP.

Adenosine—A mononucleoside consisting of adenine and D-ribose that is produced in the hydrolysis of adenosine monophosphate (AMP).

Adherence (of pesticides)—Ability to stick to a surface; retention.

Adhesion—The molecular attraction between unlike substances, e.g., water and sand particles.

Adhesive disc—*See* Holdfast.

Adhesive—A material in a pesticide formulation that increases its sticking power or adhesiveness to the treated area. Also known as a sticker.

Adhesorium—Organ developed from a resting zoospore of *Plasmodiophora* for attachment to, and penetration of, the host.

adj.—Adjective.

Adjunct (of brewing)—Any legally permitted substance, lacking nutritional properties, that is added to the fermentation.

Adjuvant—(Of pesticides) any substance added to a formulation to improve its physical or chemical properties. The term includes such materials as wetting agents, spreaders, stickers, emulsifiers, dispersing agents, foam suppressants, penetrants, correctives, etc.; (of bacteriology) a subsidiary ingredient in an inoculum that modifies the host response; (of serology) substance injected with antigens (usually mixed with them but may be given before or following the antigen) that nonspecifically enhances, or modifies, the immune response to that antigen. Thus, antibody production or the reaction of well-mediated immunity is more vigorous than would be the case were the antigen injected without adjuvant. In addition, the response may be modified qualitatively, e.g., antibody of different immunoglobulin types may be stimulated.

A-DNA—Right-handed, double helical DNA with about 11 residues per turn. Formed from B-DNA by dehydration. *See* DNA.

Adnate—(Of lamellae or tubes) broadly attached to the stipe; if lamellae, proximal end not notched (*see* Sinuate); sometimes restricted to lamellae widely joined to the stipe (*see* Adnexed); (of pellicle, scales, etc.) tightly fixed to the surface.

Adnexed (of gills or lamellae)—Narrowly attached to the stem, or only by the upper corner of the gills (*see* Adnate); an ambiguous term.

Adpressed—*See* Appressed.

ADP—*See* Adenosine diphosphate.

Adsorption—Adhesion in very thin layers of ions or molecules to the surface of a solid with which it is in contact; (of serology) nonspecific binding of molecules to solid surfaces, typically in very thin or monomolecular layers, e.g., binding of antibodies or antigens to glass, plastic, or nitrocellulose surfaces.

Adspersed—Widely distributed; scattered.

Aduncate—Hooked; curved; bent.

Adventitious septum—Septum formed in the absence of, or independent of, nuclear division; characteristic of lower fungi. *See* Septum.

Adventitious— Describing roots, tubers, flowers, buds, etc. produced in an unusual or irregular position, e.g., buds on roots, or at an unusual time of development; arising sporadically.

Adventive branching (of fruticose lichens)—Branching in an abnormal pattern.

Aecial initial—Cells of the monokaryotic mycelium that initiate formation of the aecium; site of initiation of the dikaryophase.

Aecidioid—Describing aecia, uredinia, or telia that resemble the anamorphic genus *Aecidium* in having a well-defined, short peridium and catenulate spores.

Aecidium—An anamorphic genus for O and I stages only in the rust fungi (Uredinales).

Aeciospore, aecidiospore—Dehiscent, dikaryotic spore produced in a cup-shaped aecium of a rust fungus; infects a host; nonrepeating but essential in the sexual spore cycle, often yellow to orange in mass, borne in chains in an aecium; in heteroecious rusts, incapable of infecting the host on which it was produced; applies to spores produced by the anamorphic genera *Aecidium, Caeoma, Elateraecium, Peridermium, Roestelia,* and *Uraecium*.

Aecium (pl. aecia)—"Cluster-cup" fruiting structure (sorus) of rust fungi that contains one-celled spores (aeciospores) in chains; forms after the spermagonium and before the uredinium in the macrocyclic life cycle; often cup-shaped and abbreviated as II; generalized term that applies to any of the anamorphic rust genera. *See* under Aeciospore.

Aequihymeniiferous (of hymenial development in agarics)—Having basidia that mature and shed their spores evenly over the surface of each lamella; non-*Coprinus* type. *See* Inaequihymeniiferous.

Aeration, mechanical—*See* Cultivation.

Aeration, soil—Process by which water in the soil pores is replaced by air from the atmosphere.

Aeroaquatic fungi—Fungi that grow under water and produce spores in the air above.

Aerobe (adj. **aerobic**)—A microorganism living or active only in the presence of free molecular oxygen; describing an environment or condition in which oxygen is deficient for chemical, physical, or metabolic processes. *See* Anaerobe.

Aerobic respiration—Respiration in the presence of free, gaseous oxygen.

Aerogenic—Describing an organism that produces detectable gas during the breakdown of carbohydrate.

Aerolation, aerolated (of nematodes)—The condition in which the transverse body striae enter the lateral fields.

Aerosol—Atomized particles (solids, smoke) or liquid droplets (fog) suspended in air. The term usually refers to liquids in a container under pressure, as in the case of pesticidal aerosols.

Aethalium (of myxomycetes)—Sessile fruitbody made by massing of all or part of a plasmodium.

Aetiology (British)—*See* Etiology.

Affinity—The strength of the antigen-antibody interaction; strength of bonds formed by this reversible interaction, like other reversible bimolecular equilibrium reactions, determines the rate of association between antibody (i.e., monospecific, such as a monoclonal antibody) and antigen (e.g., a hapten) versus the rate of dissociation. High-affinity antibodies have a higher rate of association with antigen and bind more antigen (e.g., 10^8 to 10^{12} mol⁻) than low-affinity antibodies, which have a lower rate of association with antigen and bind less antigen (e.g., 10^5 to 10^7 mol⁻).

Aflatoxins—Mycotoxins (or a series of toxic metabolites) produced by strains of *Aspergillus flavus* and other species of this fungus when growing on peanuts (groundnuts), maize (corn), cereals, etc.; the cause of aflatoxicosis in animals.

Agamic, agamous—Asexual.

Agar stroke—Agar slant in a test tube.

Agar, agar-agar—A non-nitrogenous gelatinous mixture of dried polysaccharides, some anionic, extracted from certain red algae (seaweeds, Rhodophyceae) and used for preparing semisolid culture media on which microorganisms are grown and studied; forms a gel with water (below about 40°C), and is usually used at a concentration of 1.5–2.0% to solidify media.

Agaric—A fungus in the Agaricales; a gill-bearing fungus such as a mushroom.

Agaricic acid—A hydroxylated tribasic acid from *Fomitopsis* (*Fomes*) *officinalis*; used to control tubercular night sweats.

Agaritine—An amino acid from *Agaricus bisporus*.

Agarose gel electrophoresis—Technique used for separating proteins or nucleic acids by passage of an electric current through the gel.

Agarose—A constituent of agar that does not contain inhibitors of virus development. It is widely used in gel electrophoresis, as the pore size is more uniform than that of agar.

Agent of disease—Organism or abiotic factor that causes disease; a pathogen.

Agent of inoculation—That which transports inoculum from its source to or into an infection court, e.g., wind, rain splash, flowing water, insects, mites, humans, tools, and equipment.

Agglutinate, agglutinated (of hyphae)—Appearing gelatinized as if glued together.

Agglutination—Clumping; specifically, a serological test in which viruses or bacteria suspended in a liquid collect into clumps whenever the suspension is treated with antiserum containing antibodies specific against these viruses or bacteria; visible clumping of particulate antigens (e.g., red blood cells, bacteria, etc.) when reacted with a specific antibody, or visible clumping of inert particles to which antibodies are adsorbed (e.g., latex microspheres) when reacted with a specific antigen. Clumping is caused by antigen-antibody bridging between adjacent cells or particles.

Agglutinin—An antibody (or compound) that causes another particular antigen (compound) to clump and settle out of suspension. *See* Antigen.

Aggregate plasmodium—Structure formed by the massing of separate myxamoebae or cells before reproduction. *See* Plasmodium.

Aggregate—(Of fruit) a cluster of fruits developed from the ovaries of a single flower; (in taxonomy, usually abbreviated "aggr.") a group of related taxa, especially species of uncertain circumscription; (of soil) many primary soil particles held in a single unit as a clod, crumb, block, or prism.

Aggregated—Crowded close together.

Aggressive strain (of a fungus)—Genotype having enhanced pathogenic capacity toward a host genotype(s). *See* Virulence.

Aggressiveness (of a plant pathogen)—Relative ability to colonize and cause damage to plants. *See* also Virulence.

Agitate—To keep a pesticide or ingredient in a medium (e.g., $CaCO_3$) in suspension and keep it from settling out or separating in the spray tank or medium.

Agonomycetales—Filamentous fungi (Hyphomycetes, Deuteromycotina) that do not form spores other than chlamydospores; may be states of ascomycetes, basidiomycetes, or deuteromycetes. *See* Mycelia Sterilia.

Agricole, agaricicolous—Living on agarics.

Agroclavine—A major alkaloidal constituent of *Claviceps fusiformis* sclerotia. *See* Ergot.

Agronomy—The art and science of crop production and soil management.

Air incinerator—An electric or fuel-fired furnace for the sterilization of microbiologically contaminated air (or other gases).

Air layering—A process or procedure causing roots to develop on an undetached aerial portion of a plant, commonly as the result of wounding or other stimulation.

Air pollution—Undesirable material(s) in the atmosphere. The most common air pollutants that affect plants, especially lichens, are ozone, sulfur dioxide, fluorides, nitrogen oxides, and particulate deposits (e.g., smoke, heavy metals, and photochemical smog).

Air porosity—The proportion of the bulk volume of soil that is filled with air at any given time or under a given condition, such as a specified moisture tension.

Air spora—*See* Spora.

Airborne—Transported from place to place through the air.

Air-dry (adj.)—Being in a state of dryness at equilibrium with the moisture content in the surrounding atmosphere.

Akaryote (of Plasmodiophoraceae)—Stage in the nuclear cycle before meiosis in which little or no chromatin is seen in the nucleus.

Akinete—Nonmotile reproductive structure (e.g., a spore); a resting cell.

Alae (of nematodes)—Expansions or projections formed by a longitudinal thickening of the cuticle. There are three types: 1) *cervical alae*—confined to the anterior region of nematodes parasitic on animals; 2) *caudal alae*—occurring in the posterior region of males in many genera; 3) *longitudinal alae*—usually four in number that extend the length of the body sublaterally.

Alate—Winged; a form in the life cycle of certain insects (e.g., aphids). *See* Apterous.

Albino, albinism—A plant or part of a plant that is white and lacking chlorophyll. Albinism is usually lethal in higher plants.

Alcohol-fast—Resistant to decolorization with alcohol.

Alepidote—Lacking scales or scurf; smooth.

-ales (suffix)—Attached to the stem of a generic name to form the name of an order.

Aleuriospore, aleurispore, aleurospore—Term formerly used for a thick-walled and pigmented but sometimes thin-walled and hyaline conidium developed from the blown-out end of a conidiogenous cell or hyphal branch from which it secedes with difficulty; has been used in various senses ("chlamydospore," "gangliospore") and now generally rejected as a confused term.

Aleurone grains—Particles of protein found in the seeds of some plants.

Aleurone—The outer layer of cells surrounding the endosperm of a cereal grain (caryopsis).

Algicole, algicolous—Living on algae.

Aliform—Wing-shaped.

Alimentary tract (of nematodes)—The tubular food-carrying passage extending from the mouth to the anus. The parts are: oral aperture, stoma (including stylet), esophagus or pharynx, intestine, prerectum, rectum, and anus.

Aliquot, aliquot part—Portion contained an exact number of times in the whole; not equivalent of "sample" because the concepts of both uniformity and representation are implicit.

Alkali soil—*See* Sodic soil.

Alkaline hydrolysis—Use of a high pH to degrade or hydrolyze a bond. DNA is not hydrolyzed at high pH, while RNA is degraded to mononucleotides.

Alkaline phosphatase—An enzyme that removes the 5' terminal phosphate groups from linear DNA molecules. It prevents the religation of plasmid vector molecules following cleavage with a restriction endonuclease.

Alkaline soil—A soil with a basic or "sweet" reaction above pH 7.0. *See* pH and Sodic soil.

Alkaloid—A nitrogenous organic compound with alkaline (basic) properties, usually bitter in taste, poisonous to animals, and produced by certain plants.

Allantoid (especially of spores)—Sausage-shaped; cylindrical, uniform in diameter, and somewhat curved with rounded ends.

Allele—One of two or more alternate forms of a gene occupying the same locus on a particular chromosome; a form of gene. Two genes that are derived from the same gene by mutation and are alternative occupants of the same chromosomal locus are described as **allelic** with respect to each other. *See* Gene.

Allelopathy (adj. **allelopathic**)—Ability of one species to inhibit or prevent the growth of the same or other species through the excretion of toxic substance(s).

Allergy—A condition of hypersensitivity to a specific substance. Spores of molds (e.g., *Alternaria, Aspergillus, Cladosporium, Penicillium,* etc.) and cereal rusts, smuts, lichens, etc. in the air are capable of inducing, by allergy, asthma and hay fever in humans.

Alliaceous—Having a taste or smell of onions, garlic, or other *Allium* spp; cepaceous; onion-shaped.

Allochronic—Occurring at different time levels, e.g., contemporary and fossil specimens.

Allochrous, allochroous—Changing from one color to another.

Allochthonous—Not indigenous; transported to the place where found. *See* Autochthonous.

Allocyst—An obovate, claviform, or pyriform chlamydospore-like structure in *Flammula gummosa*.

Allopatric—Describing forms of a species or genus occurring in different geographical regions. *See* Sympatric.

Allosteric effect—A change produced in the properties of an enzyme by the specific action of a small molecule acting at a site other than the active site.

Allotype—A paratype of the opposite sex of the specimen designated as the holotype; (in serology) genetically inherited antigenic sites present on a class of antibodies in some individuals and absent from others of the same species.

Alluvial soil—A recently developed soil from deposited soil material that exhibits essentially no horizon development or modifications.

Alpha-spore (A-spore, α-spore)—A germinative ovoid spore of the imperfect (anamorph) state of the Diaporthaceae, especially *Phomopsis*, in which they are fusoid to oblong and biguttulate. Beta-spores (B-spores) are commonly also produced. *See* Beta-spore.

Alternaric acid—Phytotoxic substance produced by *Alternaria solani*.

Alternate host—A plant, different from the principal host, on which a pathogen (e.g., a heteroecious rust) must develop to complete its life cycle. Do *not* confuse with secondary host or alternative host.

Alternate—Referring to bud or leaf arrangement in which there is one bud or one leaf at a node on a stem; i.e., not opposite or whorled.

Alternation of generations—The alternation of a spore-producing phase (sporophyte) and a gamete-producing phase (gametophyte) in the sexual life cycle of an organism. When the two generations are alike, they are said to be "homologous"; if unlike "antithetic."

Alternative host—One of a parasite's several hosts; alternative hosts are not required for completion of the developmental cycle of the parasite. Weeds and wild plants are often alternative hosts of certain pathogens and pests.

Alumino-silicates—Compounds containing aluminum, silicon, and oxygen as main constituents. Example: Microcline, $KAlSi_3O_8$.

Alutaceous—(Of color) pale tan, pale brown, or pinkish-cinnamon; (of texture) soft leathery.

Alveola, alveole, alveolus—A small surface cavity or hollow.

Alveolate—Pitted like a honeycomb; falveolate.

Amadou—The context of *Fomes fomentarius* or *Phellinus (Fomes) igniarius* fruitbodies after saltpeter ($NaNO_3$) is added; tinder; touchwood; punk or funk.

Amalgamate (v.)—To unite to form a unit.

Amanita factor B—*See* Pantherine.

Amanita factor C—*See* Ibotenic acid.

Amanitin—*See* Amatoxins.

Amatoxins—Cyclic octopeptides (including A- and B-amanitin, amanin) from *Amanita phalloides*, which is toxic to humans, and the nontoxic amanillin. *See* Phallotoxins.

Amber—A mutation that creates the stop codon UAG in the coding region of a gene, leading to the synthesis of a truncated protein.

Ambient temperature—Air temperature at a given time and place; not radiant temperature.

Ambifenestrate (of nematodes)—Describing the occurrence of two openings in the vulval cone that are separated by the vulval bridge; found in some species of cyst nematodes.

Ambrosia fungi—Fungi, e.g., yeasts (*Ambrosiozyma, Ascoidea,* and *Dipodascus* spp.), Hyphomycetes, etc. that grow in the tunnels of ambrosia or wood-boring beetles (Scolytidae) and serve as food for larvae and adults.

Amendment, physical—Any substance added to the soil for the purpose of altering its physical condition (e.g., sand, clay, peat moss, organic matter, calcined clay, lime, sulfur, gypsum, sawdust, or bark, etc.).

American Type Culture Collection—A large collection of bacteria, fungi (including yeasts), viruses, viroids, cells, etc. available on request by payment of a small fee. Located at 12301 Parklawn Drive, Rockville, MD 20852-1776 (Telex ATCCROVE 908-768, FAX 301-231-5826).

Amerospore—A one-celled spore in the Deuteromycotina with a length/width ratio of <15:1; if elongated, axis single and not curved through more than 180°; any protuberances, <1/4 spore body length. *See* Dictyospore, Didymospore, Helicospore, Scolecospore, Staurospore, etc.

Ametoecious—*See* Autoecious.

Amino acids—Organic, nitrogenous acids from which protein molecules are constructed. There are 20 common acids in living organisms, each with the basic formula NH_2-CHR-COOH.

Aminoacyl-RNA—An amino acid linked via its carboxyl group to the hydroxyl group (2' or 3') of the ribose residue at the 3' end of the RNA.

Aminoacyl-tRNA synthetase or aminoacyl-tRNA lyase—The enzyme that ligates the amin, and the nontoxic amanillino acid to its specific tRNA. There are 20 such enzymes, specific for each amino acid and its tRNA.

Amixis (adj. **amictic**)—Apomixis in haploid organisms. *See* Heterothallism.

Ammonia fungi—A chemoecological group of fungi in which reproductive structures form after the addition of ammonia, urea, etc., or alkali to the soil.

Ammonification—Formation of ammonia following the decomposition of proteinaceous organic compounds by ammonifying bacteria.

Ammonium fixation—The incorporation of ammonium ions by soil fractions in such a manner that they are relatively insoluble in water and nonexchangeable by the principle of cation exchange.

Ammonium sulfate—Salt commonly used to precipitate enzymes, proteins, and viruses without denaturation; a common type of fertilizer for plants.

Amoeboid—Not having a cell wall and changing in form like an amoeba; moving by temporary processes (pseudopodia) from the surface of the vegetative body.

Amorphous—Without definite form, shape, or organized structure.

Amph-, amphi- (prefix)—Of two kinds, sides, etc.; all around.

Amphicoelous—Concave on both sides

Amphid (pl. **amphids**)—A chemosensory organ in nematodes, occurring laterally in pairs, located in the anterior region; in the class Secernentea, amphids appear as small, pore-like openings on the lips; in the Adenophorea the openings are postlabial and may be varied in form, i.e., circular, spiral, or cyathiform. Amphids are composed of the following structures:

1. *Amphidial duct*—Channel that connects the amphidial pore to the sensilla pouch.

2. *Amphidial gland*—A microvillous organ enveloping the sensilla pouch.

3. *Amphidial nerve*—Nerve innervating the amphid.

4. *Amphidial pore*—Aperture in the cuticle through which the amphid opens to the exterior.

5. *Amphidial pouch* (amphidial pocket or chamber)—Chamber, usually just behind the amphidial pore, that is often cup-shaped in outline.

6. *Sensilla pouch*—Chamber containing nerve processes situated posterior to the amphidial pouch.

Amphidelphic (of nematodes)—A female having two ovaries, one directed anteriad and one posteriad.

Amphigenous—(Of leaves) growing on either the upper or lower surface; (of fungi) growing all around or on two sides.

Amphigynous—In Oomycetes (e.g., *Pythium, Phytophthora*), an antheridium that encircles the base or stalk of the oogonium.

Amphimictixis (adj. **amphimictic**)—Reproduction of nematodes by union of egg and sperm; true bisexual reproduction; male-female ratio in a population usually about equal.

Amphimixis—Copulation of two unrelated cells and nuclei, e.g., egg and sperm; reproduction by a sexual process. *See* Apomixis, Automixis and Pseudomixis.

Amphispore—A second, specialized type of urediniospore with a thicker and usually more deeply pigmented wall; acts as a resting spore.

Amphithallism (of fungi)—A thallus derived from one heterokaryotic spore with nuclei of compatible mating types.

Amphithecium—The thalline margin of an apothecium.

Amphitrichiate, amphitrichous—Describing an organism having one flagellum at each end (pole) of a cell.

Amphotericin A and B—Antifungal polyene antibiotics from actinomycetes (*Streptomyces* spp.). Amphotericin B (fungizone) is used in treating systemic mycoses of humans.

Amplectant—Covering; embracing.

Amplicate—Enlarged; made greater.

Amplication—An increase in the copy number of a gene or plasmid. *See* Chloramphenicol amplification.

Ampoule hypha—Swollen hypha found in some lower Basidiomycotina.

AMP—*See* Adenosine monophosphate.

Ampulla (pl. **ampullae**)—(Of nematodes) a membranous, flask-shaped sac or swelling; a widening in a canal, forming a reservoir; (of fungi) a bottle-shaped swelling or enlargement on a conidiophore, formed at the apex or at any point along its length.

Ampullate (of hyphae)—Swollen at the septa.

Ampulliform—Flask-shaped with the swollen part at the base.

Amygdaliform—Almond-shaped.

Amylase—An enzyme that hydrolyzes starch to maltose.

Amyloid (of spores, hyphal walls, ascus tips, cystidia)—Staining grayish to blackish violet in Melzer's iodine reagent; symbol J+. *See* Dextrinoid.

Anaerobe, anaerobic—An organism able to grow, or a process that occurs, in the absence of free (molecular) oxygen and usually in high carbon dioxide. *See* Facultative anaerobe and Obligate anaerobe.

Anaerobic respiration—Respiration in the absence of molecular oxygen or in the presence of reduced concentrations of free oxygen. *See* Aerobic respiration.

Anaerogenic—Describing an organism that does not produce visible gas when it breaks down carbohydrate.

Analogous, (n. **analogy**)—Showing a resemblance in form, structure, or function that is not considered as evidence of evolutionary relatedness; the opposite of homologous. *See* Homologous.

Anamnesis—Immune memory or a secondary immune response.

Anamorph—The asexual or imperfect reproductive state of a fungus; mitotic diasporic expression of a fungus.

Anaphase—A stage in mitosis in which the newly separate chromatids (daughter chromosomes) move toward opposite ends of the spindle.

Anaphylaxis—Manifestation of an acute change (e.g., immediate hypersensitivity) in a living animal from the uniting of an antibody with its antigen, which may result in the death of the animal, following the parenteral injection of the antigen. The reaction is caused by release of histamine or other vasoactive substances when antigen combines with antibody on the surfaces of cells. Anaphylaxis may be generalized or localized to the site of injection.

Anaphysis—A thread-like conidiophore that persists in apothecia of *Ephebe*.

Anastomosis, anastomosing (pl. **anastomoses**)—Union or fusion of a hypha or vessel with another, resulting in a sharing of their contents; fusion of hyphal branches or other structures to form a vein-like network.

Anatomy—The science and study of the shape and structure of organisms and their parts.

Anbury—Clubroot of crucifers.

Ancestor (in taxonomy)—A taxon from which others are thought to have descended.

Androgynous—Having the antheridium and its oogonium arising on the same hypha.

Androphore—A branch that forms antheridia.

Anemophilous (of spores)—Disseminated by air currents.

Aneuploid—Having a chromosome number that is not a multiple of the haploid (1n) set.

Angio-, -ange, -angium—Indicating a structure without an opening; cavity.

Angioange, angioangium—Cavity; structure without an opening.

Angiocarpic, angiocarpous (of a sporocarp)—Closed until the spores mature. *See* Endocarpous, Hemiangiocarpous, and Cleistocarp.

Angiosperm—A flowering plant; a plant bearing seeds that develop in an enclosed carpel.

Ang-kak—An Oriental food coloring obtained by growing *Monoascus purpureus* on polished rice.

Angle rotor, fixed-angle rotor. A centrifuge rotor in which the wells holding the tubes are drilled at an angle to both the axis of rotation and the lines of centrifugal force.

Angstrom (Å)—A unit of length equal to 1/10 millimicron (mμm) or nanometer (nm) = 1/10,000 micron or micrometer (μm) = 10^{-10} m. Commonly used for measuring wavelengths and to express dimensions of some intracellular structures of microorganisms and viruses.

Anguilluliform—Worm- or eel-shaped.

Angustate—Narrow, narrowed.

Anheliophilous—Preferring diffuse light. *See* Heliophilous.

Animalia—The kingdom of animals.

Anion—A negatively charged ion that, during electrolysis, is attracted to positively charged surfaces; contrasted with cation.

Anionic detergent—A detergent having a negatively charged surface ion, e.g., sodium deoxycholate, sodium dodecyl sulfate.

Aniso- (prefix)—Unequal.

Anisogamy, (adj. **anisogamous**)—Union of gametes of unlike form or physiology, i.e., *anisogametes*.

Anisokont—Having flagella of unequal length.

Anisometric—Describing virus particles that are not isometric, e.g., rod-shaped particles.

Anisospory—Having spores of more than one type.

Anisotropy of flow—Differences in the flow properties of a solution of macromolecules in different directions.

Annealing (v. **anneal**)—Formation of double-stranded nucleic acid molecules from single-stranded molecules with complementary base sequences.

Annellate (n. **annellations**)—With ring-like markings on the wall showing where percurrent growth has taken place repeatedly.

Annellide—A holoblastic, percurrently proliferating conidiogenous cell.

Annellidic—Describing holoblastic conidiogenesis in which the annellide (*annellophore*), by repeated percurrent proliferations, results in a basipetal sequence of conidia.

Annellophore—A conidiogenous cell with a series of ringlike scars left by succeeding blastic conidia (*annelloconidia, annellospores*).

Annual (n.)—A plant that completes its life cycle from seed and dies within one year or less; fungus fruiting body in an active, living state for one season, occasionally surviving into the beginning of the next year.

Annual ring—Single-year cylinder growth of secondary xylem added to a woody plant stem by the vascular cambium.

Annular—Ringlike; ringlike arrangement.

Annulations (of nematodes)—Transverse depressions (or striae), appearing as lines at regular intervals in the cuticle; ring or collarlike folds.

Annule (of nematodes)—The thickened interval between two transverse striae in the cuticle; ring or collarlike fold.

Annulus (pl. **annuli**)—(Of fungi) a ring; ringlike partial veil around the stipe of some agarics after the pileus (cap) has expanded; a remnant of the inner veil; hymenial veil; apical veil; (of ferns) a row of specialized cells in the wall of a sporangium; contraction of the annulus causes rupture of the sporangium and dispersal of spores.

Anoderm—Without skin or pellicle.

Anomalous—Deviating from the general rule.

Anoxybiont (adj. **anoxybiontic**)—Bacteria unable to use atmospheric oxygen for growth.

Antagonism—The phenomenon of one microorganism producing toxic metabolic products that kill, injure, or inhibit the growth of some other microorganism(s) in close proximity; a general term for counteraction between organisms or groups of organisms; (of pesticides) the action of two or more pesticides that reduces the effectiveness of one or all of the pesticide components; the opposite of *synergism*.

Anteriad (of nematodes)—Directed forward; opposed to posteriad.

Anterior—At or toward the front or head as opposed to posterior; (of lamellae) the end at the margin of the pileus; in front.

Anther mold of clover—Mycelial growth caused by *Botrytis anthophila*, teleomorph *Sclerotinia spermophila*.

Antheridiol—A sterol sex hormone of *Achyla bisexualis* that induces antheridial formation in male strains of *Achyla*.

Antheridium, antherid (pl. **antheridia**)—The male sex organ (gametangium) in fungi; a structure that produces sperms.

Antherozoid—A motile male cell or sperm.

Anther—The saclike, pollen-bearing portion (stamen) of a flower; usually borne on a filament.

Anthesis—Developmental stage in flowering at which anthers rupture and shed pollen; a state of full bloom.

Anthocyanins—Water-soluble flavanoid pigments that account for many of the usually red or blue, sometimes purple, colors in leaves, flowers, and fruits. Anthocyanins occur in the vacuole of the cell.

Anthracine—Coal-black.

Anthracnose—A leaf, stem, flower, and/or fruit disease with characteristic limited, sunken, necrotic lesions (often zonate and with a distinct margin), caused by a fungus producing nonsexual spores in an acervulus (Melanconiales). A number of these fungi also have a sexual stage. *See* Spot anthracnose.

Anthracobiontic—Obligately inhabiting burned-over areas.

Anthracophilous (of sporulation)—Favored by burned-over areas.

Anthracophobic (of sporulation)—Reduced or checked on burned-over areas.

Anthracoxenous (of incidence and growth)—Unaffected by burned areas.

Anthropophilic (mostly of dermatophytes)—Preferentially pathogenic to humans. See Zoophilic.

Anti- (prefix)—Against.

Antibiosis—Antagonistic association between two organisms or between one organism and a metabolic product of another organism, to the detriment of one of the organisms. *See* Vector resistance.

Antibiotic resistance—Resistance to the lethal effects of an antibiotic. The five main mechanisms of resistance are: 1) inactivation of the antibiotic; 2) reduction in cellular uptake or an increase in cellular excretion of the antibiotic; 3) production of an altered target protein that no longer binds the antibiotic; 4) overproduction of the target protein so that the antibiotic is "titrated out"; and 5) elaboration of some alternative enzyme or pathway that is not susceptible to the antibiotic.

Antibiotic—Damaging to life; especially a chemical compound produced by one microorganism that inhibits growth or kills other living organisms in very small amounts.

Antibody—A new or altered specific protein (modified serum immunoglobulin molecule) produced by an animal's B lymphocytes in response to an antigenic stimulus (a foreign substance); it reacts (binds) specifically with the antigen and renders it harmless. An antibody that causes lysis is called a *lysin*; those causing agglutination and precipitation are *agglutinins* and *precipitins*, respectively. *See* Anaphylaxis, Complement-fixation, ELISA, and Serology.

Anticlinal—Perpendicular to the surface; cell division perpendicular to, rather than parallel to, the surface. *See* Periclinal.

Anticodon—A sequence of three nucleotides complementary to the codon triplet in a mRNA; a group of three bases in a tRNA molecule that recognizes a codon in the mRNA molecule.

Antidesiccant—Chemical compound that minimizes water loss from a plant through transpiration.

Antidote—Substance used to counteract the effects of a poison.

Antigen (adj. **antigenic**)—Foreign protein, occasionally a complex lipid or polysaccharide and some nucleic acids, which, when introduced into living animal tissues, induces the production of specific antibodies antagonistic to the substance injected and capable of specifically binding to antigen binding sites of antibody molecules (*see* Immunogen). All immunogens are antigenic, but not all antigens are immunogenic (e.g., haptens). The capacity of an antigen to react specifically with an antibody is referred to as an *antigenic reactivity*. Antigens from specific fungi may be used for the identification of fungi or to detect infection in humans or higher animals by pathogenic fungi. *See* Immunogen.

Antigen binding capacity—A measurement of the ability of an antibody to bind antigen, based on the effects of dilution of the antibody.

Antigen binding site = Fab—That area of an antibody molecule that binds to antigen; paratope.

Antigen presenting cells (APC)—Cells involved in the immune response that initially interact with, engulf, break down, and present an antigen on the cell's surface to T cells.

Antigen-antibody reaction—The specific interaction between an antigen and an antibody that recognizes a specific structural feature of the antigen and binds to it.

Antigenic determinant—The small site on an antigen molecule to which an antibody is specifically bound. It is determined by structural complementarity between antibody (Fab) and antigen molecules. An antigen molecule may comprise more than one determinant. Attached haptens add further determinants. *See* Epitope.

Antigen-mimicking antibody—*See* Ab 2 beta.

Anti-idiotope antibody—An antibody that interacts with an idiotope of another antibody.

Antimetabolite—A substance resembling a chemical structure of a naturally occurring compound essential in a living process and that specifically antagonizes the biological action of such an essential compound.

Antimicrobial—Inhibiting microorganisms by biological or chemical agents.

Antipodals—Three small ephemeral cells within the embryo sac in an ovule; three or more nuclei at the end of the embryo sac opposite the egg nucleus (female gamete). They are produced by mitotic divisions of the megaspore and degenerate following sexual fertilization.

Antiseptic (n.)—A substance that inhibits or destroys growth and activities of bacteria and other microorganisms in living tissues.

Antiserum (pl. antisera)—The blood serum (fluid fraction of coagulated blood) of any warm-blooded animal that contains antibodies to one or more specified antigens, e.g., anti-ovalbumin.

Antiserum titer—The highest dilution of an antiserum that will react with its homologous virus. *See* Homologous reaction and Heterologous reaction.

Antisporulant—A chemical or physical agent that prevents spore production without killing vegetative growth of a fungus.

Antitoxin—An antibody capable of uniting with and neutralizing a specific toxin or cause it to flocculate.

Antrorse—Directed upward or forward; opposite of retrose.

Apandrous—Describing oospores formed in the absence of antheridia.

Aparaphysate—Without paraphyses or falsely paraphysate.

Apex (pl. apices, adj. apical)—Tip or end; uppermost point; part of root or shoot containing apical meristem.

Aphanoplasmodium—A plasmodium composed of a network of undifferentiated strands of nongranular protoplasm.

Aphelenchoid—*See* Esophagus.

Aphid—A small, sucking insect of the family Aphididae (order Homoptera) that produces honeydew and injures plants when in large populations or when serving as a virus vector.

Apical—At or near the end or tip (apex).

Apical cells (of a rust fungus)—Cells of the upper part of a pedicel and to which spores are attached.

Apical dominance—The inhibition of lateral buds on a shoot due to auxins produced by the apical bud or the apical meristem.

Apical granule—A deeply staining granule at the tip of a hypha, especially in Hymenomycetes and Gasteromycetes.

Apical growth—Growth that occurs at the tip or apex of an organ.

Apical meristem—A mass of undifferentiated cells capable of division at the tip of a root or shoot. These cells differentiate by division that allows a plant to growth in depth and height.

Apical paraphyses—Downward growing hyphae with free tips in the centrum of hypocrealean fungi. *See* Paraphyses.

Apical veil—*See* Annulus.

Apical wall building—*See* Wall building.

Apiculate (of spores)—Having a distinct, small pointed tip (apiculus or apiculum) or hilar appendage.

Apiculus (of spores)—Projection at the basal end of a spore by which it is attached to the sterigma; also referred to as the hilar appendage; a short sharp point; apicule.

Apileate—Without a pileus; resupinate.

Aplanetism—The condition of having nonmotile spores instead of zoospores.

Aplanogamete—A nonmotile gamete.

Aplanospore—A nonmotile sporangiospore; may be an abortive zoospore.

Aplerotic (of oospores of Pythiaceae)—Not filling the oogonium.

Apobasidiomycete—A gasteromycete possessing apobasidia.

Apobasidium—Basidium with nonapiculate spores borne symmetrically on the sterimata and not forcibly discharged.

Apodial—Without a stem or stipe; sessile.

Apogamy—Apomictic development of diploid cells.

Apomixis (adj. **apomictic**)—The development of sexual cells into spores, etc., without being fertilized; the asexual (vegetative) production of seedlings in the usual sexual structures of the flower but without the mingling and segregation of chromosomes. Seedling characteristics are the same as those of the maternal parent. *See* Amphimixis, Automixis, and Pseudomixis.

Apophysis—(Of nematodes) a swelling or expansion as found in the organ-Z, or Z-organ, in species of *Xiphinema*; (of fungi) a swelling or swollen filament, e.g., at the end of a sporangiophore below the sporangium in certain Mucorales (e.g., *Rhizopus* but not in *Mucor*) or the stem of some species of Basidiomycotina; the swelling at the tip of a sterigma from which the basidiospore develops and which becomes the hilar appendage.

Apoplasm, apoplast—Continuous space outside each living protoplast, including cell walls and free space. The total nonliving parts of a plant. *See* Symplasm, Symplast.

Apoplasmodial, apoplastogamous (of Acrasiales)—Describing nonfusion of the myxamoebae.

Apospory—Suppression of spore formation.

a posteriori—Reasoning from effect to cause, i.e., inductivity.

Apothecium (pl. **apothecia**)—A special sessile or stalked, disc-, cup-, or saucerlike, ascus-bearing fruitbody (ascocarp) of fungi in the Discomycetes; the open inner surface is lined with a hymenium consisting of asci and paraphyses (sterile cells). In size, apothecia vary from nearly microscopic to several centimeters in diameter and height.

a priori—Reasoning from cause to effect, i.e., deductivity.

Appendage—A process (outgrowth), organ, or limb of any sort; (of conidia) cellular, setulose, or mucilaginous ornamentation, usually filiform but sometimes infundibular or capitate.

Appendiculate—(Of a spore) a spore with one or more appendages (setulae); (of an agaric basidiocarp) having the margin of the expanded pileus fringed with toothlike fragments of the veil.

Applanate—Thin, flattened horizontally.

Apple or Nectria canker—Canker caused by *Nectria galligena*, anamorph *Cylindrocarpon heteronemum*.

Apple scab—Scab caused by *Venturia inaequalis*, anamorph *Spilocaea pomi*.

Application rate—Amount of a fertilizer, pesticide, etc. applied to a given area.

Appressed, adpressed—Closely flattened out or pressed against a surface.

Appressed-strigose (of pilear surfaces)—With coarse, parallel hyphal strands closely flattened to the surface.

Appressorium (pl. **appressoria**)—Thick-walled, swollen or flattened terminal or intercalary cell of a germ tube of a fungus, serving for attachment to the host, formed prior to penetration, often with a circular germination pore, found in anthracnose fungi and rusts. Some parasitic seed plants produce a similar structure.

Apterous—Wingless stage in the life cycle of certain insects (e.g., aphids). *See* Alate.

Apud—Among, within, in.

Aquatic fungi—Fungi living in water, especially fresh water in contrast to marine fungi. The main groups are Chytridiomycetes and Oomycetes, especially Chyridiales and Saprolegniales.

Arabinosyl adenine or ARA A—An antiviral agent that inhibits viral DNA polymerase.

Arabinosyl cytosine or ARA C—An antiviral agent that probably inhibits DNA polymerase.

Arable land—Land that is suitable for the production of crops.

Arachnoid—Covered with, or formed of, delicate hairs or fibers like a cobweb with very loosely interwoven hyphae; cobweblike; araneose, araneous.

Arborescent—Describing branched, treelike growth.

Arboricolous, arboreal—Growing on or in trees.

Arboriform hypha—Much-branched skeletal hypha of *Ganoderma*.

Arbuscle, arbuscule (adj. **arbuscular**)—Finely branched haustorial structure in cortical cells; exhibited by vesicular-arbuscular mycorrhizae. The interface of arbuscule with plant protoplast is a site of exchange of nutrients and growth-regulating chemicals.

Arch (of nematodes)—That curved portion of the perineal pattern dorsal to the lateral field or lines in females in the genus *Meloidogyne*.

Archegonium—A multicellular egg-producing gametangium.

Archicarp (of Ascomycotina)—The cell, egg, hypha, or coil, that later develops into the fruitbody, or a part of it, such as a perithecium.

Archontosome (of ascomycetous fungi)—An electron-dense body found near nuclei at all stages from crosier formation to the development of young ascospores in *Xylaria polymorpha*.

Arcuate—Strongly curved as a bow; arclike.

Ardosiaceous, ardesiaceous—Slate-colored.

Arenaceous, arenicole, arenicolous—Growing in sandy places.

Areola (pl. **areolae**) (of nematodes)—Typically, the lateral field of some nematodes delimited by transverse and longitudinal markings.

Areolate—Marked out in little areas, usually by cracks or crevices.

Areolation—A condition in which the transverse body annulation in a nematode traverses the lateral field.

Areole (of cacti)—A small pit or cavity on the surface, from which a spine may arise.

Arescent—Becoming crustose on drying.

Arid—Dry.

Armilla (pl. **armillae**)—Braceletlike frill.

Armillate—Edged; frilled; fringed.

Arrect—Stiffly upright or erect.

Arthric (of conidiogenesis)—Hyphal conidiogenesis characterized by the conversion of a preexisting determinate hyphal element into a conidium.

Arthro- (prefix)—Jointed.

Arthrocatenate—Arthroconidia formed in chains by the fragmentation of a hypha.

Arthroconidium, arthrospore-A jointed conidium composed of more than one cell produced by a thallic conidiogenous cell that can become separated; often cylindrical, truncate at each end or one end; formed by hyphal fragmentation.

Arthropod—Member of the phylum Arthropoda, which consists of invertebrate animals with articulated bodies and limbs and includes insects, arachnids, and crustaceans.

Arthrospore—An asexual spore formed by a hypha fragmenting into separate cells, e.g., in *Geotrichum*. See Arthroconidium.

Articulate—Jointed, segmented.

Artist's conk—*Ganoderma applanatum.*

Ascending, ascendent—(Of conidiophores) curving up; (of lamellae) on a conelike or unexpanded pileus.

Ascigerous—Having asci; ascus-bearing (sexual) stage in the development of an Ascomycete.

Ascigerous centrum—The specialized tissue that gives rise to asci, paraphyses, or other tissues between asci. *See* Hamathecium.

Ascitic fluid—Serous fluid from the peritoneal cavity of mice used in raising large quantities of monoclonal antibodies following injection of hybridoma cells into the peritoneal cavity.

Asco- (prefix)—Pertaining to an ascus.

Ascocarp, ascoma (pl. **ascomata**)—Many-celled sexual fruitbody (ascus-bearing organ) of an Ascomycete, e.g., apothecium, cleistothecium, hysterothecium, perithecium.

Ascoconidiophore—Phialide bearing an ascoconidium.

Ascoconidium (pl. **ascoconidia**)—A conidium formed directly from an ascospore, especially when still within the ascus.

Ascogenous—Ascus-producing or supporting.

Ascogenous hypha—A specialized, dikaryotic hypha that gives rise to one or more asci.

Ascogone, ascogonium—The initial, specialized cell or group of cells that gives rise to ascogenous hyphae and asci; the female gametangium of Ascomycotina (sac fungi).

Ascolocular ascomycetes—Ascomycetes having asci (and paraphyses) developing in cavities within a preformed stroma.

Ascoma (pl. **ascomata**)—*See* Ascocarp.

Ascomycetes (adj. **ascomycetous**)—The Ascomycotina, or sac fungi, one of the major classes of fungi, that produces sexual spores (ascospores), often eight in number, within an oval or tubular membranous sac called an *ascus*.

Ascomycotina—The subdivision of fungi containing the ascomycetes.

Ascoparaphysis—Multicellular diploid storage hypha originating from the base of the ascus in powdery mildew fungi (Erysiphaceae).

Ascophore—Ascus-producing hypha.

Ascospore—A sexually produced spore formed inside an ascus by "free cell formation" after the union of two nuclei.

Ascostroma (pl. **ascostromata**)—A stroma bearing or containing asci, as in the Loculoascomycetes; often microscopic, dark in color, and partially to wholly embedded in cavities in dead plant tissue.

Ascus (pl. **asci**)—Typically a saclike or clavate cell in which meiosis occurs, where ascospores are formed (usually eight) by "free cell formation," and borne in a fungus fruitbody (ascocarp). The spores are often forcibly discharged into the air. Nine main types have been distinguished by light microscopy: annellate, bitunicate (fissitunicate), hypodermataceous, lecanoralean (archeasceous), operculate, ostropalean, prototunicate, pseudoperculate, and verrucarioid.

Ascus crown—Crownlike, annular thickening in *Phyllachora*.

Ascus mother cell—In Ascomycotina, the binucleate cell in which karyogamy and the diploid nucleus occur and which develops into the ascus.

Ascus plug—Thickening in the apex of an ascus through which the spores are forcibly discharged.

Aseptate—Without cross walls.

Aseptic—Free of microorganisms capable of causing infection or contamination. *See* Axenic.

Asexual reproduction—Any type of reproduction not involving the union of gametes (karyogamy) and meiosis (independent of sexual processes); the imperfect (anamorph) or vegetative state of an organism; without sex organs, gametes, or sexual spores; imperfect; vegetative.

Asexual—Without sex organs or sex spores; vegetative.

Ash—The solid, noncombustible residue left after burning.

Asperate—Rough with projections, hairs, spines, warts, etc., or points.

Aspergilliform (of a sporulating structure)—Resembling a condiophore of *Aspergillus*.

Aspergillosis—Any diseases of humans or animals caused by species of *Aspergillus*, especially *A. fumigatus*.

Asperulate, asperulous—Delicately roughened or asperate.

Aspicilioid (of lecanorine apothecia)—More or less immersed in the thallus, at least when young.

A-spore—Alpha spore.

Asporogenic, asporogenous—Not forming spores.

Assimilative (n. **assimilation**)—Pertaining to growth before reproduction; nonreproductive; vegetative; the transformation of foods into living protoplasm.

Association—An assemblage of plants, usually dominated by a few species and representing a major division of a plant formation. *See* Phytosociology.

Astatocoenocytic (of nuclear behavior in Basidiomycotina)—Describing a condition in which haplont mycelium cells are coenocytic, diplont binucleate but coenocytic and lacking clamps when aeration is insufficient, basidiocarp binucleate. This is in contrast to *holocoenocytic* (haplont and diplont coenocytic, with only the developing basidium being binucleate), *heterocytic* (haplont is regularly coenocytic), and the *normal* condition when the haplont is uninucleate and the diplont is binucleate.

Asteroid—Radiating; starlike; "stellate."

Asteroid body—Usually a stellate cell of *Sporothrix schenckii* in animal tissues, resulting from an antigen-antibody complex precipitate formed on the cell wall.

Asteroseta—*See* Cystidium and Seta.

Astomate, astomous—Lacking an ostiole.

Asymmetric (of spores)—Having a flattened or concave side.

Asymmetrical—Lacking symmetry about a designated axis.

Ataxic—Lacking coordination of voluntary muscular movement.

ATCC—*See* American Type Culture Collection.

Atmospheric pollution—*See* Air pollution.

Atom—The smallest particle in which an element combines, either with itself or with other elements; the smallest quantity of matter possessing the properties of a particular element.

Atomate—Having a powdered surface.

Atomize—To reduce a liquid to fine droplets by passing it under pressure through a suitable nozzle, or by applying drops to a spinning disc.

ATP—*See* Adenosine 5′-triphosphate.

Atrophy—To progressively decline in size (of an organ or an entire body); wasting away or arrested development.

Attenuate (adj. **attenuated**)—To weaken or decrease in virulence or pathogenicity; thin, extended, gradually tapered; to draw out, narrowed, more or less to a point; to thin in consistency.

Attenuated virus strain—A selected strain of a virulent virus that does not cause the clinical signs associated with the parent virus but still replicates well enough to induce immunity.

Attenuation—The process of producing an attenuated (weakened) virus strain; reduction in virulence of a pathogen.

Atypical—Not normal, not true to type.

Aucuba—Bright yellow mosaic leaf variegation of genetic or virus origin.

Auleate—A closed fruitbody of a basidiomycete (gasteromycete basidioma) in which pleated plates of trama project into the glebal cavity from the top and sides.

Auricle—Clawlike appendages in grasses that occur in pairs at the base of the leaf blade or at the apex of the leaf sheath.

Aut-eu-form—An autoecious rust with all its spore stages.

Authentic (of specimens, cultures, etc.)—Identified by the author of the name of the taxon to which they were originally referred.

Auto- (prefix)—Self-inducing, -producing, etc.

Autobasidium—Basidium with spores borne symmetrically and forcibly discharged.

Autobody—An antibody with a paratope that reacts with an idiotope on the same antibody as well as with the antigen against which the antibody was generated. The self-binding characteristics of an autobody are inhibitable by antigens.

Autochthonous—Indigenous; organisms continuously active.

Autoclave—An apparatus that sterilizes using pressurized steam. A temperature of 121°C must be maintained for a minimum of 15 min to kill all pathogens.

Autodeliquescent (of lamellae and pileus of *Coprinus*)—Becoming liquid by *autodigestion* in which the spores are freed in drops of liquid end-products.

Autoecious, autoecism—Pertaining to certain parasitic fungi (especially rusts) that complete their entire life cycle on a single host species. *See* Heteroecious.

Autogamy—The fusion of nuclei in pairs in the female organ, without cell fusion having occurred; pollination within the same flower.

Autoimmune—A situation in which an organism, virus, antibody, etc., produces an immune response to its own tissues.

Autolysis—Disintegration of cells by the action of enzymes within them; self-digesting. *See* Lysis.

Automates—Robotic machines.

Automictixis (adj. **automictic**)—Nematode reproduction by parthenogenesis or self-fertilization; males usually rare or absent in a population.

Automixis—Self-fertilization by the fusion of two closely related sexual cells or nuclei.

Autoradiography—A technique that uses X-ray film to visualize radioactively labeled molecules or fragments of molecules; it is used to analyze the length and number of nucleic acid (DNA) molecules after they have been separated by gel electrophoresis.

Autotroph (adj. **autotrophic**)—An organism capable of growth independent of an outside source of food; plant capable of utilizing carbon dioxide or carbonates as the sole source of carbon and obtaining energy for life processes from the oxidation of inorganic elements or compounds or from radiant energy. *See* Heterotroph.

Auxin—A plant growth-regulating substance controlling cell elongation; a natural hormone that regulates plant growth particularly through cell elongation. Auxin is also required for cell division.

Auxotroph—Biochemical mutant of a phototroph that will grow only on the minimal medium after one or more specific substances are added.

Availability threshold period—The time following inoculation of a host plant in which the pathogen can first be acquired by a feeding vector.

Available nutrient—That portion of a molecule or compound in the soil that can be readily absorbed, assimilated, and utilized by growing plants.

Available water—That portion of water in a soil that can be absorbed by plant roots; soil moisture held in the soil between field capacity (FC) and permanent wilting percentage (PWP) (available water = FC − PWP).

Avellaneus—A pale brown color.

Avena test—A test for auxin activity, based on the response of living oat (*Avena*) coleoptiles to the growth-regulating compound.

Avenacein—*See* Enniatin.

Avenacin—A fungus inhibitor from oats (*Avena*).

Aversion—Inhibition of growth at the adjacent edges of colonies of microorganisms, especially in a culture of a single species. *See* Antagonism and Barrage.

Avidity—The stability of the antigen-antibody complex as governed by the affinity of the antibody for the antigen, the valency of the antibody or antigen, and the spatial arrangement of the epitope and respective paratope. An IgM antibody (i.e., 10 paratopes potentially interacting with 10 epitopes) is capable of forming a more stable antigen-antibody complex than an IgG antibody (two paratopes, for example, with the same epitope affinities as IgM).

Avirulent—Non-pathogenic; lacking virulence; unable to cause disease.

Awl-shaped—Fairly long, narrow, and pointed.

Awn—Bristlelike structure at the apex of the outer bract of some cereal and grass flowers.

Axenic or gnotobiotic culture—Growth of a single species of organism without another organism being present; germfree; pure culture; uncontaminated.

Axeny—Inhospitality; "passive" as opposed to "active" resistance of a host plant to a pathogen.

Axial—Belonging to, around, in the direction of, or along an axis.

Axillary—Pertaining to buds, branches, or meristems that occur in the axil of a leaf.

Axil—Upper angle between the petiole of a leaf (or spikelet) and the stem axis.

Axoneme—The main core of a flagellum composed of two central microtubules surrounded by nine double microtubules.

Azonate, azonous—Uniform in color, without distinct zones, furrows or concentric markings.

Azotodesmic—Nitrogen-fixing.

Azygospore—A parthenogenetic zygospore; characteristic of many Mucorales; also formed by some vesicular-arbuscular mycorrhizal fungi (family Endogonaceae).

B

"b"—Body length of a nematode divided by the distance from the anterior end to the junction of esophagus and intestine; often given as a range; a part of deMan's formula.

B cells—Antibody-producing cells found mainly in the blood, lymph nodes, and spleen of the body.

B horizon—A soil layer of varying thickness (usually beneath the A horizon) characterized by an accumulation of silicate clays, iron and aluminum oxides, and humus, alone or in combination, and/or a blocky or prismatic structure.

Baccate—Soft throughout like a berry.

Baccatin—An actibacterial wilt toxin formed by *Gibberella baccata*, anamorph *Fusarium lateritium*.

Bacillar, bacilliform—Shaped like a short, blunt, thick, cylindrical rod, rounded on the ends; bacillus-shaped.

Bacillary layer (of nematodes)—The internal surface covering of the intestinal cells consisting of microvilli that increase the surface area.

Bacillus—A short, rod-shaped, spore-forming, aerobic bacterium; a genus of the family Bacillaceae.

Backcross—In breeding, a cross of a hybrid with one of the parents or with a genetically equivalent organism. In genetics, a cross of a hybrid with a homozygous recessive.

-bacter (suffix)—Used to form many generic names for bacteria.

Bacteremia—The presence of bacteria in the blood stream.

Bacteria—One-celled, prokaryotic organisms. *See* Bacterium.

Bactericidal—Capable of killing bacteria.

Bactericide—A chemical or physical agent that kills bacteria.

Bacteriocin—Protein produced by some bacteria that are able to kill strains of the same or closely related bacterial species.

Bacteriology—The science or study of bacteria.

Bacteriolysin—An agent or substance that causes disintegration of bacteria.

Bacteriophage (phage)—A virus that infects, replicates and causes lysis (death) of bacterial cells.

Bacteriostatic—A chemical or physical agent that inhibits the growth of, but does not kill, bacteria.

Bacterium (pl. bacteria, adj. bacterial)—Microscopic (0.25–2.0 µm), prokaryotic, generally one-celled organism that lacks a nuclear membrane, usually lacks chlorophyll, exists mostly as a parasite or saprophyte, and increases by binary fission (division of a mother cell into two like daughter cells). Some bacteria attack plants and cause such diseases as crown gall, wilts, cankers, rots, leaf spots, blights, and other diseases.

Bactobiont—*See* Photobiont.

Baermann funnel extraction technique—A method of isolating nematodes from soil, screening residue, plant tissue, or other matter in which the material is placed on porous paper or cloth supported in a water-filled funnel; the nematodes by their own action move out of the soil layer or tissue, through the cloth or paper layer, settle, and are drawn off from tubing attached to the funnel stem; there are many modifications.

Bagasse—The dry refuse of sugarcane (*Saccharium officinarum*) after the juice has been expressed; often used as a mulch.

Bakery mold—*Chrysonilia sitophila*, teleomorph *Neurospora sitophila*.

Balanced salt solutions—Solutions having a composition that maintains a balance of the requirements of the cells for which they are providing nutrients. They also control the osmolarity of the nutrient solution.

Balanoid—Acorn-shaped.

Ballistospore—A forcibly discharged basidiospore.

Band application—Application of a spray, dust, or granules in a continuous restricted area, such as along a crop row, rather than over the entire field or area (*broadcast*).

BAP—The Biotechnology Programme of the Commission of European Communities.

Bar—Unit of pressure used to express water potential (1 atm = 1.013 bar; equal to 1 million dyn/cm^2).

Barbate—Bearded; having one or more tufts or groups of hairs.

Bare-root—Describing plants from which the soil surrounding the roots has been removed.

Bark—(Of woody plants) the aggregation of tissues outside the cambium; (of fungi) a compact membrane burst by enlargement of tissues within and renewed at the suture of dehiscence.

Barm—Froth or foam on the surface of fermenting malt liquors; brewer's or baker's yeast.

Barrage—Space between two mycelia that have an aversion or mutual repulsion for each other.

Barrel, U.S.—31.5 gal = 7056 in.^{3} = 0.11924 m^3.

Basal—Related to, located at, or forming near the base of a structure.

Basal body (of Blastocladiaceae)—1) The lower portion of the thallus attached to the substratum by rhizoids; 2) *see* Blepharoplast.

Basal bulb (of nematodes)—A muscular and/or glandular swelling of the esophagus at its posterior end, abutting the intestine. A valvular structure may be present or absent.

Basal cell—(Of rust fungi) a cell that produces several spores; (of the crozier of the Discomycetes) the stalk cell with one nucleus that supports the "loop" or "dome cell" and the "tip" or "ultimate cell" in order.

Basal flanges (of nematodes)—The posterior knoblike expansions at the base of the odontostylet.

Basal frill (of a spore)—The apical part of an annellide or annellophore, conidiophore, or basal cell that is carried away when the conidium detaches.

Basal gland lobe (of nematodes)—Basal glandular region of the esophagus, overlapping the intestine.

Basal knobs (of nematodes)—*See* Stylet knobs.

Basal region of esophagus (of nematodes)—Posterior portion of the esophagus; may be variable in shape, i.e., a distinct bulb, a lobe overlapping the intestine, cylindrical, etc.

Basal ring (of nematodes)—The posterior circular base of the cephalic framework.

Basal whorl—Group of leaves attached at the same point at the base of a plant.

Basauxic (of conidiophores)—Growing and elongating at the base.

Base—An alkaline, usually nitrogenous organic compound; used particularly for the purine (adenine or guanine) or pyrimidine (uramine, thymine, or cytosine) bases forming the structure of RNA and DNA in cells, viruses, and viroids. *See* Nucleic acid.

Base analogue—A compound resembling one of the natural bases of RNA or DNA that is incorporated into newly synthesized nucleic acid by substituting for the normal "base." Example: 5-fluorouracil, which substitutes for uracil.

Base pair—A pair of nucleotides (nitrogen bases) held together by hydrogen bonding and found in double-stranded nucleic acids. DNA contains the base pairs adenine with thymine and guanine with cytosine. *See* Nucleic acid.

Base ratio—The ratio of adenine (A) and thymine (T) to guanine (G) and cytosine (C) in the deoxyribonucleic acid (DNA) in an organism.

Base sequence—The order in which the purine and pyrimidine molecules A, G, C, and U (T) occur along the polynucleotide chain of nucleic acids.

Basic dye—A dye consisting of a basic organic grouping of atoms (cations), the active staining part, combined with an acid, usually inorganic; has affinity for nucleic acids.

Basidiocarp, basidioma, basidiome, carpophore—Sexual fruitbody (sporocarp, conk, mushroom, fructification, etc.) of a basidiomycete on or in which basidia and basidiospores develop.

Basidiograph—A straight-line graph obtained by plotting the ratio of the length (l) to the width (w) against the length of the basidia of a species of agaric.

Basidiole—A basidium-like hymenial element that lacks sterigmata due to being young or permanently sterile; best restricted to immature basidia.

Basidioma (pl. **basidiomata**)—The basidiospore-bearing fruitbody of the Basidiomycotina = basidiocarp, basidiome, carpophore, hymenophore, sporophore.

Basidiomycetes—The class containing fungi characterized by septate mycelium, often with clamp connections, that forms sexual spores (basidiospores, usually four) on a club-shaped basidium; one of the Basidiomycotina, a major group of higher fungi.

Basidiomycotina—The subdivision of fungi containing the basidiomycetes.

Basidiosorus—A sorus comprised of basidia.

Basidiospore—Haploid (1n) sexual spore (typically a ballistospore) produced on a basidium following karyogamy and meiosis; also called sporidium in rust and smut fungi.

Basidium (pl. **basidia**, adj. **basidial**)—Specialized cell or organ, often club-shaped, in which karyogamy and meiosis occur, followed by production of externally borne basidiospores (generally four) that are haploid. There are a number of types of basidia.

Basifugal—developing from the base up; acropetal.

Basinym, basionym, basonym (in nomenclature)—The name- or epiphet-bearing synonym on which a new name or new combination is based.

Basipetal—Toward the bottom; producing structures (such as conidia) in succession from the apex towards the base, making the apical part the oldest. *See* Acropetal.

Basipetally—Successively from apex to base.

Basket fungi—*Clathrus* species.

Basocatenate (of conidia)—Formed in chains, with the youngest conidium at the basal or proximal end of the chain.

Bauch test for smuts—A microscopic test for determining whether monosporidial lines are *compatible* (a mixed culture has aerial white mycelium) or *incompatible* (absence of aerial mycelium).

Beaded—(Of a lamella) having a row of droplets exuded from the edge; (in bacteriology) separate or semiconfluent colonies, in stab or stroke culture, along the line of inoculation.

Beak (of a perithecium or pycnidium)—An elongated neck through which the spores are discharged. *See* Rostrum.

Beard moss—Species of *Alectoria, Bryoria, Ramalina,* and *Usnea*.

Beefsteak (or liver) fungus—The edible wood rot fungus *Fistulina hepatica*, cause of brown oak.

Beeli formulae—Numerical designations for coding the characteristics of *Meliola* species.

Beetle—Any insect of the order Coleoptera characterized by elytra (thickened outer wings), chewing mouthparts, and complete metamorphosis.

Behind (of lamellae)—End nearest the stipe.

Bentonite—A native clay (hydrated aluminum silicate) of the montmorillonite type having the property of viscous suspensions or gels in water. Used as a carrier in dusts; has good sticking properties.

Berry—A simple fleshy fruit formed from a single ovary; the ovary wall fleshy and including one or more carpels and seeds.

Beta-spore (B-spore)—Elongate asexual, hyaline, usually hamate, one-celled spore produced in members of Diaporthaceae (especially *Phomopsis, Melanconium,* and *Coryneum*) and not known to germinate. *See* Alpha-spore.

Betulicole, betulicolous—Growing on birch (*Betula*).

Bi- (prefix)—Two; twice; twofold.

Biallelic (of an incompatibility system)—Having two alleles per locus. *See* Multiallelic.

Biatorine (of lichen apothecia)—Of the lecideine type, pale or more or less pigmented and soft in consistency.

Bibulous (of the surface of a pileus)—Capable of absorbing water.

Bicampanulate—Like two bells arranged mouth to mouth.

Biciliate—Having two cilia.

Biconic (of spores)—Cone-shaped at each end.

Biconvex—Convex on both sides.

Bicornate—Two-horned; two-branched.

Biennial—A plant that develops seed at the end of its second year of growth and then dies; a plant that completes its life cycle within two years.

Bifarious—In two rows or series; distichous.

Bifid—(Of nematodes) divided into two equal lobes or parts by a median cleft; (of fungi) having a split, cleft or division near the middle; forked.

Biflagellate—Have two flagella.

Bifurcate—Divided into two branches; two-forked.

Bifusiform—Elongate with a narrow central isthmus.

Biguttulate—Having two globules or vacuoles.

Bilabiate—Two-lipped.

Bilaminate—Two-layered; applies to spore walls often with an outer pale layer and an inner pigmented layer.

Bilateral symmetry—The condition whereby the two halves of the body (of a nematode) are exact counterparts one of the other (mirror images if divided in one plane); bilaterally symmetrical organisms are characterized by a ventral and dorsal side that are marked by openings; the condition of having distinct and similar right and left sides as in flowers of sweet pea and snapdragon.

Binary fission—Division of a microbial cell transversely into two (usually equal) elements and preceded by division of the genetic material into two parts.

Binary symmetry (of virology)—Combination of cubic and helical symmetry.

Binary system—A system of naming in which binomial nomenclature is used for species.

Binate—In two parts.

Binding or ligative hyphae—Highly-branched, thick-walled, aseptate, interwoven, narrow, often coralloid hyphae in basidiocarps of species with a trimitical hyphal system.

Binding site—Region of a protein in which binding occurs with a compound (e.g., an enzyme or its substrate) whose structure is changed by the activity of the protein; also, the site on a cell to which virus particles adsorb before entry.

Binomial nomenclature or system—The scientific method of naming organisms.

Binomial, binominal—*See* Latin binomial.

Binucleate phase—The dikaryophase.

Binucleate—Containing two nuclei.

Bio- (prefix)—Pertaining to life.

Bioassay—A test involving the response of a living cell or organism to an artificial stimulus; quantitative estimation of biologically active substances by the extent of their actions under standardized conditions on living organisms; the test organism is often used to measure the relative infectivity of a pathogen or toxicity of a substance; determination of the amount of a virus by measuring its biological activity, e.g., infectivity for the host.

Biocide—A compound that is toxic to all forms of life.

Biodegradable—Describing materials readily decomposed by microorganisms such as bacteria and fungi.

Biogenous, biophagous, biophilous—Living on another living organism; parasitic.

Biologic form or race—Physiologic race.

Biological management or control—Disease or pest control through counterbalance of microorganisms and other natural components of the environment; control of pests (bacteria, fungi, insects, mites, nematodes, rodents, weeds, etc.) by means of living predators, parasites, disease-producing organisms, competitive microorganisms, and decomposing plant material, which reduce the population of the pathogen.

Biology—The science or study dealing with living organisms or their fossils.

Bioluminescence—Emission of light by living organisms.

Biomass—Volume (quantity, weight, etc.) of organisms (or living material) in a particular environment (e.g., bacteria, fungi, nematodes, etc., in soil); sometimes extended to the quantity of organic matter in a material (e.g., domestic refuse).

Biome—A large, natural assemblage of associated plants and animals extending over large regions of the earth's surface.

Biorational pesticides—Pest and disease control agents of biological origin including bacteria, fungi, nematodes, viruses, and biochemical products of natural origin (or identical to a natural product).

Bios—A mixture of aneurin (thiamine, vitamin B1), "biotin," and other substances in yeasts that, when added to culture media, give a better growth of yeasts.

Biostat—A substance that causes organisms to stop growing.

Biosynthesis—Production of a chemical compound by a living organism.

Biota—The biological population; all living things in an area.

Biotaxonomy—Classification of living things.

Biotechnology—The use of genetically engineered microorganisms and/or modern techniques and processes with biological systems for industrial production; also, technology concerned with machines in relation to human needs.

Biotic (syn. **animate**)—Of or pertaining to living organisms.

Biotin—A small, water-soluble macromolecule (vitamin H in the B complex or coenzyme R) used as a nonradioactive reporter group for labeling nucleic acid probes.

Biotroph (adj. **biotrophic**)—A parasite that obtains its energy from living host cells, commonly an obligate parasite. *See* Necrotroph.

Biotype—Subspecies group of organisms that differ in one or a few biochemical, physiologic, or behavioral properties; a group of individuals with a like genetic makeup; a subdivision of a pathologic race; a geographical isolate or a clone of the same vector species. *See* Physiologic race.

Bipartite—Divided into two; two-parted.

Bipolar—(Of a spore) at the two ends or poles; (of heterothallic basidiomycetes) with incompatibility controlled by one pair of alleles; a single dikaryon gives rise to two mating types; also referred to as unifactorial. *See* Tetrapolar.

Birch canker—Siberian chaga fungus; sterile basidiocarps of *Inonotus obliquus*.

Birch canker fungus—*Piptoporus (Polyporus) betulinus*.

Bird's nest fungi—The Nidulariales. *See* Splash cup.

Birefringence—Double refraction. Polarization of light when passing through crystals and other substances. Such substances appear bright when viewed with cross polaroid filters in a microscope.

Biserial, biseriate—In two series or rows.

Bisexual—Describing an organism that produces both eggs and sperms, and a flower that bears both stamens and pistils.

Bitubular—Having a tube within a tube.

Bitunicate—A two-walled ascus in which the inner wall is elastic and expands greatly beyond the rigid outer wall at the time of spore liberation.

Biuncinate—Two-hooked.

Biuret—A compound ($H_2NCONHCONH_2, H_2O$) formed by the thermal decomposition of urea (H_2NCONH_2) that is phytotoxic to many plants. Therefore, "biuret-free" urea is used for fertilization of crops.

Bivalvate (of spores)—Lens-shaped with a hyaline rim.

Biverticillate (of a penicillus)—Branching at two levels as in some species of *Penicillium*, i.e., having metulae-bearing phialides.

Black dot (of potato and tomato)—Disease caused by *Colletotrichum coccodes*.

Black knot (of plum and cherry, *Prunus* spp.)—Disease caused by *Dibotryon morbosum*, synonym *Aposporina morbosa*.

Black line—*See* Zone lines.

Black mildews—Genera and species of Meliolales.

Black mold—Especially *Aspergillus niger* and *Rhizopus stolonifer*, which are common laboratory contaminants.

Black or sooty blotch of clovers—Disease caused by *Cymadothea trifolii*.

Black pustule (of *Ribes* spp.)—Disease caused by *Dothiora ribesia*, anamorph *Aureobasidium* sp.

Black root rot (of many plants)—Disease caused by *Thielaviopsis basicola*, syn-anamorph *Chalara elegans*.

Black scurf (of potato)—Disease caused by *Rhizoctonia solani*, teleomorph *Thanatephorus cucumeris*.

Black slime disease (of bulbs)—Caused by *Sclerotinia bulborum*.

Black spot (of rose)—Disease caused by *Diplocarpon rosae*, anamorph *Marssonina rosae*.

Black stem rust (of cereals and grasses)—Disease caused by *Puccinia graminis* and its *formae speciales*.

Black tree lichen—*Bryoria fremontii*.

Black yeasts—*See* Yeasts.

Blackfellows' bread, native bread—The sclerotium of *Mylitta australis* of the Australian *Polyporus mylittae*.

Blackleg (of beet)—Disease caused by *Phoma betae* and *Pythium* spp; of *Pelargonium* (*Pythium* spp.).

Black-tip and leaf spot (of banana)—Disease caused by *Deightoniella torulosa*.

Bladder plum—*See* Pocket plum.

Blade—The flat or expanded part of a leaf or petal.

Blast, blasting—Sudden death of young buds, flowers, or fruits. Also, a failure to produce flowers, fruit, or seeds.

Blastic conidium ontogeny—One of the two basic types of conidiogenesis characterized by the marked enlargement (blown out) of a recognizable conidial initial *before* the initial is delimited by a septum. The conidium is formed from part of a cell.

Blastidium—A lichen propagule formed by yeastlike budding of thalli.

Blastocatenate (of blastoconidia)—Formed in chains with the youngest at the apical or distal end of the chain.

Blastoconidium—*See* Enteroblastic conidiogenesis and Holoblastic conidium ontogeny.

Blastomycin S—An antifungal antibiotic from *Streptomyces griseochromogenes* that is used against rice blast caused by *Pyricularia oryzae*.

Blastomycin—An antigen made from *Blastomyces dermatidis*.

Blastomycosis—Disease of humans caused by *Blastomyces dermatitidis*, teleomorph *Ajellomyces dermatitidis*; any mycotic disease in humans having budding (sprouting) cells in the parasitized tissues.

Blastospore—A spore developed by budding or blowing out of the wall of an existing initial or cell and enlarging before the initial is delimited by a septum. *See* Enteroblastic and Holoblastic.

Bleached—White to straw-colored; used to describe areas of necrotic tissue.

Bleeding—Sap flow from a wound.

Blematogen, blematogen layer—Undifferentiated tissue in agarics that becomes the universal veil.

Blend—A combination of two or more cultures or cultivars of a single plant species.

Blepharoplast (of zoospores)—The basal body or granule (*kinetosome*) from which arise the longitudinal fibers that make up the axoneme of a flagellum; joined to the nucleus by a *rhizoplast*.

Blewits, blewitt, blue leg, bluette—The edible *Lepista saeva*.

Blight—General term for sudden, severe withering and/or killing of leaves, flowers, shoots, fruit, or the entire plant. Usually young growing tissues are attacked. May be coupled with the name of the host part attacked, e.g., bud blight, leaf or needle blight, tip blight, shoot or twig blight, flower blight.

Blind shoot—Flower bud that fails to form and/or develop.

Blind virus passage—Inoculation of material, from an animal or cell culture, that shows no evidence of infection, into a test animal or cell culture, usually with the aim of growing and identifying an infective agent.

Blister rust—A disease of five-needled pines (*Pinus*) caused by *Cronartium ribicola*, alternate host *Ribes* spp.

Bloat—Excessive accumulation of gases in the rumen of some animals.

Blood plasma—The fluid portion of blood minus all blood corpuscles.

Blood serum—The fluid expressed from clotted blood or clotted blood plasma.

Blotch—An irregular necrotic area on a leaf, shoot, stem, or fruit; usually superficial.

Blotting—A technique to transfer DNA, RNA, or protein to an immobilizing matrix, e.g., nitrocellulose, nylon fibers, or diazobenzyloxymethyl paper. *See* Northern blotting, Southern blotting, and Western blotting.

Blue mold—Disease caused by *Penicillium expansum* (of apple), *P. italicum* (of *Citrus*), *Peronospora tabacina* (syn. *P. hyoscyami*) (of tobacco).

Blunt or flush end—Describing DNA fragments generated by certain restriction endonucleases that are perfectly base-paired for their entire length. There are no single-stranded regions after cleavage with the enzyme.

Blusher—The edible *Amanita rubescens*.

BOD—Biological oxygen demand. A measurement of the oxygen required to stabilize the organic matter in sewage or water supplies.

Body pores (of nematodes)—A series of minute, slightly submedial or lateral depressions (or openings) traversing each side of the body in the superfamily Dorylaimoidea.

Bole—The trunk or main stem of a tree.

Bolete—One of the Boletales; fleshy, mushroom-like fungi with a tubular hymenophore.

Boletinoid (of hymenophores)—Having a structure intermediate between pores and gills (lamellae).

Bolting—Rapid production of flower stalks in some herbaceous plants after sufficient chilling or a favorable photoperiod.

Bombysine—Silklike.

Boom—Several nozzles joined together by sections of pipe or tubing to apply pesticides or fertilizer over a wide area at one time.

Booster dose—A dose of immunogen given several days, weeks, months, or years after the initial immunizing dose to stimulate continued production of antibody.

Boot—Sheath of the uppermost leaf that encloses the head of a cereal or grass plant.

Booted—Sheathed by the volva or universal veil; "peronate."

Bordeaux mixture—A spray developed about 1880 near Bordeaux, France, to combat downy mildew (*Plasmopara viticola*) of grape (*Vitis*). It is still used somewhat for controlling a wide range of diseases. A common mixture is 4-4-50 (4 lb of copper sulfate, 4 lb of hydrated lime, and 50 U.S. gal of water [1 imperial gal = 1.2 U.S. gal]). The copper sulfate is dissolved in about a third of the water; lime is added to the rest of the water; and the two liquids then mixed. Application should start and finish as soon as possible to avoid plugging of nozzles. Homemade mixtures are desirable, but prepared mixtures are available. Bordeaux mixture leaves a conspicuous residue on plant surfaces and may be injurious to copper-sensitive plants in cold, wet weather.

Boreal—Northern.

Borer—Insect or insect larva that forms tunnels within the wood of trees or stems of herbaceous plants.

Botany—The science or study of plants, including their characteristics, functions, life cycles, and habits.

Botryo-aleuriospore—One of an apical cluster of aleuriospores formed basipetally from the conidiogenous cells.

Botryo-blastospore—One of a cluster of conidia borne on the swollen apex (*ampulla*) of a conidiogenous cell, either singly or in chains.

Botryose—Clustered or grouped like grapes; racemose.

Botrytis (gray mold)—Genus name for a widespread fungus (*B. cinerea*) that causes blight of many flowers and damping-off of seedlings. A dense gray mold grows on fading flowers and other affected parts in damp weather. The mold is composed largely of microscopic, grapelike clusters of spores (conidia).

Botuliform—Cylindrical with rounded ends; sausage-shaped; allantoid.

Bouillon—Meat broth used in or as a culture medium.

Bovine serum albumin (BSA)—The major protein constituent of bovine serum.

Brace root, anchor root—A type of adventitious root that grows from above-ground parts of the stem and serves to support some plants, e.g., maize or corn.

Brachy- (prefix)—Short.

Brachymeiosis—The third division, once claimed to occur in the ascus.

Bracket (of fungi)—A shelflike basidiocarp.

Bract—A reduced leaf associated with a flower or inflorescence; modified leaf from the axil of which a flower arises.

Bramble—A cane bush (e.g., raspberry, blackberry [*Rubus*]), usually with spines. Fruit is an aggregate (commonly referred to as a berry).

Branch gap—An interruption in the vascular tissue of a stem at the point where a branch trace arises.

Branch trace—Vascular tissue that passes from a stem into a branch.

Brand spore—Urediniospore; smut spore.

Bread mold—*Monilia sitophila, Aspergillus* spp., *Rhizopus* spp., specifically *R. stolonifer, Mucor* spp., etc.

Breaking—Disease symptom usually caused by a virus; addition or loss of flower color to create a variegated pattern. *See* Flower break.

Breeder seed—Seed (or vegetative propagating material, e.g., potato) increased by the originating, or sponsoring, plant breeder or institution and used as the source to increase foundation seed.

Brevicollate—Short-necked.

Bridging hypha—A special branch hypha joining two other hyphae.

Bright-field microscope (light microscope)—A microscope that uses transmitted light.

Bristle—Seta.

Brittle (in bacteriology)—Fragile growth easily broken with an inoculating needle.

Broadcast—To scatter seed or fertilizers uniformly over the soil surface rather than placing in rows.

Broadcast application—Application of a spray, dust, or granules uniformly over an entire area. *See* Band application.

Broadleaf or broad-leaved—Any plant with a flat leaf and netted veins other than a grasslike plant.

Broad-spectrum pesticide—A chemical that controls a wide range of pests.

Bromthymol blue—A pH indicator with a range of 6.0–7.6.

Bronchomycosis—*See* Mycosis.

Brood—Eggs and young animals (especially insects); a generation of an insect species.

Broom cells (of agarics)—Cells with apical appendages giving a broomlike appearance on the pileus or the edges of lamellae.

Brooming—Profuse branching of woody stems from a single stem position; cluster of stunted, slender twigs in crowded bunches from a stem with shortened internodes (a witches' broom).

Broomrape—A whitish to yellowish, annual, leafless herb (family Orobanchaceae, the broomrape family) growing as a parasite on the roots of another plant.

Broth—A liquid nutrient culture medium, especially one containing meat extract.

Brown oak—Oak wood stained by *Fistulina hepatica.*

Brown rot (of wood)—Decay resulting from selective removal of cellulose and hemicellulose, leaving a brown, amorphous residue that usually cracks into cubical blocks and consists largely of slightly modified lignin.

Brown rot fungi (of fruit)—Species of *Monilinia (Sclerotinia)* causing fruit rots and other damage to fruit trees, especially *Prunus* spp.

Brownian motion—A characteristic "dancing motion," exhibited by finely divided particles and bacteria in suspension, due to the bombardment by the molecules of fluid.

Brushing—Using riverbank brush for constructing a barrier for sheltering crop plants from wind and for modifying the plant's microenvironment in order to accelerate growth.

Bryicolous—Fungi growing on bryophytes (Bryophyta = mosses and liverworts).

B-spore (Beta-spore) A lineate, sometimes curved spore. *See* Alpha-spore.

Buccal aperture (of nematodes)—*See* Oral aperture.

Buccal capsule (of nematodes)—*See* Stoma.

Buckeye rot—Disease of tomato fruits caused by *Phytophthora nicotianae* var. parasitica.

Buckle—Clamp connection.

Bud—A terminal or axillary structure on a stem consisting of a small mass of meristematic tissue, generally covered wholly or in part by modified scale leaves.

Bud scale—A specialized protective leaf of a bud.

Bud scar—A scar left on a twig by the falling away of a bud or a group of bud scales.

Bud sport—A mutation arising in a bud and producing a genetically different shoot. Includes changes due to gene mutation, somatic reduction, chromosome deletion, or polyploidy.

Bud trace—The microscopic vascular connection between stem and bud.

Budbreak—Resumption of active growth by resting buds.

Budding—A special type of grafting using a single bud as a scion; method of vegetative (asexual) reproduction of plants by implanting a stem bud of a desired species or cultivar onto the rootstock of another plant, often a different species or cultivar; (of fungi) the development of a spore from a small outgrowth (bud) from a parent fungal cell; a method of asexual reproduction in unicellular fungi or in spores; (of viruses) method of release of enveloped virus particles from the cells in which they have grown.

Buff—A pale tan to creamy gray or yellow color.

Buffer—A mixture of an acid and a base in a solution capable of maintaining hydrogen-ion concentration and thereby avoiding rapid changes in acidity or alkalinity of a solution.

Bug (true bug)—Any insect of the order Hemiptera (especially suborder Heteroptera), characterized in part by piercing-sucking mouthparts, a triangular scutellum, two pairs of wings, and gradual metamorphosis. It is incorrect to apply this term to all insects or creatures that resemble insects.

Bulb—A short, flattened, usually globose or disc-shaped, underground, perennial, storage organ composed of concentric layers of overlapping fleshy scale leaves attached to a stem plate at the base; essentially a subterranean bud.

Bulbil—A number of cells aggregated into a small sclerotium-like structure but lacking a distinct rind layer.

Bulbillate (of a stipe)—Having a small or obscure bulb at the base.

Bulbillosis (of Agaricales)—Condition in which basidiocarp sporulation is suppressed and the basidial function is assumed by bulbils, as in *Rhacophyllus*.

Bulblet—Small, immature bulb produced on the stem in the axils of leaves, or below the soil line or at the base of an older, "mother" bulb.

Bulboid—*See* Esophagus.

Bulbous—Bulb-like; a stipe enlarged at the base.

Bulk density (soil)—The mass of a known volume (including air space) of soil. The soil volume is determined in place, then dried in an oven to constant weight at 105°C. Bulk density (D_B) = oven dry weight of soil/volume of soil.

Bulla (pl. **bullae**)—A dark colored, irregularly shaped structure in nematodes located within the vulval cone of cysts of some *Heterodera* species near the underbridge or fenestra.

Bullate, bulliform—Having bubble- or blisterlike swellings; (of a pileus) having a rounded knob at the center.

Bunch-type growth—Plant development at or near the soil surface without production of rhizomes or stolons.

Bundle scars—Scars left where conducting strands (vascular bundles), passing from the stem into the petiole, were broken during leaf fall.

Bunt—A wheat and grass disease caused by *Tilletia caries* and *T. laevis*; stinking smut. *Dwarf bunt* is caused by *Tilletia contraversa*.

Bunt ball—Smut sorus that replaces a cereal or grass kernel but is covered by plant tissue at maturity.

Buoyant density—The density that a virus or other macromolecule possesses when suspended in an aqueous solution of a heavy metal salt, calcium chloride, or a sugar such as sucrose; the density at which the macromolecule is in equilibrium and neither floats nor sinks.

Burgundy truffle—*Tuber uncinatum*. *See* Truffle.

Burn—The condition in which the cells of a host plant become reddish to dark brown and collapse.

Bursa (pl. **bursae**)—Paired lateral cuticular fin- or winglike extensions (caudal alae) at or near the posterior end of some male nematodes used to clasp the female during copulation.

Bursiculate, bursiform—Bag-, purse-, or pouchlike.

Bushel, U.S.—4 pecks = 32 qt = 64 pt = 128 cups = 1.2445 ft^3 = 35.2383 liters = 0.304785 barrel = approximately 1/20 cu yard.

Butt rot—Decay in the heartwood of the basal part of a living tree.

Button—The "head" of an immature mushroom.

Butyrous (of bacteriology)—Describing growth of a butterlike consistency.

Byssisede—Seated on a cottony subiculum.

Byssoid, byssine (of subiculum, context, or marginal tissue of basidiocarps)—Soft, cottony, made up of delicate threads; floccose.

C

"c"—Body length of nematodes divided by the tail length, i.e., the distance from the anus or cloaca to the terminus; usually given as a range; a part of deMan's formula.

°C—Celsius (formerly Centigrade); unit of temperature between boiling and freezing points of water at a standard pressure. *See* Celsius and Fahrenheit.

C horizon—A soil layer beneath the B layer that is relatively little affected by biological activity and pedogenesis and is lacking in properties diagnostic of an A or B horizon; parent material from which upper horizons developed.

C region = Fc.—Carboxyterminal portion of immunoglobulin H or L chains that is identical for a given class or subclass.

C_3 cycle—The Calvin-Benson cycle of photosynthesis in which the first products after CO_2 fixation are three-carbon molecules.

C_4 cycle—The Hatch-Slack cycle of photosynthesis, in which the first products after CO_2 fixation are four-carbon molecules.

Cacomatoid—Refers to aecia, uredinia, or telia that resemble the anamorphic genus *Caeoma* in having catenulate spores but no peridium.

Cadavericole—An organism living on corpses.

Caducous (of spores, leaves, etc.)—Tending to shed; falling off or dropping readily; transitory; deciduous; perishable, e.g., deciduous leaves, sporangia of *Phytophthora infestans*.

Caecum (of nematodes)—A cavity open at one end, as the blind end of a lumen or duct (refers mostly to an outpocketing of the intestine at the anterior end).

Caeoma (pl. **caeomata**)—An aecium without peridial walls and with or without paraphyses, e.g., *Caeoma*.

Caeomoid—With no well-defined peridium.

Caerulous, coeruleous—Sky blue.

Caesar's mushroom—The edible *Amanita caesarea*.

Caespitose, caespitous, cespitose—Aggregated in dense groups, clumps, or tufts like grass and frequently arising from a common stroma; gregarious; (of stipate basidiocarps) developing in clusters; stems usually more or less united at the base.

Caespitulus (pl. caespituli)—Tuft of spores.

Calcarate—Having a projection or spur.

Calcareous, calciferous—Containing calcium carbonate (lime) or lime compounds; (of color) chalk-white; chalky.

Calcareous soil—Soil containing sufficient calcium and/or magnesium carbonate to effervesce visibly when treated with cold 0.1 normal (0.1*N*) hydrochloric acid.

Calceiform, calciform, calceolate—Shoe- or slipper-shaped.

Calcicole (adj. **calcicolous**)—Organism growing on a substrate rich in calcium; especially of lichens on limestone or chalky rocks or soils.

Calciform, calycular—Cup-shaped.

Calcined clay—Clay minerals, such as montmorillonite and attapulgite that have been fired at high temperatures to obtain absorbent, stable, granular particles; used as an amendment in soil modification.

Calcium pectate—An organic calcium compound found in the middle lamella between plant cells and serving as an intercellular cement.

Calico—Distinctive yellow leaf pattern associated with certain plants infected with a virus.

Callorites—Fossil fungi.

Callose (n. and adj.)—Hard or thick, sometimes rough; a carbohydrate component of plant cell walls often forming over sieve plates and in calcified cell walls; a glucan formed in response to injury.

Callosities (of fungi)—Wall thickenings associated with the penetration of hyperparasites. *See* Papillae.

Callus (adj. **callous**)—Tissue overgrowth (mass of large, thin-walled, undifferentiated cells) around a wound or canker, formed in response to injury, and in tissue culture, developing from cambium or other parenchyma cells with meristematic potential.

Calorie (gram calorie)—Unit for measuring energy; defined as the heat necessary to raise the temperature of 1 g of water from 14.5 to 15.5°C at standard pressure; 1 kilocalorie (kcal) raises the temperature of 1 kg of water from 0 to 1°C. Thus, 1 kcal = 1,000 cal.

Calvacin—Non-diffusible, mucoprotein antibiotic from *Langermannia gigantea* that is active against tumors in mice, rats, and hamsters.

Calvescent—Becoming bare or bald.

Calvous—Naked, bare.

Calyciform, calyular—Cup-shaped.

Calyculus (of Myxomycetes)—Cup- or calyxlike structure at the base of the sporangium.

Calyptra—Cap or hood.

Calyx (pl. **calyxes**)—Outermost whorl of flower parts; sepals collectively.

Cambium, cork—A lateral meristem that produces cork.

Cambium, vascular—A one- or two-celled meristematic layer between the phloem and xylem that produces both these tissues and results in secondary diameter growth. If the cambium is destroyed, as may occur in banding trees, the plant dies.

Campanulate, campanuliform, campanaceous—Bell-shaped.

Campestroid (of agarics)—Having a pileus with a diameter:stipe ratio of 1 or >1. *See* Placomycetoid.

Canaliculate—Having a longitudinal groove.

Cancellate—Reticulate; like a network.

Candelabrum (pl. **candelabrums**)—Clustered group of basidia formed by repeated branching of a single subhymenial generative hypha.

Candicant, candidous—Shining-white.

Candicidin—An antifungal and antibacterial antibiotic produced by a strain of *Streptomyces griseus*.

Candidiasis, candidosis—A cosmopolitan disease of humans and animals caused by species of the fungus *Candida*, especially *C. albicans*; moniliasis.

Candle-snuff fungus—*Xylaria hypoxylon*.

Canescent, canous—Becoming gray or hoary.

Cane—The externally woody, internally pithy, usually flexible stem of certain plants, e.g., brambles, roses, and vines; (of nematodes) a thickening with a lack of ornamentation of the posterior cuticle.

Caninoid venation (of lichens)—A condition in which the hyphal strands are separated to the tips of the lobes.

Canker—A definite, localized, usually dry, dead, often discolored, sunken, raised, or cracked area (necrotic lesion) on a stem, twig, branch, limb, or trunk surrounded by living tissues. Cankers may girdle affected parts, resulting in a dieback starting from the tip. Usually differs from *rot* in having a definite line of demarcation. Cankers may be caused by pathogens or by injuries of various types.

Canker, annual—A canker that enlarges only once and does so within an interval briefer than the growth cycle of the plant, usually less than one year.

Canker, diffuse—A canker that enlarges without characteristic shape or noticeable callus formation at its margins.

Canker, perennial—A canker that enlarges during more than one year.

Canker, target—A canker that includes concentric ridges of callus.

Canopy—Expanded leafy top of a plant or plants.

Cap—(Of fungi) structural part of higher basidiomycete fungi that supports the hymenium of basidia. (*See* Pileus.) (Of viruses) a sequence of methylated base at the 5′ terminus of an eukaryotic mRNA molecule joined in the opposite orientation, i.e., 5′ to 5′ instead of 5′ to 3′. The cap interacts with various proteins involved with the initiation of translation.

Cap cell (of nematodes)—Apical cell at the distal end of female and male gonads; the primordial germ cell.

Capillaceous, capilliform—Hair- or thread-like; filiform.

Capillary water—Water retained in the spaces among and on the surfaces of soil particles after drainage with a tension greater than 60 cm of water.

Capillitium (pl. **capillitia**)—Mass of sterile, threadlike structures (tubes or fibers) present among the spores in the fruitbodies of many Myxomycetes and Gasteromycetes.

Capitate hyphopodium—*See* Hyphopodium.

Capitate-fastigiate (of macrolichens)—Having a thallus cortex of parallel, erect hyphae with swollen, pigmented, apical cells at the tip.

Capitate—Having a well-formed head, usually of conidia; (of hypha, cystidium, or seta) having a bulb-like swelling or knob at the apex and resembling a drumstick.

Capitellum (pl. **capitellitia**)—A little head of any sort.

Capitulum—(Of nematodes) a sclerotized guiding piece for the gubernaculum on the ventral cloacal surface of some Hoplolaimae; (of fungi) a stalked, globose, apical lichen apothecium, as in the Caliciales. *See* Mazaedium.

Capnetic—Describing organisms that grow or grow better under conditions of increased carbon dioxide (CO_2).

Capping or granylation—Addition of the methylated base to the primary transcript in the nucleus while it is being spliced and polyadenylated. *See* Cap.

Capsid polypeptide—The protein-forming part of the capsid structure of a virus particle.

Capsid—The protein coat of a virion forming a closed shell or tube that contains the nucleic acid (either DNA or RNA) and consists of protein subunits or capsomeres. The capsid and nucleic acid form the nucleocapsid.

Capsidiol—A phytoalexin from bell or sweet pepper (*Capsicum frutescens*).

Capsomer(e)—Also called a protein subunit; a small protein molecule that is the structural and chemical unit of the protein coat (capsid) of a virus. It is built up from varying numbers of protein subunits (polypeptide chains).

Capsulate—Enclosed in a capsule.

Capsule—A hyaline, gelatinous, relatively thick layer of mucopolysaccharides (and occasionally protein) that surrounds some kinds of bacterial and yeast cells; a simple, dehiscent fruit of the pod type, composed of two or more carpels, that is dry at maturity, usually opening to disperse seeds in a dry condition; the spore case of a moss or liverwort; (of nematodes) a membrane or saclike form enclosing a structure or organ.

Carbamate insecticide—A synthetic compound derived from carbamic acid. Carbamate insecticides are contact killers with relatively short-lived effects. Examples: Carbaryl and bendiocarb.

Carbohydrases—Enzymes that digest (hydrolyze) carbohydrates.

Carbohydrate—A compound (foodstuff) comprised of carbon, hydrogen, and oxygen (CH_2O) with the last two frequently in a 2 to 1 ratio; various chemical compounds of carbon, hydrogen, and oxygen, such as sugars, starches, and celluloses.

Carbon dioxide fixation—The addition of H^+ to CO_2 to yield a chemically stable carbohydrate. The H^+ is contributed by NADPH, the reduced (hydrogen-rich) form of $NADP^+$, produced in the noncyclic phase of the light reactions of photosynthesis. The H^+ comes originally from the photolysis of water.

Carbonaceous, carbonous—Hard, black, brittle and easily broken, resembling charcoal or cinders.

Carbonicole, carbonicolous—Living on burned-over soil; pyrophilous.

Carboniferous period—A period of the Paleozoic era, characterized by the formation of great coal beds.

Carbon-nitrogen ratio—The ratio of the weight of organic carbon to the weight of total nitrogen in a soil or in organic material.

Carboxymethylcellulose—A cellulose derivative used to separate proteins by ion exchange chromatography.

Carboxyphilic—Liking or favored by carbon dioxide.

Carcinogen (adj. **carcinogenic**)—A substance or agent capable of producing or inciting cancer in experimental animals or known to do so in humans.

Cardia (of nematodes)—A valvular apparatus connecting the esophagus and intestine (sometimes called the cardiac valve or esophago-intestinal valve) that functions to prevent regurgitation..

Cardiac glands (of nematodes)—Three glandular bodies located at the base of the esophagus.

Cardinal temperatures—The minimum, optimum, and maximum temperatures for growth or germination of an organism.

Carinate—Furnished with a keel; boat-shaped.

Cariose, carious—Decayed or decaying.

Carioso-cancellate—Becoming latticed by decay.

Carminophilic (of basidia)—Becoming densely granular after treatment with aceto-carmine stain.

Carna-5 RNA (syn. **cucumber mosaic virus RNA 5**)—A satellite RNA of CMV dependent on the remainder of the CMV genome for its own replication, but not essential for the replication of the CMV particle. *See* Genome and Satellite.

Carnose, carnous—Fleshy-appearing.

Carnulose—Somewhat fleshy.

Carotenes—Yellow pigments found in plant cells, precursors of vitamin A. Alpha, beta, and gamma carotenes are converted into vitamin A in the animal body.

Carotenoids—A large class of related polyene compounds that includes carotenes and xanthophylls; mostly with C_{40}, yellow, red, or (rarely) colorless.

Carotiform—Carrot-shaped.

Carpel—A simple pistil, or a member of a compound pistil; the ovule-bearing structure of a flower in angiosperms; often regarded as a single, modified, seed-bearing leaf.

Carpogenous—Living on fruit.

Carpogonium—Female gametangium of a red alga, sometimes of fungi (e.g., powdery mildew fungi, Erysiphaceae).

Carpophore—Stalk of a sporocarp, *or* a sterile, sporocarp-like body of unknown function; sometimes equals basidioma.

Carpophoroid (of agarics)—Sterile carpophore-like body of unknown function.

Carrier—Plant, animal, or human harboring an infectious disease agent (e.g., virus, bacterium) but not showing marked symptoms. A carrier plant can be a source of infection to others. An insect contaminated externally with an infectious agent (e.g., bacterium, fungus, virus, nematode) sometimes is called a carrier. *See* Vector. Also, a liquid or solid inert material (e.g., dust, clay, oil, water, air) added to the pesticidal compound to prepare a proper formulation and to improve its physical dispersion, spread, and adsorption, or absorption.

Cartilaginous—Firm and tough, like cartilage, not fibrous. Cartilaginous stems break cleanly with a snap, when bent sharply.

Cartilaginous layer—Sometimes applied to the sterome in *Cladonia* and the chondroid axis in *Usnea*.

Caruncle—A spongy structure at one end of a seed such as a castor bean (*Ricinus communis*).

Caryallagia (of reproduction)—Not having nuclear change, as in clone development.

Caryallagic (of reproduction)—Having nuclear change.

Caryogamy (karyogamy)—*See* Karyogamy.

Caryopsis—Small, one-seeded, dry, indehiscent fruit with the seed coat and a thin pericarp completely united and adhering to the seed; the "seed" (grain) or fruit of grasses (Poaceae).

Caryo-—*See* Karyo-.

Casparian strip—A secondary thickening that develops on the radial and end walls of some endodermal cells.

Cassideous—Helmet-shaped.

Casting—Premature loss of abscised leaves or twigs.

Castings, earthworm (wormcasts)—Soil and plant remains excreted by earthworms that are deposited at the soil surface or in the burrow; forms a relatively stable soil granule that can be objectionable in some places (e.g., a golf green).

Catabolism—The dissimilation, or breakdown, of complex organic molecules. A part of the total process of metabolism.

Catahymenium—A hymenium in which hyphidia are the first-formed elements and the basidia are embedded at various levels, elongate to reach the surface, and then form a palisade.

Catalase—Enzyme of qualitative significance in plants and some bacteria, converting hydrogen peroxide into water and oxygen.

Catalyst—Any substance that accelerates a chemical reaction without entering or being used up in the reaction itself.

Cataphyses—Pseudoparaphyses. *See* Hamathecium.

Catathecium, catothecium—A flattened ascoma with a wall more or less radial in structure and with a basal plate. *See* Thyriothecium.

Category—The equivalent of taxonomic rank in a hierarchical system. The species or the genus is a category while a taxon is a taxonomic group.

Catenate, catenulate—Developing in chains, or in an end-to-end series. *See* Arthrocatenate, Basocatenate, and Blastocatenate.

Catenuliform—Chainlike.

Caterpillar—The "worm-like" stage (larva) of a moth or butterfly.

Caterpillar fungi—Species of *Cordyceps* on larvae of caterpillars and other insect larvae and pupae.

Cation—A positively charged ion; in contrast to anion.

Cation exchange capacity (base-exchange capacity)—A measure of the total amount of exchangeable cations that a soil can hold; expressed in meq/100 g soil at pH 7.

Cationic detergent—A detergent having positively charged surface ions such as molecules containing a quaternary ammonium ion with a group of 12–24 carbon atoms attached to the nitrogen atom in the cation. Example: Cetyl-trimethyl ammonium bromide or CTAB.

Catkin—A type of flower cluster (spike), generally bearing either female (pistillate) flowers or male (staminate) flowers; conelike male or female fruit of angiosperms.

Catothecium—*See* Catathecium.

Cat's ear—Basidiocarp of *Clitopilus passeckerianus*, an invader of mushroom beds.

Cauda—Tail or tail-like appendage.

Caudal—Pertaining to or located near the posterior region or tail.

Caudal alae (bursa)—*See* Alae and Bursa.

Caudal glands (of nematodes)—Three to five hypodermal glands located in the tail and emptying subterminally or terminally through a pore, the spinneret. These glands occur in some of the Adenophorea; they secrete a fluid that hardens to form an attachment thread in water.

Caudal supplements (of nematodes)—Papillate glandular structures on the ventral surface of male Adenophorea in the caudal region; may serve the same purpose as the bursa.

Caudalid (of nematodes)—A narrow, clear, bandlike structure below the cuticle, located slightly anteriad of the anus in some Tylenchida (similar to hemizonid or hemizonion); may correspond to the paired anolumbar commissures linking the preanal ganglion to the lumbar ganglia.

Caudate—Having a tail.

Caulescent—Having a stem; becoming stemmed (stalked).

Caulicole, caulicolous—Living on herbaceous stems.

Caulocystidium—*See* Cystidium.

Cauloplane—The stem surface.

Causal agent (or organism)—The organism or agent (bacterium, fungus, mycoplasma, nematode, virus, viroid, etc.) that incites a given disease or injury.

Cavernose, cavernous—Having hollows or cavities.

Cavernula (pl. **cavenulae**)—Cavity.

cc—Cubic centimeter (now expressed cm^3); equivalent to 1.0 ml of water under standard conditions = 0.06102 $in.^3$ = 1,000 ml^3 = 0.0000011 m^3.

cDNA—Abbreviation for complementary DNA that is synthesized from a messenger RNA template.

Cecidium—Plant gall, usually caused by an animal (*zoocecidium*), especially an insect or, more rarely, a fungus (*mycocecidium*).

Cedar apple—A popular term given to the hard, brown, rust gall produced on junipers by the cedar-apple rust fungus (*Gymnosporangium juniperi-virginianae*).

Cell—The structural and functional unit of all plant and animal life. The living organism may have from one cell (bacteria) to hundreds of billions of cells (a large tree). The essential feature of a cell is its living protoplasm limited by a membrane. It is surrounded in plant cells by a wall.

Cell culture—A culture in a liquid or soft gel medium *in vitro*. *See* Organ culture, Tissue culture.

Cell division—A process whereby cells reproduce.

Cell fusion—The formation by fusion of cell membranes of multinucleate giant cells (*syncytia*).

Cell line—A culture of cells that can be subcultured indefinitely.

Cell manipulation—*See* Genetic engineering.

Cell membrane—The outer membrane of the cytoplasm, next to the cell wall, that regulates the flow of material into and out of the cell.

Cell plate—A thin partition formed between daughter nuclei in a cell undergoing *cytokinesis*; the precursor of the cell wall, formed as cytokinesis starts during cell division. It develops in the region of the equatorial plate and arises from membranes in the cytoplasm.

Cell sap—The liquid in the vacuoles of plant cells.

Cell wall—Protective, resistant, but permeable polysaccharide structure secreted externally to the cell membrane in plants and certain other organisms; bacteria and fungi typically have cell walls with other components; the barrier that develops between nuclei during mitosis.

Cellar fungus—*Coniophora puteana* or *Rhinocladiella ellisii*.

Cellular (appendage)—An appendage, at least when immature, with cellular contents and a cell wall, later frequently lacking contents; originates as an outgrowth of the cell and maintains cellular continuity, although sometimes delimited by a septum.

Cellular immunity—Immunity ascribed to various cellular functions other than those that produce antibody.

Cellular response—That part of the immune response that involves the interaction between cells and an antigen.

Cellulase—An extracellular enzyme capable of breaking the cellulose molecule into fragments of lesser weight, or ultimately into basic units (i.e., glucose).

Cellulin—A chitin-glucan complex that occurs as granules in the cells and plugs of fungi.

Cellulin plugs—Plugs that occur at hyphal constrictions in the Leptomitales.

Cellulolysis adequacy index (of a fungus)—An estimate, derived from dividing the rate of cellulolysis by the mycelial growth rate on an agar plate, as to whether the rate of cellulose decomposition is adequate to supply the needs for saprophytic survival.

Cellulolytic (of certain enzymes)—Able to digest cellulose.

Cellulolytic fungi—Fungi able to utilize cellulose-containing substances (including plant cellulose, paper, cloth, etc.).

Cellulose—A complex polysaccharide that is composed of hundreds of beta-glucose molecules (polymers) linked in an unbranched chain and that makes up 40–55% by weight of plant cell walls.

Ceno- (prefix)—*See* Coeno-.

Cenozoic—The geologic era extending from about 65 million years ago to the present time.

Center of origin—A geographical area in which a species is thought to have evolved through natural selection from its ancestors.

Centimorgan—A unit of measurement of recombination frequency; one centimorgan is equal to a 0.01 chance that a genetic locus will be separated from a marker because of recombination in a single generation. In humans, a centimorgan equals, on average, one million base pairs.

Central—Occupying a midposition, e.g., a spore in the center of a sporangium.

Central body (of ascomycetes)—Cell structure, or central apparatus, from which astral rays emanate and initiate a cleavage of the cytoplasm.

Centric, central (of a stipe)—At the center of the pileus; (of oogonium of Saprolegniaceae) having one or two layers of fat droplets surrounding the central cytoplasm; *subcentric*, having the cytoplasm surrounded by a layer of droplets on one side with two or three layers on the other side; *excentric*, having one large drop or a lunate row of droplets on one side.

Centrifugal—From the center outward; around the margin.

Centrifuge—An apparatus used to separate or remove particulate matter (including viruses), suspended in a liquid, by centrifugal force.

Centriole—An organelle to which spindle tubes are attached to chromosomes during nuclear division. They are typically present in animal cells and absent in plant cells; may be present in other organisms.

Centripetal—Toward the center.

Centromere—The constricted portion of a chromosome to which, in mitosis, the chromosomal fiber is attached.

Centrum—The structures within an ascoma, i.e., asci and hamathecium.

Cep—The edible *Boletus edulis*.

Cepaceous—*See* Alliaceous.

CEPH—Abbreviation for Centre d'Etude du Polymorphism Humaine.

Cephalic (of nematodes)—Pertaining to or located in the head region.

Cephalic framework (of nematodes)—The sclerotized framework in the head region that provides rigidity. It is subcuticular and axially forms a stylet guide in stylet-bearing forms.

Cephalids (hypodermal commissures)—Two highly refractive structures (posterior and anterior) situated in the cephalic region of a nematode and extending in a complete circle around the body, possibly part of the nervous system.

Cephalodium (pl. **cephalodia**)—Delimited internal region or a warty, squamulose, or fruticose structure on the surface of a lichen thallus containing an alga (usually blue-green) different to that characteristic of the remaining thallus (usually green alga).

Cephalosporins—Antibacterial antibiotics from *Emericellopsis minimum*, anamorph *Acremonium* sp.

Ceraceous, cereous—Waxy.

Ceranoid—Having hornlike branches.

Cereal—Grass grown primarily for its edible seed.

Cereal forage—Cereal crop harvested when immature for hay, silage, green chop, or pasturage.

Cerebriform, cerebroid, cerebrose—With brainlike convolutions or folds.

Cernuous—Drooping; hanging down; nodding.

Certification of seeds, transplants, cuttings, other plant parts, or nursery stock—Seeds, plant parts or plant material, the progeny of foundation, registered or certified seed, or other plant material produced and sold under inspection control to maintain genetic (varietal) identity and purity and freedom from harmful diseases, insect and mite pests, and weed seeds. It is approved and certified by an official certifying agency.

Cervical (of nematodes)—Located in the neck region.

Cesium chloride density gradient centrifugation—Method for separating viruses or macromolecules according to their density. Sedimentation ceases when the molecules reach the position in the gradient that is the same as their own buoyant density. *See* Isopycnic density.

Cespitose—*See* Caespitose.

CFT—Abbreviation for complement fixation test.

CFU—Abbreviation for colony-forming unit. The progeny of bacteria or fungi resulting from the deposit of a single cell or group of cells onto a nutrient medium; the number of colonies formed per unit of volume or weight of a cell or spore suspension.

CH_{50} unit—The amount of serum (dilution) required to inhibit lysis of antibody-coated red blood cells by 50%, in a standard assay for hemolytic complement.

Chaff—Nonseed portion of a mature grass or cereal head.

Chaga fungus—*See* Birch canker.

Chain (unit of length) = 100 uniform links = 66 ft (Gunter's or surveyor's chain) or 100 ft (engineer's chain).

Chains (of bacteriology)—Four or more cells attached end to end.

Chalky (of basidiocarp tissue)—Crumbly and brittle in consistency.

Channel (phialide)—Zone at the phialide apex in which the area occupied by the protoplast is usually of much smaller diameter and surrounded by periclinal thickening. The conidial primordium emerges through the channel to produce a new conidium.

Chantarelle—The edible *Cantharellus cibarius*.

Character—The expression of a gene in the phenotype.

Chartaceous—Papery.

Cheesy—Soft and easily cut, uniform in texture.

Cheilocyst, cheilocystidium—*See* Cystidium.

Cheilostome (of nematodes)—The anterior portion of the stoma.

Cheiroid, chiroid—Roughly the shape of a hand with the fingers together.

Chelating (or sequestering) agent—A large organic compound such as sodium ethylenediaminetetraacetic acid (EDTA) that attracts and tightly binds with specific bivalent (divalent) and trivalent metallic cations. Often used to correct nutrient deficiencies, to inhibit biological interactions that require bivalent ions, to assist in virus purification, and for other purposes.

Chemostat—A device for maintaining a bacterial culture in the log phase of growth.

Chemosyndrome—Biogenetically meaningful set of major and minor natural metabolic products produced by a species.

Chemosynthesis—A process of food manufacture in certain bacteria that utilizes energy derived from chemical reactions, such as the oxidation of sulfur, ammonia, etc.

Chemotaxis, chemotropism—Movement or growth of organisms in response to a chemical stimulus; often a reaction in relation to food material.

Chemotaxonomy—Taxonomy using chemical characteristics (biochemical systematics).

Chemotherapy—Treatment of plant disease by chemicals (chemotherapeutants) absorbed and translocated internally. The chemical agent has a toxic effect directly or indirectly on the pathogen without undue injury to the host plant or affecting host-cell metabolism.

Chemotroph—An organism that uses energy from inorganic chemicals, such as iron and sulfur, in the manufacture of carbohydrates.

Chemotype—Group of chemically differentiated individuals of a species of unknown or of no taxonomic importance.

Cherry scab—Disease caused by *Fusicladium cerasi*, teleomorph *Venturia cerasi*.

Chestnut brown—A dark reddish brown.

Chilling injury—Direct and/or indirect injury to plants or plant parts from exposure to low, but above-freezing, temperatures (as high as 9°C or 48°F).

Chimera—A plant composed of two or more genetically different tissues. Includes *periclinal chimera*, in which one tissue lies over another as a glove fits over a hand; *mericlinal chimera*, where the outer tissue does not completely cover the inner tissue; and *sectorial chimera*, in which the tissues lie side by side.

Chinese mushroom—*See* Straw mushroom.

Chionophilous—*See* Nitrophilous.

Chip bud—Method of grafting a small piece of scion, bearing a single bud, onto the stock.

Chiroid—*See* Cheiroid.

Chisel (subsoil)—A tillage implement with one or more cultivator-type shanks to which are attached knifelike units that shatter or break up hard, compact layers, usually in the subsoil.

Chitases—Enzymes produced by bacteria and fungi able to hydrolyze chitosan.

Chitin—A complex, nitrogen-containing polysaccharide in many fungal cell walls and animal exoskeletons; lacking in nematodes, except in the egg shell.

Chitinoclastic—Chitin-decomposing.

Chitosan—Partially deacetylated form of chitin characteristic of Zygomycetes.

Chitosome—A microscopic spheroidal structure, 40–70 nm in diameter, found in many fungi.

Chlamydocyst (of Blastocladiaceae)—Two-celled resting zoosporangium within a hypha.

Chlamydospore—A thick- or double-walled asexual spore formed directly from a vegetative hyphal cell (terminal or intercalary), or by transformation of a conidial cell, that functions as a resistant or overwintering stage.

Chloramphenicol acetyl transferase (CAT) gene—The CAT gene product catalyzes the acetylation of chloramphenicol, disrupting its antibiotic activity. It is commonly used as a marker in genetic cloning experiments.

Chloramphenicol amplification—Increasing the copy number of relaxed control plasmids by inoculating *Escherichia coli* in the presence of the protein-synthesis inhibitor chloramphenicol. In the absence of protein synthesis, the replication of the *E. coli* chromosome stops but that of the plasmid continues; a useful method for increasing the properties of plasmid DNA in extracts of *E. coli* cells.

Chlorinated hydrocarbon insecticide—A synthetic pesticide that contains hydrogen, carbon, oxygen, and chlorine. These compounds are persistent insecticides that kill insects mainly by contact. They are insoluble in water and are decomposed by alkaline materials and high temperatures. Examples: Aldrin, chlordane, DDT, dieldrin, heptachlor, lindane, methoxychlor, and toxaphene. Most of these materials are not sold at present in industrialized countries.

Chlorinous—Slightly greenish yellow, approximately hyaline.

Chlorophycophilous (of fungi)—Lichenized with a green photobiont.

Chlorophyll (adj. **chlorophyllous**)—The green, light-sensitive pigments found chiefly in the chloroplasts of leaves and other green parts of higher plants, that absorbs the light energy used in the process called photosynthesis.

Chloroplast—Specialized cytoplasmic organelle (plastid) in plant cells that contains chlorophyll and is the site of photosynthesis.

Chlorosis (adj. **chlorotic**)—Paling, yellowing, or whitening of normally green tissue characterized by the partial to complete destruction of chlorophyll. May be due to a virus, the lack of or unavailability of some element (e.g., iron, manganese, zinc, nitrogen, boron, magnesium), lack of oxygen in a waterlogged soil, alkali injury, or some other factor. *See* Yellowing.

Chocolate spot—Disease of *Vicia* spp. and other legumes caused by *Botrytis fabae* and *B. cinerea*.

Choke—Disease of grasses caused by *Epichloë typhina*, anamorph *Acremonium typhinum*.

Chondriosomes—A generic term for small cytoplasmic structures including mitochondria.

Chondroid axis—Cartilaginous axis filling the central portion of the medulla of *Neuropogon* and *Usnea*.

Christie-Perry extraction technique—A method for isolation of nematodes from soil, combining the wet sieving and Baermann funnel techniques.

Chrom-, chromo- (prefix)—Color.

Chromatid—Shortened and thickened chromatin formed during the prophase stage of mitosis and meiosis. Two chromatids are formed for each chromosome during mitosis and meiosis.

Chromatin—Deeply staining nuclear material of which hereditary determiners are composed.

Chromatin body—Bacterial nuclear material.

Chromatography—A physicochemical analytical procedure by which metabolic and other chemical products are separated by distribution into two or more phases. In *gas chromatography*, separation occurs in a gaseous phase; in *high-pressure liquid chromatography* (HPLC), the solvent is under pressure; in *paper chromatography*, separation occurs on filter paper; in *thin-layer chromatography*, (TLC) separation takes place in thin layers on glass, aluminum, or plastic plates.

Chromatophore—Pigment-containing body; specifically applied to chlorophyll-bearing granules in bacteria.

Chromogen—Stain-producing organism.

Chromogenic, chromogenous (n. **chromogenesis**)—Able to produce color or pigment by microorganisms under suitable conditions of temperature, light, certain nutrients, and the presence or absence of oxygen.

Chromomycosis, chromoblastomycosis—Skin disease in humans caused by *Phialophora* spp.; dermatitis verrucosa.

Chromophilous—Deeply staining.

Chromoplast—Yellowish or red cytoplasmic body containing carotene and xanthophyll.

Chromosomal fiber—A minute strand in mitosis that connects a chromosome with the spindle apparatus.

Chromosomes—Self-replicating, strandlike bodies within cell nuclei in eukaryotes, formed from chromatin and bearing an aggregate of genes (hereditary determiners); a store of genetic information composed of protein and DNA. One chromosome of each pair is inherited from each parent. In bacteria, the entire genome is contained within one double-stranded, circular DNA molecule.

Chronic—Slow-developing, persistent, or recurring symptoms that appear over a long period of time; an infection that lingers, often without symptoms; pertaining to a condition that is of long duration. *See* Acute.

Chronic toxicity (poisoning)—A prolonged exposure to a pesticide from small, repeated dosages over a period of time that may result in injury or death.

Chryseous—Golden yellow.

Chrysocystidium (pl. chrysocystidia)—*See* Cystidium.

Chytrid—Microscopic, usually aquatic, often nonmycelial fungus member of the Chytridiales.

Cicatricose—Having longitudinal ridges.

Cicatrized (of conidiogenous cells, conidiophores)—Bearing thickened scars.

Ciliate, ciliolate—Having protoplasmic filaments (cilia) similar to flagella, but shorter and more numerous on the cell; fringed or edged with fine hairs.

Ciliatulate—Slightly ciliate.

Cilium (pl. cilia)—Hairlike, protoplasmic appendage like a flagellum that propels certain types of unicellular organisms (especially protozoa, zoospores, gametes, and bacteria) through water; also, a hairlike outgrowth from certain cells (e.g., from the edge of an apothecium or lichen thallus). *See* Flagellum.

-cillin (suffix for penicillins)—Derivatives of carboxy-6-amino-penicillanic acid.

Cincinnate, cincinnal—Curled; rolled around.

Cineraceous, cinereous—Ashy-gray; dirty white.

Cingulate—Edged all around.

Cinnabar, cinnabarine—A reddish orange (vermillion red) color.

Cinnamon—A light yellowish brown color.

Circadian—Pertaining to a day (e.g., a 24-h rhythm). *See* Diel and Diurnal.

Circinate—Coiled inwards; twisted round; ring-shaped; (of setae) curved or hooked at the apex like a shepherd's crook.

Circulative virus—Virus acquired by a vector through its mouthparts, accumulates internally in the salivary glands, then passes through the vector's tissues, and is introduced into other plants via the mouthparts of the vector.

Circum- (prefix)—All around; round about.

Circumcinct—Having a band around the middle.

Circumesophageal commissure, circumentric ring (of nematodes)—*See* Nerve ring.

Circumfenestrate (of nematodes)—Describing the condition in some species of *Heterodera* in which a vulval bridge across the vulval cone is not present, producing only a single opening.

Circumscissile—Opening or cracking in a circle or equatorial line.

Circumscription—Statement that defines the limits of a taxon, and shows by implication how it differs from similar taxa.

Cirrate, cirrhate, cirrhose, cirrose—Rolled around (curled) or becoming that way; curly.

Cirrhus, cirrus (pl. **cirrhi** or **cirri**)—A curled, tendril-like mass of exuded spores held together by a slimy matrix as it issues from an ostiole; also termed a "spore horn"; elaborate cephalic appendages found in certain nematodes.

Cisternal ring (of fungi)—Ringlike arrangement of the endoplasmic reticulum that appears to bud and give rise to vesicles.

Cistron—The basic unit of genetic information. It usually refers to a gene or the coding region for a protein.

Citreoviridin—Polyene toxin produced by *Penicillium citreonigrum* (syn. *P. citreoviride*), the cause of cardiac beri-beri in humans.

Citrinin—Toxic yellow pigment (a phenolic carboxylic acid) formed by *Penicillium citrinum*, *P. viridicatum*, etc., causing nephrotoxicosis in pigs.

Citrus scab—Disease caused by *Elsinoë fawcettii*, anamorph *Sphaceloma fawcettii*.

Cladistic—Indicates the degree of relatedness, as shown by the pathways or phyletic lines, by which taxa are linked.

Cladode—A stem with leaflike form; on certain cacti, a joint or stem segment.

Clamp, clamp connection—A bridge- or buckle-like hyphal connection around the septa of a hypha; characteristic of dikaryotic mycelium of many Basidiomycotina; associated with conjugate division of nuclei in the dikaryotic cell; nodose septum; bypass hypha.

Class—A taxonomic category made up of closely related orders; (of soil) a group of soils having a definite range in a particular property such as acidity, degree of slope, texture, structure, land-use capability, degree of erosion, or drainage.

Classification—The systematic arrangement of names for organisms into categories on the basis of characteristics. Broad groupings (e.g., division, subdivision, class, subclass, order, suborder, etc.) are made on the basis of general characteristics and subdivisions (e.g., family, subfamily, tribe, subtribe, genus, subgenus, etc.) on the basis of more detailed differences in specific properties. The relative order of groups is governed by the International Code of Botanical Nomenclature.

Clathrate, clathroid—Latticed; like a network.

Clava—Clublike fruiting structure, such as produced by *Cordyceps*.

Clavate, claviform—Club-shaped; narrowing in the direction of the base; (of an agaric stipe) narrowing to the apex.

Clavatin, clavacin, claviformin—*See* Patulin.

Clavillose—Clubbed; markedly club-shaped.

Clavine alkaloids—A group of ergoline alkaloids in sclerotia of *Claviceps*; also synthesized by other fungi, including *Aspergillus fumigatus* and *Penicillium chermesinun*.

Clavulate—Somewhat club-shaped.

Clay—Complex, colloidal, inorganic fraction of soil, consisting largely of aluminosilicates; clay particles are usually negatively charged and absorb to positively charged ions; soil particles less than 0.002 mm in equivalent diameter; soil material containing more than 40% clay, less than 45% sand, and less than 40% silt.

Claypan—A compact, slowly permeable layer of varying thickness and depth in the subsoil having a much higher clay content than the overlying material. Claypans are usually hard when dry, plastic and sticky when wet.

Cleavage—The cutting of nucleic acid or protein, usually enzymatically, at specific sites. *See* Restriction endonuclease, Protease.

Cleistocarp, cleistothecium (pl. **cleistothecia**, adj. **cleistocarpous**)—A more or less spherical, entirely closed ascocarp, lacking a definite hymenium and ostiole, that ruptures at maturity to release its spores; typical of powdery mildew fungi (Erysiphaceae).

Climacteric—The period in the development of some plant parts involving a series of biochemical changes associated with the natural respiratory rise and autocatalytic ethylene production.

Climax community—A relatively permanent plant community that maintains itself with little change in a given region so long as there are no major changes in environmental conditions.

Cline—A gradation in a measurable character or characters in a population.

Cloaca—A common duct or chamber in male nematodes formed by the junction of the digestive and reproductive systems; the hind gut.

Clod—A compact, coherent mass of soil produced artificially, usually by tillage operations, especially when performed on soils either too wet or too dry.

Clone, noun or verb (syn. **colony**)—An aggregate of individual organisms produced asexually (vegetatively), originating from one sexually produced individual (e.g., rooted cuttings) or from a mutation; any plant propagated vegetatively is considered to be a genetic duplicate of its parent; (in virology) a population of recombinant DNA molecules all carrying the same inserted sequence; (in microbiology) a colony of microorganisms containing a specific DNA fragment inserted into a vector; a population of cells or organisms of identical genotype; use of *in vitro* recombination techniques to insert a particular DNA sequence into a vector.

Close (of gills)—Narrow space between the gills.

Clove—Segment of a garlic, shallot, or other small bulb used for propagation.

Club fungi—Members of the Clavariaceae; has been used to describe basidiomycetes in general.

Clubroot—Disease of crucifers (Brassicaceae) caused by *Plasmodiophora brassicae.*

Cluster analysis—Mathematical treatment of character comparisons used in the identification of taxonomic groups from similarity values. *See* Similarity index.

Cluster-cup—One kind of aecium.

Clypeate—Having a clypeus.

Clypeiform—Shield-shaped.

Clypeus (pl. **clypei**)—A superficial, shield-shaped stromatic tissue, usually dark brown or black, around an ostiole overlying one or more embedded perithecia or pycnidia.

cm—Centimeter = 10 mm = 0.01 m = 0.39 in. = 0.010936 yard.

CMI AAB—Initials of the Commonwealth Mycological Institute/Association for Applied Biologists in the United Kingdom.

Co- (prefix)—Together, with.

Coacervate, coacervulate—Massed or heaped together.

Coadnate—United, cohering, connate.

Coagulase—An enzyme that causes coagulation of blood plasma.

Coagulation system—A system of 12 proteins in serum that results in the formation of fibrin and blood clotting.

Coagulation—Formation of a clot or gelatinous mass.

Coalesce (adj. **coalescent**)—to grow or join together into one body or spot; overlap; merge.

Coarctate—Crushed together; constricted; crowded.

Coat protein—The protective layer(s) surrounding the viral nucleic acid.

Coccidioidin—An antigen prepared from *Coccidioides immitis*; used especially for skin testing. *See* Spherulin.

Coccoid—Sphere-shaped.

Coccus (pl. **cocci**)—A spherical-shaped bacterial cell.

Cochleariform—Spoon-shaped.

Cochleate—Shell-shaped; twisted like a shell.

Cochliobolin—*See* Ophiobolin

Cocoon—A silken case constructed by an insect larva (e.g., caterpillar) to protect the pupal stage.

Coding capacity—The amount of protein that a given DNA or RNA sequence can in theory encode.

Coding sequence—The process by which nucleotides within a certain area of RNA or DNA determine the sequence of amino acids in the synthesis of a particular protein.

Codon—A sequence of three adjacent nucleotides (in a nucleic acid) that codes for a specific amino acid or the initiation or termination of a polypeptide chain.

Coelomycetes—A class of the Deuteromycotina (Fungi Imperfecti) producing conidia in pycnidial, pycnothyrial, acervular, cupulate, or stromatic conidiomata. *See* Hyphomycetes.

Coeno- (prefix)—Living together, e.g., multinucleate.

Coenoapocyte—Cell temporarily or secondarily multinucleate.

Coenocentrum (of oomycetes)—Small deeply staining body at the center of the multinucleate oosphere to which the egg nucleus goes.

Coenocyte (adj. **coenocytic**)—A cell or an aseptate hypha with several to many nuclei; nonseptate; referring to the fact that nuclei are embedded in the cytoplasm without being separated by cross-walls—that is, the nuclei lie in a common matrix; a multinucleate mass of protoplasm.

Coenogametangium (pl. **coenogametangia**)—Multinucleate differentiated sac that produces gametes.

Coenogametes (of lower fungi)—Multinucleate gametangia that fuse to form a *coenozygote*.

Coenosyncytium—Multinucleate structure resulting from the fusion of several protoplasts.

Coenzyme—A substance, usually nonprotein and of low molecular weight, necessary for the action of certain enzymes.

Coevolution (of fungi)—Evolution of obligately parasitic or symbiotic fungi together with their hosts.

Cofactor—Additional (nonprotein) components required by an enzyme for action.

Coffeate—Coffeelike; shaped like a coffee bean.

Cognate—Related.

Cognomen—Nickname or surname.

Cohabitant—Living together.

Cohesion—Holding together; a force holding a solid or liquid together, owing to attraction between like molecules.

Colchicine—An alkaloid, derived from the autumn crocus (*Colchium autumnale*) used specifically to inhibit the spindle mechanism during cell division and thus cause a doubling of chromosome number.

Cold frame—An unheated plant bed on the ground enclosed by side walls that are usually 12–24 in. high, covered with glass or plastic, and useful for growing and protecting young plants in early spring. The heat comes from sunlight.

Cold water insoluble nitrogen (CWIN)—Fertilizer nitrogen not soluble in water at 25°C.

Cold water soluble nitrogen (CWSN)—Fertilizer nitrogen soluble in water at 25°C.

Coleoptile—Ephemeral, nonpigmented tissue sheathing the first true leaf in seedlings of certain monocots. It protects the plumule as it emerges through the soil.

Coleorhiza—Sheath that surrounds the radicle of the grass embryo and through which the young developing root emerges.

Colicins, colicines (adj. **colicinogenic**)—Antibacterial substances (endotoxins) produced by some bacteria (usually coliforms, e.g., *Escherichia coli*) active against other bacteria; each colicin has its distinctive characteristics in the antibacterial spectrum it prevents production of, and immunity to. A colicin is usually encoded by genes on a plasmid termed a Col factor. These plasmids form the basis of a number of cloning vectors, e.g., pBR322.

Coliform—General term for fermentative Gram-negative rods that inhabit the intestinal tract of humans and other animals.

Collabent—Collapsing; falling in.

Collar—A light-colored band at the junction of the leaf blade and sheath on the outside of a grass leaf.

Collar rot—Rotting of the stem at or about soil level.

Collarette—A small, expanded collar- or cup-shaped structure terminating or placed laterally on a phialide; remains of the outer conidiogenous cell wall left on the conidiogenous cell after secession of the first conidium.

Collariate—Having a collar; collared; attached to a collar.

Collateral host—A secondary, alternative, or subordinate host.

Collenchyma (adj. **collenchymatous**)—The tissue in elongating soft stems and certain other parts of plants that helps support the plant; composed of elongated, parenchymatous living cells with unevenly thickened primary walls.

Colliculose, colliculous—Having rounded swellings or "hillocks"; blistered.

Colloid—State in which a finely divided solid remains suspended in a liquid. Colloidal systems are usually more stable than emulsions or suspensions, are generally electrically charged and usually turbid; (of soil) organic and inorganic matter with very small particle size and a correspondingly large surface area per unit of mass.

Collulum—Neck of a phialide or annellide.

Colonization—The period following infection during which a pathogen becomes established in its host. *See* Infection.

Colonize—To establish an infection within a host or part of a host.

Colony—A macroscopically visible group of individuals (e.g., bacteria and yeasts), usually of the same species, living in close association (often on a solid culture medium); in mycelial fungi, the term usually refers to a group of hyphae growing out of a single point and forming a round or globose thallus.

Color break—*See* Flower break and Breaking.

Colorant—A paintlike material, usually a dye or pigment, applied to brown warm-season turfgrasses that are in winter dormancy, brown cool-season grasses that are in summer dormancy, or turfs that have been discolored by environmental stress, turfgrass pests, or the abuses of humans. Its purpose is to maintain a favorable green appearance.

Columella (pl. **columellae**)—A persistent, sterile, central, branched or unbranched axis within a sporangium or other mature fructification; often an extension of the stalk; (of nematodes) four columns of cells that are part of the oviduct.

Comate—Shaggy with hairs.

Combining site—The site on the antibody molecule that binds specifically with its corresponding antigenic determinant; present on the Fab portion of IgG. The heavy (H) and light (L) chains are both involved and the combining site is formed from contiguous areas of the variable regions of both chains, involving five to 15 amino acids.

Comixed, commixt—Intermingled; mixed with.

Commensal, commensalism—A form of symbiosis (e.g., an alga and fungus in lichens) in which an organism lives in close proximity or within another organism without causing harm to either; one or neither may be benefited. *See* Mutualism.

Commercial pesticide applicator—A person who applies pesticides (for hire, not for hire, or public) to land or commodities.

Commissure (latero-ventral, rectal)—A bundle of nematode nerve fibers connecting ganglia; (of fungi) a seam or closing joint.

Common (coined) name—A generic name given to the active ingredient of a fungicide, nematicide, or other pesticide by a recognized committee [e.g., American Standards Association (ASA), Sectional Committee on Common Names for Pest Control Chemicals, the United States Standards Organization (USSO), the International Standards Organization (ISO), etc.].

Common or field mushroom—*Agaricus campestris.*

Community—Any phytosociological taxon; all the plant populations within a given habitat. Usually the populations are considered to be interdependent.

Community pot—A pot to which seedlings or "plantlets" are first transferred before being potted or set as individuals.

Comose—Having hairs in groups or tufts; hairy.

Compaginate—Joined tightly together; packed closely.

Companion cells—Elongated specialized cells adjacent to sieve tubes in phloem (inner bark) tissue of most plants.

Companion crop—A crop sown with another crop and harvested separately. Small-grain cereal crops are often sown with forage crops (grasses or legumes) and harvested in early summer, allowing the forage crop to continue to grow (e.g., oats sown as a companion crop with red clover (*Trifolium*) or alfalfa or lucerne (*Medicago*).

Compartmentalization—Isolation of a specific tissue area by host barrier tissues.

Compatible—Describes different types of plants that set fruit when cross-pollinated, or make a successful graft union when intergrafted; (of mating types, strains, etc.)

able to be cross-mated or cross-fertile; (of pesticides) two or more compounds able to be mixed together without deleterious effects; an interaction between a plant and a pathogen resulting in disease. *See* Incompatible.

Compensation point (light)—The light intensity at which the rates of photosynthesis and respiration are equal.

Competitive inhibition—In serology, process in which a population of molecules inhibits reaction of a second molecule population by competing for common binding sites on molecules of a third population, e.g., antigen A competing with antigen B for the antigen-binding site of an antibody.

Complanate—Flattened; compressed; smooth.

Complement—A normal thermolabile protein constituent of blood serum that reacts nonspecifically with antigen-antibody complexes; a system of at least 13 serum proteins that are activated by enzymatic cleavages and aggregations to produce components with biological activity.

Complement fixation—A sensitive test for antigen-antibody reactions that depends on binding (consumption) of complement by antigen-antibody complexes.

Complementary base sequence—A nucleic acid sequence that is able to form a perfectly hydrogen-bonded duplex with the one to which it is complementary.

Complementary determining region (CDR)—An antigenic site composed of amino acids arranged in a specific conformation. *See* Hypervariable region.

Complementary DNA (cDNA)—DNA synthesized by reverse transcription from an RNA template; a single-stranded DNA molecule that is complementary in base sequence to the single strand from which it was transcribed. If the cDNA is made double-stranded and cloned it is described as a cDNA clone.

Complementary RNA (cRNA)—A single-stranded RNA molecule that is complementary in base sequence to the single strand from which it was transcribed. Most single-stranded RNA (ssRNA) viruses use complementary RNA as intermediates in replication.

Complementary strand—A double-stranded nucleic acid molecule complementary in base sequence to the single strand from which it was transcribed.

Complementation group—A group of viruses that have mutations for the same codon and that cannot complement each other.

Complementation test—A test to determine whether two virus mutants are defective in the same cistron.

Complementation—Process occurring when a virus is assisted by another virus (or a strain of the same virus) to replicate; repairing a gene defect by the presence of another, functional copy *in trans*; the process by which one genome provides functions that another genome lacks. There are two types of complementation between viruses: *intergenic*, in which mutants defective in different genes assist one another, and *intragenic*, in which mutants defective in the same gene produce a functional gene product.

Complement-fixation test (CFT)—A sensitive test in which the antigen-antibody reaction can be detected and quantified; often used for comparing different or related antigens, e.g., viruses.

Complete fertilizer—Any fertilizer containing the three basic elements often deficient in soil for optimum plant growth: nitrogen (N), phosphorus (P), and potassium (K).

Complete flower—A flower that has all the flower parts (sepals, petals, stamens, and pistil[s]) attached to a receptacle.

Complex—Sometimes used to designate a group of closely related species.

Complicate—Bent or folded upon itself.

Compost—A mixture of organic residues and soil that have been piled, mixed, moistened, and allowed to decompose biologically. Mineral fertilizers are sometimes added.

Compound—Composed of a number of similar parts aggregated into a whole.

Compound fruit—*See* Multiple fruit.

Compound leaf—Leaf blade composed of two or more distinct leaflets.

Compound middle lamella—A collective term for the middle lamella and the primary walls of two adjacent cells.

Compound oosphere—*See* Oosphere.

Compound pistil—A pistil composed of two or more partially or wholly fused carpels.

Compressed (of a stipe, spores, etc.)—Flattened transversely or lengthwise.

Concatenate—In chains; catenulate.

Concave—Having a surface that curves inward; round-depressed like a bowl or basin.

Concentration—The amount of actual pesticide or active ingredient contained in a formulation or mixture; the amount of a chemical compound in a given volume or weight of diluent. Expressed as percent, pounds per gallon, etc.

Concentric (of zonation, etc.)—One or more circles within one another with a common center. A frequent symptom of numerous diseases caused by fungi, viruses, and bacteria. *See* Ringspot.

Concentric bodies—Ultrastructural bodies found in the mycobionts of certain lichens and in other fungi, e.g., *Rhopographus, Sphaerotheca, Cercospora*, etc.

Conceptacle—A cavity within certain algae within which antheridia or oogonia are produced; (of fungi) any hollow structure producing spores or spermatia.

Conchate, conchiform—Like a bivalve shell.

Concolorous—Describing cells of even pigmentation; of the same color as a previously described structure.

Concrescent—Growing or joining together, fusing.

Concrete (of fungi)—Joined by growth.

Conducive soil—Soil in which disease readily occurs (as opposed to suppressive soil).

Conducting hyphae—Specialized hyphae characteristic of *Stereum* and *Xylolobus* with contents that are globular to granular, ranging from hyaline to brown. *See* Pseudocystidium and Hypha.

Cone—A specialized woody, usually elongated, seed-bearing organ of a conifer consisting of a central stem, woody scales, bracts (often not visible) and seeds; (of nematodes) the posterior region of a female cyst of the genus *Heterodera*.

Conferted—Crowded; near together.

Confervoid—Composed of loose filaments or cells.

Confluent (of fruitbodies)—Coming together; fusing with growth; becoming joined by hyphae, cells, or stromatic tissue; (flesh of a stipe) continuous with the trama of the pileus.

Congeneric, conspecific—Pertaining to one of two or more genera or species considered to be one taxon.

Congested—Crowded; very near together.

Conglobate—Massed into a ball; (of the bases of stipes) together forming a fleshy mass.

Conglutinate—Glued together, especially of paraphyses in certain fungal groups.

Conic, conical, conoid—More or less cone-shaped.

Conico-truncate—Having the shape of a truncated cone.

Conidiogenesis—Process by which conidia are formed from conidiogenous cells.

Conidiogenous cell—Any fungus cell that directly produces one or more conidia.

Conidiogenous locus—Place on a conidiogenous cell where a conidium arises.

Conidiole—A small conidium, especially one on another; a secondary conidium.

Conidioma (pl. **conidiomata**)—Any organized hyphal structure (or asexual fruitbody) bearing or containing conidia, e.g., acervulus, coremium, pycnidium, sporodochium, synnema, etc. *See* Conidiophore.

Conidiophore (pl. **conidiophores**)—A specialized, simple or branched hyphal cell or group of cells bearing conidiogenous cells that produce conidia. *See* Conidiogenous cell.

Conidium initial—A cell or part of a cell from which a conidium develops.

Conidium ontogeny—Conidiogenesis.

Conidium, conidiospore (pl **conidia**)—Any asexual, nonmotile spore (except intercalary chlamydospores or sporangiospores) that develops externally or is liberated from a conidiogenous cell; the term is generally restricted to members of the Ascomycotina, Basidiomycotina, and Deuteromycotina.

Coniferous—Having cones.

Conjugate—Joined; in twos; (of serology) the product of joining two or more dissimilar molecules by covalent bonds. In immunological contexts, one is usually a protein and the other a hapten or a label such as fluorescein, ferritin, or enzyme.

Conjugate nuclear division—The simultaneous division of the two paired nuclei (*conjugate nuclei*) in a dikaryon, giving rise to four daughter nuclei; these generally become separated by a septum into two cells, with the sister nuclei migrating into different daughter cells.

Conjugation—A process of sexual reproduction involving the fusion of gametes that are morphologically similar; (in bacteria) directional transfer of DNA from a donor cell to a recipient cell; the transfer of a plasmid from one cell to another.

Conjugation tube—Short hyphal element that replaces basidiospores in the life cycle of some smut fungi to allow nuclear transfer between promycelial cells; a tube between two copulating cells as in oomycetes.

Conk—Fruitbody (basidiocarp, basidioma, sporophore) usually of a polypore wood-rotting fungus (family Polyporaceae) formed on tree stumps, branches, trunks, or occasionally on lumber. Usually they are spongy to hard, become large when mature, and persist for one or more years.

Connate—Congenitally or firmly united by growth.

Connective—*See* Disjunctor.

Connective hyphae—Hyphae of the connective tissue of the context.

Connivent—Touching but not joined; (of a pileus margin) touching the stipe.

Consortium—A form of symbiosis.

Conspecific—Individuals belonging to the same species. *See* Congeneric.

Constant region (in serology)—Portions of either the heavy (H) or light (L) chains of immunoglobulins that exhibit conserved amino acid sequences.

Constipate—Crowded together.

Constricted (conidia)—Having the periclinal wall indented to varying degrees where it meets transverse septa.

Constriction (of nematodes)—Condition in which the junction of the lip and cervical regions is delimited by a sharply pronounced groove.

Constrictive resistance—Genetically controlled, inherited resistance.

Contact inhibition—Inhibition of growth and/or movement when cells, especially in tissue-culture systems, come into contact.

Contact movement—Turgor movements, chiefly of leaves and floral parts, that result from contact stimuli.

Contact pesticide—A pesticide that kills on contact.

Contagious—Spreading from one to another.

Contaminant—A substance or microorganism that renders an otherwise pure culture system unclean or unsuitable for the intended purpose. Bacteria, fungi (e.g., *Alternaria, Aspergillus, Mucor, Penicillium, Rhizopus*), and mites are common contaminants in cultures of fungi in petri dishes, etc.

Contaminated (n. **contamination**)—Bearing, or intermixed with, a pathogen as spores on seeds, fungi in soil; entry of undesirable organisms into some material or object; (of cultures) not pure.

Context—The sterile inner hyphal tissue of the cap or pileus that supports the hymenophore (hymenium) of pileate basidiocarps (not tubes, gills, etc.).

Contigs—Groups of clones representing overlapping or contiguous regions in a genome.

Contiguous—Touching; joining.

Contingent—Touching.

Continuous (spores, hyphae, etc.)—Lacking septa; (of a stipe) with the tissue of the pileus or peridium.

Continuous cell line—Cells with uniform morphology and capable of indefinite propagation *in vitro*.

Continuous flow centrifugation—Centrifugation in a rotor with a fluid seal that allows the continuous flow of a sample in and out of the rotor while it is rotating at high speed.

Continuous or contiguous epitope—An antigenic site composed of two or more groups of peptides linked to one other.

Continuous resistance—A response involving a gradient from severe infection to extreme resistance in a segregating population.

Contorted—Twisted.

Contour length—The length of a nucleic acid molecule as measured by electron microscopy.

Control, of plant diseases—Prevention, retardation, or alleviation of disease. There are four principal methods: *Exclusion*—Keeping the pathogen away from a disease-free area through quarantines, embargoes, and disinfection of plants and plant parts; *Eradication*—Destruction (roguing) of infected plants or plant parts or killing of the pathogen or agent on or in the host; *Protection*—Application of a chemical or physical barrier to prevent entrance of the pathogen; and *Immunization*—Production of genetically resistant or immune plant cultivars, chemotherapy, or other treatment to inactivate or nullify the effect of the pathogen within the plant. Some scientists recognize a fifth method, *Avoidance*, which involves choosing cultural practices (e.g., sanitation, rotation, choice of planting site and date, propagating and planting only disease-free material, etc.) that avoid disease.

Controlled atmosphere (CA) storage—A process of fruit storage in which the CO_2 level is raised (to 1–3%) and the O_2 level is lowered (to 2–3%).

Convergent evolution—Evolution of unrelated or distantly related groups of organisms along similar lines, resulting in the development of similar traits or features in the unrelated groups.

Conversion—A term often applied to interactions between a temperate phage and a prokaryote host cell in which new properties, which have no obvious relation to the replication cycle of the phage, are conferred on the host by the genome of the prophage. These may include changes in colony morphology or pigmentation, modification of the antigenic properties of the host, and effects on toxin production.

Convex—Having a surface or line that curves outward; regularly rounded or regularly bulging, rising, or swelling into a raised, rounded form; (of a pileus) regularly rounded, broadly obtuse. *See* Concave.

Convexo-expanded—Having the edge bent over.

Convexo-plane—Convex when young and flat after expansion when mature.

Convolute—Rolled up longitudinally.

Convoluted—Describing a conidioma in which one or more cavities are of irregular shape and imperfectly separated.

Coolplate—Temperature-controlled plate on which cultures under a light source are maintained without an undesirable rise in temperature.

Cool-season turfgrass—Turfgrass species adapted to favorable or optimum growth during cool portions (15.5–24°C) of the growing season.

Coombs' test (in serology)—*In vitro* agglutination test for non-agglutinating antibodies to red blood cells, employing anti-Ig. *Direct Coombs' test*: Anti-IG is added to test red cells; agglutination indicates that cells were coated with antibody. *Indirect Coombs' test*: Serum is added to red blood cells, the cells are washed, and anti-Ig is added; agglutination indicates that serum contains antibody to red blood cells.

Coprogen—A growth factor in dung required by species of *Pilobolus*.

Coprophilous fungi—Fungi living on dung; funicolous fungi.

Copulants—Copulating structures of like form.

Copulation—The fusion of sexual elements; conjugation; the act of sexual union between a male and female leading to fusion of gametes. Various types of copulation include gametangial, heterogamic, isogamic, and planogamic (which see).

Coral spot—A stem disease of many angiospermous woody plants caused by *Nectria cinnabarina*, anamorph *Tubercularia vulgaris*.

Coral spot fungi—The Clavariaceae.

Coralliform, coralloid—Coral-like in form; much branched; especially basidiomata of *Clavaria*.

Corbicula (pl. **corbiculae**)—Protective structure forming a stroma around the telia of certain rusts; paraphyses; pseudoparaphyses.

Cord or **chord** (of nematodes)—Longitudinal internal thickenings of the hypodermis (dorsal, two lateral, ventral).

Cordate—Heart-shaped; usually used to describe leaves with a pair of rounded basal lobes.

Cordon—A microscopic ropelike strand of intertwined hyphae found in the subiculum or marginal tissues of resupinate basidiocarps.

Core—The central part of a virus particle (virion) enclosed by a capsid and comprising protein and the viral nucleic acid genome; internal body enclosed in a capsid or envelope; is or contains a nucleoprotein complex. Also the inner tube of contractile phage tails.

Core aerification—A method of turf cultivation in which soil cores are removed by hollow tines or spoons to control soil compaction and to aid in the penetration and distribution of pesticides and water.

Coremioid—Broomlike.

Coremium (pl. **coremia**)—An asexual fungal fruitbody consisting of a headlike cluster or fascicle of intertwined, erect conidiophores bearing conidia at their tips; also termed a synnema (which see).

Coriacellate—Somewhat leathery in texture.

Coriaceo-membranaceous—Describing a leathery membrane.

Coriaceous, corious—Leathery in texture, flexible when fresh; tough; fungal hyphae tightly interwoven.

Coriaceous-ceraceous—Describing a condition in which the hymenium and/or subhymenium is waxy, the remainder coriaceous; or between coriaceous and ceraceous.

Cork—An external, secondary, protective, suberized tissue, usually on stem or root surfaces of woody plants, impermeable to water and gases. It is often formed by cork cambium (*phellogen*) in response to wounding or infection.

Cork cambium—Meristematic tissue that is formed in woody plants, usually from certain cells of the cortex, and that produces cork cells and phelloderm cells.

Corky—Firm but not hard; of a density similar to cork.

Corm—A short, enlarged, solid, often globose, fleshy base of an underground stem that stores food and contains undeveloped buds with a few thin, scalelike leaves. Examples are crocus, freesia, and gladiolus. *See* Bulb.

Cormel—A small or secondary, usually hard-shelled, corm produced around the base of the "mother" corm.

Corneous—Hornlike in texture; a substance like horn.

Corniculate—Having horn-like projections.

Corniform (appendage)—Horn-shaped; when branched, with several branches on one side.

Cornute—Horned; horn-shaped; (of aecia) *see Roestelioid.*

Corolla—The petals, collectively, of a flower.

Coronate—Crowned.

Corpus—The anterior, usually cylindric, part of a nematode's esophagus, the basal portion of which may be swollen to form the metacorpus; the main body of an organism.

Correlated species (of Uredinales, rust fungi)—A species derived by reduction (of life cycle or morphology) from a parent heteroecious macrocyclic species, or the parent species itself.

Correlation—The mutual interaction of plant parts and processes.

Cortex (adj. **cortical**)—The primary, soft, mostly parenchymatous tissue between the epidermis and the phloem in stems and the pericycle in roots; (of fungi) a more or less thick outer covering; periderm; the upper and lower or outer layers of a lichen thallus; (of nematodes) the outermost layer (cortical) of the cuticle.

Cortical haustoria (of mistletoes)—*See* Cortical strands.

Cortical strands—Of mistletoes, radiating strands of mistletoe tissue that grow through cortex and secondary phloem of the parasitized tree stem.

Corticate—Woody limbs or trunks possessing bark; having a cortex.

Corticole (adj. **corticolous**)—An organism living on bark.

Cortina (of agarics)—A partial veil, frequently weblike, covering the mature gills.

Cortinarins—Fluorescent, cyclic, toxic decapeptides produced by species of *Cortinarius*.

Corymb—Flat-topped type of inflorescence, the main axis of which is elongated; the pedicels of the older flowers, arranged so the flowers are all approximately in one plane.

Corymbose—Arranged in clusters.

Coryneform bacteria—General term for Gram-positive, non-sporeforming rods that show clump formation rather than chains; some are club-shaped.

Coryneform—Club-shaped.

Cos site—The "sticky" ends of certain DNA phage molecules.

Coscinoid—A long, filamentous, brown, pitted conducting element in *Linderomyces*.

Cosinocystidium—A cystidium projecting as a pseudocystidium.

Cosmid—An artificial plasmid that combines transduction and replication properties of a phage and a plasmid and is used as a cloning vehicle for genetic engineering; lamba and one or more selectable markers; a plasmid vector that contains the cos site of phage.

Cosmid vectors—Plasmids that also contain specific sequences from the bacterial phage lamba; cosmids are designed for cloning large fragments (typically 40,000 base pairs) of eukaryotic DNA.

Costa—A ridge or rib.

Costate—Veined, fluted, or ribbed.

Cotyledon—The seed leaf; one in the monocotyledonae, two in the dicotyledonae; primary embryonic leaf; a food-digesting and food-storing part of an embryo.

Cotyliform—Plate- or wheel-like with an upturned edge.

Cotype—A strain (specimen) of a normal species by an author who failed to designate a holotype; equivalent to syntype.

Counts per minute (CPM)—A measure of the radioactivity of a sample; the basic output of a radioactivity counter (does not allow for quenching of counts by components in the sample).

Covalent bonding—Sharing of electrons between adjacent atoms in a way that reduces the total electronic energy. Classically, two shared electrons constitute a covalent bond. This is the bonding that holds atoms together in molecules, in contrast to the electrostatic attractions that act between ions, dipoles, and induced dipoles. Covalent bonds, unlike these electrostatic forces, show characteristic bond lengths and bond angles responsible for molecular shape, and atoms so bonded cannot be separated by physicochemical techniques such as chromatography and electrophoresis.

Covalently closed circular DNA (cccDNA)—A form of double-stranded DNA in which both strands are circular without free ends. *See* Supercoiled DNA.

Cover crop—A close-growing, natural or introduced crop grown primarily to improve and maintain soil structure, add organic matter, and prevent soil erosion.

Coverage—The degree of spread or distribution of a pesticide over a discontinuous surface such as leaves, stems, fruits, and seeds.

Covered smut—A smut in which the spore mass (teliospores) is within the sorus for a time, commonly after the sorus becomes free of the host. The teliospores may be retained because the pericarp does not break down. *See* Smut.

Covered smut of barley, oats, rye, and several grasses—Disease caused by *Ustilago hordei*. *See also* Bunt, a *covered smut*.

Covered smut of sorghum— Disease caused by *Sporisorium sorghi*.

cpe—*See* Cytopathic effect.

cpm—*See* Counts per minute.

Cramp balls—Fruitbodies of *Daldinia concentrica*.

Cratera, crateriform—Crater- or cup-shaped.

Crawler—Newly hatched or first-stage insect (i.e., a whitefly) that is still able to crawl.

Crazy top—Disease symptom manifested by twisting, proliferation, and distortion of upper plant parts caused by a downy mildew fungus, e.g., *Sclerophthora macrospora*.

Crenate—Having the edge scalloped; round-toothed.

Crenation, crenature—One of a series of rounded projections forming the edge of an object or structure; used particularly to describe the tail of a nematode, where body annulation continues to the terminus (crenate vs. smooth tail terminus).

Crenellate, crenulate—Very finely crenate.

Cresote or kerosene fungus—*Hormoconis resinae*, teleomorph *Amorphotheca resinae*.

Cretaceous—Chalklike.

Cribose, cribriform—Having a sievelike network.

Criconematoid (of nematodes)—Resembling the genus *Criconema* and related forms (plant-parasitic). *See* Esophagus.

Crispate—Finely curled, twisted, or crinkled.

Cristae—Internal membrane extensions in mitochrondia; site of ATP production during aerobic metabolism.

Cristate—Crested.

Cristiform—Have the form of a crest.

cRNA—*See* Complementary RNA.

Crop residue—Portion of crop plants remaining after harvest.

Crop rotation—Growing different crop plants in a systematic sequence to help control insects, diseases, and weeds, improve soil structure and fertility, and decrease erosion.

Crop tolerance—The ability of a crop to endure treatment with a chemical or injurious environmental factor with minor adverse effect; also the amount of a pesticide that is legally allowable on or in a crop at harvest time.

Cross immunoelectrophoresis—A technique in which a mixture of antigens is first separated by electrophoresis in an agarose gel and then electrophoresed at right angles into a gel containing antibodies.

Cross section—The surface exposed when a plant stem is cut horizontally and the majority of the cells are cut transversely.

Cross-hybridization—Hybridization between complementary nucleic acids from different sources.

Crossing over—The exchange of matching segments of chromatids of two homologous chromosomes during meiosis.

Cross-pollination—Transfer of pollen from a stamen to the stigma of a flower of another plant, except for clones in which the plants must be of two different clones.

Cross-protection—The condition whereby a normally susceptible host is infected with an avirulent pathogen (usually a virus) and thereby becomes resistant to infection by a second, usually related, virulent pathogen in the same host.

Cross-reacting antigen—Antigen capable of combining with antibody produced in response to a different antigen; may cross-react due to sharing of determinants by the two antigens or because the antigenic determinants of each, although not identical, are closely enough related stereochemically to combine with antibody against one of them; antigen of identical structure in two strains of bacteria, so that antibody produced against one strain will react with the other.

Cross-reactive idiotope (CRI)—An idiotope, also known as a public idiotope or IdX, that is shared by two or more antibodies that may or may not share similar antigenic specificities.

Crown—Compacted series of nodes from which shoots and roots arise at the base of the culm (stem) of cereals and forage species from which tillers or branches arise. In woody plants, the root-stem junction; upper portions of a tree, bearing leaves, flowers, and fruit.

Crown gall—A tumor-like enlargement of roots or stems caused by *Agrobacterium tumefaciens*.

Crown rot (Southern blight)—A disease caused by a fungus, *Sclerotium rolfsii*, teleomorph *Athelia rolfsii*, that attacks hundreds of different ornamentals, vegetables, and field crops in warm, moist weather in warmer regions. White wefts of mycelium spread fanwise up the stem from the crown and also out into soil.

Crown rust—A disease of oats and grasses caused by *Puccinia coronata*.

Crozier—The hook of an ascogenous hypha before ascus development; ascus crook; the hooked or curved apical portion of a blighted stem, e.g., an apple (*Malus*) stem blighted due to infection by the fire blight bacterium.

Crozier formation—Process of ascus development from coiled tips ("hooks") of ascigerous hyphae.

Cruciate, cruciform—Cross-shaped; (of nuclear division in *Plasmodiophora*) having the chromosomes in a ring around a dumbbell-shaped nucleolus.

Cruciately septate (of basidia)—Having vertical, right-angled septa that divide the basidium into four cells, characteristic of the Tremellaceae.

Crucifers—Members of the cabbage family (Brassicaceae) including cabbage, broccoli, brussels sprouts, cauliflower, horseradish, rape, kohlrabi, radish, turnip, alyssum, honesty, and stock.

Cruciform division—*See* Promitosis.

Crumb (soils)—A soft, porous, irregular, natural unit of structure from 1 to 5 mm in diameter.

Crumbly—Breaking up readily into small pieces.

Crust—(Of fungi) general term for a hard surface layer, especially of a sporocarp; (of soil) surface layer ranging in thickness from a few millimeters to a few centimeters. It is more compact, hard, and brittle when dry than the soil beneath it.

Crustose, crustaceous—Forming a thin, hard, compact brittle crust; closely appressed (fixed) to the substratum and with a crusty appearance.

Crypta—Sleevelike formation around a true root (especially evergreens) in warm-to-hot regions, developed by certain agarics.

Cryptic—Hidden or inconspicuous.

Cryptobiology—Study of life and living things at low temperatures.

Cryptococcosis—A disease of humans and animals caused by the fungus *Cryptoccocus neoformans*, teleomorph *Filobasidiella neoformans*; torulosis.

Cryptogram—A descriptive code summarizing some of the main properties of a virus or viroid; term now seldom used.

Cryptotope—An antigenic site that is hidden, "buried" inside the antigen, or unavailable to an antibody until the molecule bearing the antigenic site is structurally altered. *See* Epitope.

Ctenoid—Comb-like.

cu ft—Cubic foot = 0.80356 bushel = 1728 in.^{3} = 0.037037 cu yard = 0.028317 m^3

= 7.4805 U.S. gal = 6.229 British or Imperial gal = 28.317 liter = 29.922 qt (liquid) = 25.714 qt (dry).

cu ft of dry soil (approximate) = sandy (90 lb), loamy (80 lb), clayey (75 lb).

cu ft water—Cubic foot water = 62.43 lb (1 lb water = 27.68 in.^{3} = 0.1198 U.S. gal = 0.01602 ft^3.

cu in.—Cubic inch = 16.38716 cm^3 = 0.0005787 ft^3 = 0.004329 U.S. gal.

cu m—Cubic meter = 1.30794 cu yard = 35.3144 ft^3 = 28.3776 bushels = 264.173 U.S. gal = 1056.7 qt (liquid) = 2113.4 pt (liquid) = 61,023 cu in.^{3}.

cu yd—Cubic yard = 27 ft^3 = 46,656 in.^{3} = 764.559 liter = 292 U.S. gal = 168.176 British gal = 1616 pt (liquid) = 807.9 qt (liquid) = 21.694 bushels = 0.764559 m^3.

Cubic phages—Viruses isolated from prokaryotes, whose capsids have cubic (icosahedral) symmetry.

Cubic viruses—Spherical, isometric, or polyhedral viruses with many-sided triangular symmetry. There are three types: *tetrahedral*, which has four threefold axes and three twofold axes; *octahedral*, which has three fourfold axes, four threefold axes, and six twofold axes; and *icosahedral*, which has 12 fivefold axes, 10 threefold axes, and 15 twofold axes. Many isometric viruses have icosahedral symmetry.

Cubing—Process of forming hay into high-density cubes to facilitate transportation, storage, and feeding.

Cucullate—Hood-, cowl-, or hat-shaped.

Cucurbits (adj. **cucurbitaceous**)—Members of the cucumber family (Cucurbitaceae) including cucumbers, melons, squashes, pumpkins, gourds, and watermelon.

Cull—Nonmerchantable volume in sawtimber due to decay or other factors.

Culm—Stem of grasses, cereals, and bamboos; usually hollow except at the swollen nodes.

Culmicole, culmicolous—Growing on stems, especially grass (Poaceae) stems.

Culmomarasmin—Wilt toxin of *Fusarium culmorum*.

Cultivar (cv.)—A cultivated variety; assemblage of closely related plants of common origin within a species that differ from other cultivars in certain minor details (e.g., form, color, flower, or fruit) that, when reproduced, sexually or asexually, retain their distinguishing features. *See* Variety.

Cultivation—Refers to preparing and working the soil to raise crops; (applied to turf) working of the soil and/or thatch without destruction of the turf, e.g., coring, slicing, spiking, or other means.

Culture—(v.) To artificially grow and propagate microorganisms or plant tissue on a prepared nutrient medium. (n.) Growth of one organism or a group of organisms artificially maintained on such food material. *See* Enrichment culture and Pure culture.

Culture collection—The microbiological equivalent of a botanical herbarium and a zoological museum together with their gardens of living plants and animals; a repository (such as the ATCC) of cultures of characterized bacteria, fungi, viruses, viroids, cells, and other organisms.

Culture medium—The prepared solid or liquid medium on which microorganisms or plant cells are grown; the medium usually contains various inorganic salts, sugars, amino acids, and antibiotics.

Culture plate—A shallow, circular, covered dish of thin glass or plastic used in growing bacteria and fungi in pure culture in the laboratory.

Cumulate—Massed or heaped up together.

Cuneate, cuneiform—Thinner at one end than the other; wedge-shaped.

Cup—Unit of measure = 16 tablespoons = 8 fl oz = 0.5 pt = 236.6 cm^3 or 1 ml^3.

Cup fungus—A discomycete, especially ascocarps of Leotiales and Pezizales.

Cup lichen—A species of *Cladonia* with podetia swollen into globlet-shaped scyphi.

Cupulae, cupulate, cupuliform—Cup-shaped, deeper than saucer-shaped; (of the margin of some resupinate basidiocarps) with isolated, shallow, cuplike depressions; (of aecia) having a cuplike or cylindric peridium with a recurved border.

Cupule—A small cup.

Cure—To prepare crops for storage by drying. Dry onions, other bulbs, corms, tubers, sweet potatoes, and hay crops are examples. Dehydration of fruits for storage is not considered curing.

Curing (of bacterial cultures)—Loss of lysogenic phage, thus converting the culture to a nonlysogenic state.

Curl—The distortion, puffing, and crinkling of a leaf resulting from the unequal growth of its two sides.

Curling factor—*See* Griseofulvin.

Curly-top—A common virus disease in western United States of bean, flax, melons, spinach, sugar beet, tomato, and 150 other plant species. Plants are stunted with curled and mottled leaves.

Cuspidate (e.g., of a pileus or cystidium)—Topped with a well-marked sharp outgrowth or point; toothed.

Cuticle (adj. **cuticular**)—The thin, waxy, protective, discontinuous, noncellular membrane (composed of "platelets") over most epidermal cells of higher plants, consisting mostly of wax and cutin. The cuticle is broken only by natural openings (stomates and lenticels), that prevents excessive water loss; (of fungi) a layer of specialized hyphae on the pilear or stipe surface of a basidiocarp; (of nematodes) the outermost noncellular covering.

Cutin—A clear or transparent waxy material related to cellulose, very impermeable to water, that comprises the inner layer of the cuticle of plants.

Cutinolytic (of certain enzymes)—Able to digest cutin.

Cutis (of basidiocarps)—Agglutinated zone of certain wood-decay fungi between the cortex and the tomentum (if present) that is darker than the cortex; the outer layer consisting of compressed hyphae parallel to the surface.

Cutting—A vegetative plant part (stem, leaf or root) placed in a rooting medium for the purpose of growing another plant, identical to the parent.

cv. (cultivar)—*See* Cultivar.

cwt = 100 lb = 45.45 kg.

Cyanescent—Turning blue.

Cyanobiont—*See* Photobiont.

Cyanophage(s)—Viruses isolated from blue-green algae (cyanobacteria) morphologically similar to many bacteriophages.

Cyanophilous—Becoming bright blue by readily absorbing cotton blue, gentian violet, or other blue dye.

Cyanophycophilous—Lichenized fungi with a blue-green photobiont.

Cyathiform—(Of amphid of nematodes) pocket-like or cup-shaped, but with a flaring margin; (of fungi) cup-shaped with a flaring margin.

Cybrid—The product of the fusion of a cell with a cytoplast (a cell that has no nucleus).

Cyclic AMP—A compound derived from ATP by the action of the enzyme adenyl cyclase.

Cylindrical—Round in cross section and of equal width throughout its length; (of nematodes) *see* Esophagus.

Cylindrospora—Having cylindrical spores.

Cyloheximide (actidione)—An antifungal and antibacterial antibiotic isolated from certain strains of *Streptomyces griseus*.

Cymbiform—Boat-shaped; navicular; broadly reniform.

Cyme—Type of inflorescence with a broad, more or less flat-topped determinate flower cluster, with the central flower at the apex; the first to mature and open.

Cymoid—Resembling a cyme.

Cymose—Having a determinate centrifugal cluster.

Cyphella (pl. **cyphellae**)—A roundish to ovate break, generally in the lower cortex of a lichen thallus. In section, it appears as a cup-shaped structure lined with a layer of loosely connected, often globular cells formed from the medulla.

Cyst—(In fungi) resting structure formed by a zoospore; a sac, especially a resting spore- or sporangium-like structure; resting or dormant phase of a cell or multicellular structure usually enveloped in a protective membrane or shell-like enclosure; (in rust fungi) sterile, colorless, hygroscopic cells that subtend some teliospores; (of nematodes) the egg-containing carcass or oxidized cuticle of a dead adult female of the genera *Globodera* and *Heterodera*. *See* Macrocyst, Microcyst, and Spore cyst.

Cystidiate—Having cystidia, often discernible with a 10× hand lens.

Cystidiole (of hymenomycetes)—A simple hymenial cell of about the same diameter as the basidia but remaining sterile and projecting a little beyond the hymenial surface.

Cystidium (pl. **cystidida**)—Sterile, generally hyaline or light-colored, often swollen or otherwise morphologically distinctive hyphal cell or body, projecting from the hymenium (or other part) of certain basidiomycetes, that lacks refractive contents. There are numerous types classified as to origin, position, form, or contents: *Hymenial-* (or *tramal-*) *cystidium* originates from hymenial or tramal hyphae; *pseudocystidium* is derived from a conducting element, filamentous to fusoid with oily contents, embedded but not projecting; *coscinocystidium*, see Coscinoid; *skeletocystidium* is the apical part of a skeletal hypha that is frequently more or less inflated and projects into or through the hymenium (false seta); *macrocystidium* arises deep in the trama of Lactario-Russulae; *hyphocystidium* is hyphalike, derived from generative hyphae; *pileo-* or *dermatocystidium* is found on the pileus surface, at the edge (*cheilo-*), on the side, (*pleuro-*), or within (*endo-*) a lamella; on the stipe (*caulocystidium*). *Leptocystidium* is smooth and thin-walled; *lamprocystidium* is thick-walled with or without encrustation; *setiform lampro-* is awl-shaped with a pigmented wall; *asteroseta* is a radially branched lamprocystidium; *microsclerid* is a versiform, endolamprocystidium; *lyocystidium* is cylindrical to conical and very thick-walled, abruptly thin-walled at the apex, not encrusted and colorless; *monilioid gloeocystidium* (torulose gloeo-); moniliform paraphysis; pseudophysis; *schizocystidium* is monilioid, frequently with a beaded apex; *gloeocystidium* is thin-walled, usually irregular with hyaline or yellowish contents and highly refractile; *chrysocystidium* is like lepto- but with highly staining contents; and *oleocystidium* has an oily resinous exudate. *See* Hyphidium and Seta.

Cystochroic—Having pigment in cell vacuoles.

Cystosorus (of Chytridiales)—A group of united cysts or resting spores.

Cystospore (of Chytridiales)—An encysted zoospore formed at the exit of the zoosporangium that germinates to produce a new zoospore or planospore (as in Achlya); (of Amoebidiales) spores released from an amoeboid cell..

Cyto- (prefix)—Referring to a cell.

Cytochlasin B—*See* Phomin.

Cytochlasins—Related fungal metabolites (e.g., from "*Helminthosporium,*" *Metarhizium, Phoma, Xyliariaceae, Zygosporium*) that inhibit cytokinesis, resulting in multinucleate cells.

Cytochroic—Having pigment in the cytoplasm.

Cytochrome—One of a group of reversible oxidation-reduction carriers (electron-transport proteins), in mitochrondrial oxidation and in photosynthetic electron transport.

Cytokinesis—Cytoplasmic division by cell plate formation, usually following mitosis.

Cytokinins—Class of hormones important in many growth responses of plants, including nucleic acid and protein metabolism, cell division and enlargement, bud and root formation, breaking seed dormancy, flower photoperiod, parthenocarpy, and delaying senescence.

Cytology—Science or study of cells and their components, structure, organic processes, and functions.

Cytolysis—Breaking up or solution of the cell wall; a dissolving action on cells.

Cytopathic effect (cpe)—Changes in the microscopic appearance of cultured cells often seen following virus infection. The changes may be in cell morphology, e.g., rounding up, cell fusion, or the production of inclusion bodies or other intracellular structures.

Cytoplasm—All living substance (protoplasm) of a cell except the nucleus; consists of a complex protein matrix or gel including essential membranes and cellular organelles.

Cytoplasmic membranes—Membranes on the surface of cytoplasm and between cytoplasm and vacuoles.

Cytosine—One of the nitrogenous bases found in DNA and RNA.

D

Dacryoid—Having a round end, the other more or less pointed; pear- or tear-shaped.

Dactyline, dactyloid—Fingerlike.

Daedaleoid, daedaloid (of pores)—Elongate, irregularly lobed and sinuous in outline; labyrinthiform.

Dalton—Unit of atomic size equal to the mass of a proton or hydrogen ion.

Damage threshold—The lowest pest population density at which damage occurs.

Damping-off (adj. **damped-off**)—A disease that may result in decay or death of seeds or seedlings at germination in soil (*preemergence damping-off*); most evident in young emerged seedlings that suddenly wilt, topple over, and die from a rotting at the stem base (*postemergence damping-off*), generally by common soil- and seedborne fungi (e.g, *Fusarium, Phytophthora, Pythium,* and *Rhizoctonia*) and also by a few bacteria.

Dangeardien, dangeardium—Collective term for asci and basidia; structures in which diploid nuclei are formed, undergo meiosis, and form haploid spores.

DAP—Diaminopimelic acid.

Dark mildew—One of the Meliolales or Capnodiales; the fungus is seen as a dark growth on the surface of the host; same as sooty mold.

Dark-field illumination—A method of microscope illumination in which the illuminating beam is a hollow cone of light formed by an opaque stop at the center of the condenser large enough to prevent direct light from entering the objective. The specimen is seen with light scattered or diffracted by it.

Dark-field microscopy—Microscopic examination in which the microscopic field is dark and any objects such as organisms are brightly illuminated.

Database—Repository of information.

Dauer larva (of nematodes)—A quiescent stage entered by the larva, often while enclosed in the cast cuticle of the previous stage; usually for a phoretic relationship with a vector host.

Dauer mycelium—Dark, thick-walled hyphae formed by *Verticillium albo-atrum*.

Daughter cells—Newly formed cells resulting from the division of a parent cell.

Daylength—Number of effective hours of daylight in each 24-h cycle.

Day-neutral plants—Plants capable of flowering under either short or long daylengths.

DBM paper—Diazobenzyloxymethyl paper.

Deaminase—An enzyme involved in the removal of an amino group from a molecule in which ammonia is liberated.

Deamination—Removal of an amino group, especially from an amino acid.

Decarbolase—An enzyme that liberates carbon dioxide from the carboxyl group of a molecule, e.g., amino acid.

Decarboxylation—Removal of a carboxyl group, -COOH.

Decay—Disintegration or decomposition of plant tissue or other substrates by bacteria, fungi, and possibly other microorganisms.

Deciduous—Describing plants that drop their leaves every fall, or within a year of the time of their production, as compared to evergreens, which retain their leaves for two years or longer; (of fungus spores, etc.) falling away at maturity; shed, either with or without a fragment of the pedicel or sporophore. *See* Persistent.

Declinate—Bent or curved down or forward.

Decline—Reduced vigor of perennial plantings as a result of chronic symptoms of disease; the gradual reduction in health and vigor of a plant or planting that is in the process of slowly dying.

Declivate, declivous—Sloping.

Decolorate—Colorless.

Decompose (n. **decomposition**)—Degradation into simpler compounds; rotting of colonized plant tissue, usually by microorganisms.

Decoration test—*See* Immunosorbent electron microscopy (ISEM).

Decorticate, decorticated (of dead wood)—Woody limbs or trunks without bark; having no cortex.

Decumbent—Resting on substratum with ends turned up; sprawling, prostrate, but usually with young growth ascending.

Decurrent (of lamellae)—Extending down the stipe or stem in stipitate basidiocarps.

Decurved (of the pileus edge)—Bent down.

Dedifferentiation—The process by which matured cells, chiefly parenchyma cells, revert to a meristematic condition.

Dediploidization—*See* Diplo-.

Deer balls (Lycoperdon nuts or harts' truffles)—Fruitbodies (ascomata) of *Elaphomyces*.

Defective interfering (DI) particles—Virus particles that lack part of the genome nucleic acid of the standard virus and that interfere with the replication of the standard virus.

Defective phage—Phage virus particles or subviral particles unable to multiply in host cells.

Defective virus—A virus that lacks part of its genome, or some function, and is thus unable to replicate fully.

Deficiency disease—A disease (or disorder) resulting from lack of one or more essential elements.

Definition (in taxonomy)—A list of characters of a taxonomic unit, particularly of those characters that distinguish the unit from other (often closely similar) units.

Defoliant—A chemical or method of treatment that causes only the leaves of a plant to drop off or abscise.

Defoliate—To lose or become stripped of leaves.

Degeneracy of the code (in virology)—The code is degenerate because 64 codons specify 20 amino acids. Most amino acids are coded for by several codons that differ in the identity of the third base; thus changes in nucleic acid sequence do not necessarily result in changes in amino acid sequence. *See* Genetic code, Wobble hypothesis.

DEGO—*See* Dorsal esophageal gland outlet or orifice.

Dehisce (n. **dehiscence,** adj. **dehiscent, dehiscing**)—Opening spontaneously when mature by pores, breaking into parts, or splitting along definite seams or sutures; opening up and shedding seed; typical of dry, nonfleshy fruits, e.g., pods or capsules; (of asci or fruitbodies) opening when mature by pores or being broken into parts.

Dehiscence papilla (of Blastocladiaceae)—Small rounded projection on the surface of an undischarged zoosporangium that becomes converted to an exit tube.

Dehulled seed—Seed from which pods, glumes, or other outer coverings have been removed.

Dehydrate (n. **dehydration**)—To reduce water content; to become dry.

Dehydrogenase—An enzyme that effects oxidation of a substrate by removing hydrogen from it.

Dehydrogenation—Removal of hydrogen, as from a molecule.

Deirids or **cervical papillae** (of nematodes)—Paired, porelike organs located in the lateral fields in the vicinity of the nerve ring of many of the Tylenchida; believed to be sensory in nature.

Delayed dormant spray—A spray applied mostly to fruit crops and shade trees in late winter or early spring when the new green tips are $1/8$ to $1/4$ in. out.

Deletion mutant (in virology)—A mutant generated by the loss of one or more nucleotides from a coding sequence.

Delignification—Chemical, usually enzymatic, removal of lignin from xylem, leaving a cellulosic residue.

Deliquescent (of lamellae)—Dissolving or liquefying after maturing.

Deltoid—Triangular in shape.

DeMan's values, formula, or indices (of nematodes)—A series of measurements and ratios that expresses body length, relative body width, relative lengths of the esophagus and tail, position of the vulva, etc. See "a", "b", "c", "L", "o", "T", and "V."

Dematiaceous—Pigmented more or less darkly. See Moniliaceous.

Demicyclic—Life cycle of rust fungi comprised of spermogonia, aecia, and telia but lacking uredinia (characteristic of many species of *Gymnosporangium*).

Denaturation—The destruction of secondary and tertiary structure of a protein, nucleic acid, or virus by physical or chemical means.

Denatured protein—A protein that has been altered by treatment with a physical or chemical agent to change its properties.

Dendriform, dendritic, dendroid—With repeated, irregular, treelike branching; treelike.

Dendrochin—An antifungal antibiotic from *Dendrodochium toxicum* that is toxic to farm animals.

Dendrohyphidium—*See* Hyphidium.

Dendroid—*See* Dendriform.

Dendrophyses—Generally branched paraphyses covered throughout with spines.

Denigrate—Blackened.

Denitrification—Conversion of nitrogenous compounds (nitrates and nitrites) in soil into gaseous nitrogen or nitrogen oxides by denitrifying bacteria.

Density (buoyant density)—The density at which macromolecules will band in an isopycnic gradient.

Density gradient—A gradient of a solute in a solvent used to support macromolecules during their fractionation. Usually applied to the separation of macromolecular species by centrifugation or electrophoresis.

Density-gradient centrifugation—A method of centrifugation in which the particles (or components) of a partially purified virus are separated into layers according to their differing buoyant densities.

Dentate, dentoid—Toothed; tooth-shaped projections.

Denticle, denticula (pl. **denticulae**)—A cylindrical or tapered, small tooth-like projection, especially one on which a spore (conidium) is borne; a term usually used in sympodial conidiogenesis; (of nematodes) prickles located in the stoma of some species.

Denticles (of nematodes)—Minute teeth, projections, or prickles.

Denticulate (conidiogenous cell, conidiophore)—Bearing denticles; having small teeth; minutely toothed.

Denuded (of fungi)—Uncovered or glabrous by loss of scales, etc.

Deoxyribonuclease (DNase)—An enzyme that degrades DNA.

Deoxyribonucleic acid (DNA)—The gene bearing material of each plant and animal cell; the principal constituent of chromatin. *See* DNA.

Deoxyribose—A five-carbon sugar, one of the components of DNA. It has one oxygen atom less than its parent sugar ribose.

Depauperate—Undeveloped or poorly developed due to unfavorable conditions.

Dependent—Hanging down.

Dependent transmission—Transmission of a virus (by aphids) that occurs only when the vector feeds on a source plant jointly infected by a second virus. The second virus is referred to as a *helper virus*, and the virus not transmissible on its own is called the *dependent virus*.

Deplanate—Flattened; flat.

Deposit or spray residue—The quantity of a pesticide deposited per unit area of plant, plant part, or other surface at any given time. *Initial deposit* is the amount deposited initially.

Depressed—Sunken; (of a pileus) having the middle part lower than the edge; (of lamellae) sinuate.

Depressor ani—The H-shaped muscle of nematodes that serves to dilate the rectum and elevate the posterior lip of the anus, thus permitting defecation.

Derm, dermium (of basidiomata)—An outer layer in which hyphae are perpendicular to the surface. *See* Cortex, Hymeniderm, Palisoderm, and Trichoderm.

Dermal—Pertaining to the skin.

Dermal toxicity—Ability of a compound when absorbed through the skin of animals or humans to produce symptoms of poisoning.

Dermato- (prefix)—Skin.

Dermatocyst, dermatocystidium—*See* Cystidium.

Dermatomycosis—*See* Mycosis.

Dermatophyte—A fungus that parasitizes keratinized tissue (hair, skin, nails) of humans and animals causing *dermatophytosis* (pl. *dermatophytoses*), e.g., tinea, ringworm, and athlete's foot. These fungi are typically Hymenophytes with teleomorphs in the Arthrodermataceae (onygenales), and have commonly been treated as a special group of "Dermatophytes," or Ringworm Fungi. The main genera include *Epidermophyton, Microsporum,* and *Trichophyton*. They are distinguished mainly by characteristic microconidia.

Dermatophytid—A pustular allergic eruption of the skin at a distance from a primary infection caused by a dermatophyte.

Desalination—Removal of salts from saline soil or water.

Desert—A formation (or biome) characterized by scanty rainfall and usually sparse vegetation with ability to withstand drought and conserve moisture.

Desiccant—Substance that accelerates drying of plant or other tissues.

Desiccate (n. **desiccation**)—To dry up; to remove water; a means of preservation.

Detergent—A substance used for cleaning.

Determinant (in virology)—Antigenic site or epitope.

Determinate—Not continuing indefinitely at the tip of an axis, e.g., having cessation of growth of a structure when a fixed size or shape is attained; clearly marked; definite; (of conidiophores) growth ceasing with the production of terminal conidia; (of flowering) uniform flowering of plant species within certain time limits, allowing most of the fruit to ripen about the same time.

Detersile (of villosity)—Removable to leave the surface bare.

Detoxification—The inactivation or destruction of a toxin by alteration, binding, or breakdown of the toxic molecule to nontoxic (noninhibitory) products.

Deuteroconidium (of dermatophytes)—Sporelike cell from the division of a hemispore (protoconidium).

Deuterogamy—Any secondary process of fertilization or secondary pairing that replaces or compensates for the suppression of fusion of gametes as in certain fungi, macroalgae, and phanerogams.

Deuteromycotina—Subdivision of asexual (anamorphic, imperfect) fungi. Sexual states of these fungi, where known, are in the Ascomycotina and Basidiomycotina. *See* Fungi Imperfecti.

Devalidated (of names)—Names that would have been validly published under the Code, except for the operation of Article 13. Such names may be "revalidated" or "taken up" by post-starting-point authors.

Devil's cigar—*Urnula geaster*.

Devil's snuffboxes—Puffballs.

Dextran—A polysaccharide (glucose polymer) formed outside the cell by enzymatic action of certain microorganisms.

Dextrinoid (of spores, hyphae, etc.)—Turning yellowish brown, red, or bright reddish brown in Melzer's iodine reagent; pseudoamyloid. *See* Amyloid.

Dextrose—Glucose; a hexose sugar.

DI particle = Defective interfering particle.

Diagnosis (pl. **diagnoses**)—Identification of the nature and cause of a disease problem; a shortened, Latin version of a taxonomic description of a species or other taxon.

Diagnostic—A distinguishing characteristic important for identification of disease or other condition.

Dialysis—Separation of soluble substances from colloids and high-molecular-weight solutes by diffusion through a semipermeable membrane.

Diaphanous—Transparent or nearly so.

Diaphyllous—Across the leaf.

Diaporthaceous—Formed as in the genus *Diaporthe*.

Diaporthin—An antibacterial wilt toxin from *Cryphonectria parasitica* (syn. *Endothia parasitica*).

Diaspore—Any unit of dissemination, e.g., a spore, mycelial fragment, sclerotium, etc.; (of lichens) especially applied to vegetative propagules.

Diastase—Enzyme that brings about the hydrolysis of starch with the formation of glucose.

Diastatic action—Conversion of starch into simpler carbohydrates, such as dextrins and sugars, by means of enzyme(s), e.g., diastase.

Diatomaceous earth—A geologic deposit of fine, grayish, siliceous material consisting chiefly of the dead, empty valves of diatoms; used as a diluent and conditioning agent in dust formulations.

Diazobenzyloxymethyl (DBM) paper—An activated paper used to bind nucleic acid covalently for hybridization procedures; especially useful for binding RNA in Northern blotting experiments.

Dicaryon (dikaryon)—*See* Dikaryon.

Dichogamy—Maturation of male and female flowers at different times, ensuring cross-pollination as in maple (*Acer*) and walnuts (*Carya*).

Dichohyphidium, dichophysis—*See* Hyphidium.

Dichotomous (n. **dichotomy**)—Dividing into two more or less equal arms (parts); type of branching in which the main axis or stem forks repeatedly into two more or less equal arms; (of branching hyphae) branching by repeated equal forking.

Dichotomous key—Natural key with two or more choices to consider at any point leading to the next choice and, eventually, the name of the taxon; supposedly reflecting a degree of phylogenetic relationship between taxa.

Diclinous—In oomycetes, having the oogonium and its antheridium on different hyphae. *See* Androgynous.

Dicotyledon, dicot (pl. **dicotyledonae**, adj. **dicotyledonous**)—A flowering plant having two cotyledons (seed leaves) in an embryo in contrast to monocotyledons (e.g., grasses and cereals). *See* Monocotyledoneae.

Dictydine granules—*See* Plasmodic granules.

Dictyochlamydospore—A nondeciduous, multicelled chlamydospore composed of an outer wall separable from the walls of the component cells that are rather easily separated from each other, as in some *Phoma* species.

Dictyoporospore—Deciduous, multicelled porospore the cells of which are firmly united and not enclosed by an outer wall. *See* Porospore.

Dictyosomes (of fungi)—More or less spherical vesicles associated with the edges of the membrane-bound sacs (cisternae) that make up the Golgi apparatus in oomycetes and other fungi, as seen by electron microscopy.

Dictyospore (adj. **dictyosporous**)—Spore with both longitudinal and transverse septa (e.g., *Alternaria*); a muriform spore.

Didelphic (of nematodes)—A female having two complete genital tubes or ovaries.

Didymospore (adj. **didymosporous**)—Spore differing from an amerospore in having one transverse septum; a two-celled spore. *See* Amerospore.

Dieback (v. **die back**)—Progressive death of shoots, branches, or roots generally starting at the tip. Dieback may be due to cankers, stem or root rots, insect borers, nematodes, winter injury, deficiency or excess of moisture or nutrients, some other factor, or a stress complex.

Diel—A 24-h periodicity. *See* Circadian and Diurnal.

Diethylaminoethylcellulose (DEAE-cellulose)—An ion exchange medium used in chromatography of proteins and nucleic acids.

Differential centrifugation—Cycles of low-speed (to clarify) and high-speed (to sediment) centrifugation or spinning used in the purification of a virus.

Differential hosts—Special species, cultivars, inbreds, or isogenic lines of plants varying in susceptibility to a given disease agent, such that their distinctive symptoms to the agent facilitate its identification, especially as to physiologic races. *See* Indicator plant.

Differentially permeable—Referring to membranes that allow certain substances to pass through and that retard or prevent the passage of others.

Differentiation—Development of an organism from one cell to many cells together with modification of the new cells, tissues, or organs for the performance of definite functions; maturation.

Diffluent—Readily dissolving or breaking up in water.

Diffract (of a pileus surface)—Cracked or broken into small areas; areolate.

Diffuse—Widely or loosely spreading without a distinct margin.

Diffuse (fibrous) root system—A root system in which there is no single main root larger than other roots.

Diffuse wall building—*See* Wall building.

Diffuse-porous wood—A type of wood in which the vessels are more or less uniform in size and distribution throughout each growth ring.

Diffusion—The spreading of a substance (molecules) throughout available space from high to low concentrations of that substance, as a result of molecular motion.

Diffusion coefficient—A measure of the rate at which a solute moves along a gradient from a higher to a lower concentration.

Diffusion pressure—Pressure exerted by the molecules of a diffusing substance.

Digestion—The breakdown or transformation of insoluble or complex foods into soluble or simpler substances, through the action of enzymes.

Digitate—Having lobes or fingerlike processes radiating from a common center.

Digonic (of nematodes)—Pertaining to hermaphroditic reproduction in which sperm and eggs are produced in different gonads of the same individual.

Dihybrid cross—A cross between organisms differing in two characters.

Dikaryon (adj. **dikaryotic**)—A cell or strain of fungal cells with two genetically distinct haploid nuclei (n+n); a specialized binucleate condition common in the Basidiomycotina and Ascomycotina.

Dikaryoparaphysis—*See* Hyphidium.

Dikaryotization—Conversion of a homokaryon into a dikaryon, usually by the fusion of two compatible homokaryons.

Dilacerate—Torn apart; lacerated.

Diluent (of pesticides)—An essentially inert or nonreacting gas, liquid, or solid material used to dilute the concentration of the active ingredient in a formulation to the desired concentration. *See* Carrier.

Dilution—The process of increasing the proportion of solvent or diluent to solute or particulate matter, e.g., bacterial cells.

Dilution end point—The last point at which infectivity or other activity remains in a progressive dilution series; the extent to which sap from a virus-infected plant can be diluted with water or phosphate buffer before its infectivity is lost on a mechanically inoculated plant; the greatest dilution of an antibody that gives a measurable reaction with an antigen in a serological test.

Dilution, serial—Successive dilution of a specimen, e.g., one containing bacterial cells. A 1:10 dilution equals 1 ml of specimen plus 9 ml of diluent (e.g., sterile water); a 1:100 dilution equals 1 ml of a 1:10 dilution plus 9 ml of diluent, etc.

Dimerous (of basidia)—Having a constriction between the probasidium and the metabasidium.

Dimethyl sulfoxide (DMSO)—A solvent used 1) for dissolving both inorganic and organic chemicals; 2) for substances being applied to cells in tissue culture; 3) in the preparation of cells for storage in liquid nitrogen; 4) to strand-separate double-stranded nucleic acids.

Dimidiate—Having one half very much smaller than the other; (of a perithecium) having the outer wall covering only the top part; half structure; (of pileate basidiocarps) without the stalk and semicircular in outline as viewed from above; (of lamellae) stretching only half way to the stipe.

Dimitic (of hyphal structure)—Having two different types of hyphae, generative and skeletal.

Dimorphic, dimorphism—Having two distinct shapes or forms (e.g., yeast and mycelial); producing two types of zoospores; some male and female nematodes.

Dimple—*See* Nassacc.

Dioecious aphid—An aphid whose life cycle alternates between a primary and secondary host plant. *See* Monoecious aphid.

Dioecism (adj. **dioecious**)—Having the staminate and pistillate flowers on different individuals; sexually distinct; (of fungi) condition in which the male and female sex structures are on different thalli; (of higher plants) having either staminate or pistillate flowers, but not both, e.g., asparagus, holly, and date palms. *See* Monoecism and Heterothallism.

Diorchic—Describing a male nematode having two complete genital tubes or testes.

Diorchidioid—Resembling the rust fungus *Diorchidium*; two-celled teliospores, with the septum mostly vertical or diagonal; most often applied to species of *Puccinia*.

Diphenylamine—A chemical used for the colorimetric determination of DNA (RNA does *not* give a color reaction). *See* Orcinol.

Diphtheritic—Diptheria-like.

Diphycophilous—Lichenized fungi with both a green and blue-green photobiont.

Diplanetism (adj. **diplanetic**) (of zoospores in the oomycetes)—Condition in which there are two motile stages of zoospores with a resting stage in between.

Diplo- (prefix)—Two; twice; double.

Diplobiontic (of an organism)—Having free-living haploid and diploid generations.

Diplococci—Cocci occurring in pairs.

Diploconidium—A binucleate conidium.

Diplodioecious—*See* Heterophytic.

Diplogasteroid—*See* Esophagus.

Diploid—Applied to the 2n condition, the chromosomes being twice as many in the nucleus of a cell as in the haploid condition (1n); characteristic of the sporophyte generation.

Diploidization—Process by which a haploid cell (or mycelium) is converted into a diploid cell (mycelium) having conjugate nuclei.

Diplokaryon—*See* Synkaryon.

Diplophase—The phase of the life history of a fungus or other organism in which the cells are diploid (2n).

Diplostichous—In two lines, rows, or series.

Diplostromatic—Fungus with both an ecto- and endostroma.

Direct (of fruitbody growth)—With cell enlargement occurring simultaneously with cell division. *See* Indirect.

Disaccharide—A sugar composed of two monosaccharides.

Disarticulation—Breaking or falling away from the plant.

Disbudding—Pinching off of unwanted buds to control branching or to produce larger flowers from remaining buds.

Disc—(Of a pileus) central part of the top surface; (of discomycetes) round, platelike or curved, spore-producing part of a fruit body.

Disc fungi—The Discomycetes.

Disciform, discoid—Flat and circular; resembling a disk or plate, usually with a central attachment.

Discocarp—An ascocarp in which the hymenium lies exposed when the asci and ascospores are mature; an apothecium.

Discolorous—Of a different color.

Discomycetes—The cup fungi, a class of Ascomycotina, with asci produced in apothecia.

Discontinuous or discontiguous epitope—An antigenic site that arises when two separate peptide chains are brought together or when a single peptide chain is folded.

Discontinuous resistance—A response involving distinctive, clear-cut symptoms in a segregating population; often controlled by a single dominant gene.

Discothecium—An ascostroma resembling an apothecium but bearing cylindrical bitunicate asci; differs from a hysterothecium by the weathering away of the covering layer.

Discrete—Separate; not joining; distinct; not in clusters; (of a conidiogenous cell) separate from the main axis or branches of a conidiophore and often having a distinctive shape.

Discrete body—A nonfunctional cleistothecial initial of a dermatophyte in culture; pseudocleistothecium.

Disease complex—A plant disease caused by the interaction of two or more pathogens; often manifested by a greater-than-normal array of symptoms.

Disease cycle—The sequence of events involved in disease development, including the stages of development of the pathogen and the effect of the disease on the host; the chain of events that occurs between the time of infection and the final expression of disease.

Disease gradient—Change in incidence of a disease with increasing distance from the source of infection.

Disease pyramid—A concept describing the four factors needed simultaneously for a plant disease to occur: a susceptible host, presence of a virulent pathogen, a favorable environment for infection, and time for the disease to develop.

Disease range—The geographic distribution of a disease. *See* Range.

Disease tolerance—Capacity of a plant to maintain fairly normal vigor, without excessive injury or loss in yield, even though a pathogen is established within the plant. *See* Tolerance.

Disease triangle—A concept describing the simultaneous occurrence of a pathogen, a susceptible host, and a favorable environment such that a disease may develop. *See* Disease pyramid.

Disease, plant—Any disturbance of a plant that interferes with its normal growth and development (e.g., structure and function), economic value, or aesthetic quality and leads to development of symptoms. A continuously, often progressively affected condition in which any part of a plant is abnormal (e.g., structure, function, or economic value) or that interferes with the normal activity of the plant's cells or organs. Injury, in contrast, results from a momentary damage. *See* Disorder and Injury.

Disinfectant—A chemical or physical agent that kills pathogenic microorganisms (fungi, bacteria, nematodes) once a plant, or any of its parts, has become infected or infested; same as germicide.

Disinfection—Freeing a diseased plant or plant parts from infection; or the destruction of a disease agent or disease-inducing organism in the immediate environment of the host plant.

Disinfestant—A chemical or physical agent that removes, kills, or inactivates disease-causing organisms *before* they can cause infection. It may be applied on the surface of a seed, other plant part, tools, or in the soil.

Disinfestation—Killing or inactivating disease organisms before they can cause infection on the surface of seeds or other plant parts, tools, or in soil.

Disintegrations per minute (dpm)—A direct measurement of radioactivity; the actual number of radioactive disintegrations in the sample. It is estimated from the counts per minute measured in a radioactivity counter and the quenching of counts in the sample.

Disjunctor—Thin-walled cell or projection (a disjointer) between spores in a chain; a connective.

Disk—*See* Disc.

Disorder—A harmful nonpathogenic deviation from normal growth; abiotic disease. *See* Disease.

Dispergent—Scattering; spreading.

Dispersal—*See* Dissemination.

Dispersal spore—A spore disseminated by wind, water, insects, or other agent; diaspore.

Disperse—1) To break up compound particles, such as aggregates, into the individual component particles; 2) to distribute or suspend fine particles, such as clay, in or throughout a dispersion medium.

Dispore—One of the spores of a two-spored basidium as opposed to a *tetraspore*, one of the spores of a four-spored basidium.

Dissected—Deeply cut into numerous segments, as a leaf.

Dissemination (dispersal)—In relation to plant diseases, the transfer (spread) or transport of infectious material (inoculum) to healthy plant tissue by wind, water, insects, animals, machinery, humans, or other means. *See* Agent of inoculation.

Dissepiment—A partition or wall, e.g., end wall of one of the united tubes that compose the hymenophore of polypores.

Dissociation (of bacteriology)—Variation of organisms from the parent, particularly in colony form, but in other characteristics as well.

Distal—Remote from the point of attachment or origin; away from the center of the body; terminal. *See* Proximal.

Distichate, distichous—In two lines or rows.

Distichously—In a two-ranked arrangement.

Distoseptate (of conidial septation)—Having each individual cell surrounded by a sac-like wall distinct from the outer wall as in "*Helminthosporium.*" *See* Euseptate.

Distribution—Spread of a pathogen to areas outside of its previous geographical range; "geographical distribution" is synonymous with "range."

Disulfide bonds—Chemical bonds between sulfhydryl-containing amino acids that bind polypeptide chains together; a bond that forms when two sulfhydryl (-SH) groups of cysteine side chains of a protein are close together and are oxidized by the same reagent.

Dithiothreitol (DTT), or Cleland's reagent $(CHOHCH_2SH)_2$—A very useful reagent for maintaining sulfhydryl groups in a reduced state, thus disrupting disulfide bonds. It has little tendency to be oxidized by air.

Diurnal—Occurring daily; or during daylight hours as opposed to nocturnal. *See* Circadian and Diel.

Divaricate—Spreading; divergent at right angles.

Diverticillate—Having two whorls.

Diverticule, diverticulum—A pouchlike side branch, as on mycelium of *Pythium*.

Division—A taxonomic category made up of closely related classes, used in classification of plants and fungi; a propagation method in which underground stems or roots are cut into pieces and replanted.

DMSO—Dimethyl sulfoxide.

DNA (deoxyribonucleic acid)—A molecule occurring in nuclei of plant and animal cells and composed of repeating subunits of nucleotides containing deoxyribose (a five-carbon sugar), phosphoric acid, and one of four nitrogenous bases (adenine, cytosine, guanine, and thymine). Every inherited characteristic has its origin in the code of an individual's complement of DNA. The linear chain of deoxyribonucleotides is double-stranded in some DNA viruses (dsDNA) and single-stranded in others (ssDNA). dsDNA is held together by bonds between base pairs of nucleotides (adenosine, guanosine, cytidine, and thymidine); bonding occurs only between adenosine and thymidine and between guanosine and cytidine. Therefore, it is possible to determine the sequence of either strand from that of its partner. Also in dsDNA, the two strands are antiparallel in a double helix. Purine bases of one strand form hydrogen bonds with pyrimidine bases of the other strand. In A-DNA the base pairs are slanted with respect to the helix axis and are pulled away from the center. In B-DNA the helix adopts a right-handed conformation with each chain making a complete turn every 10 bases. In Z-DNA the double helix winds in a left-handed manner with the purine and pyrimidine bases alternating, giving it a zig-zag formation, hence the name Z-DNA. *See* Nucleic acid.

DNA cloning—A means of isolating individual fragments from a mixture and multiplying each to produce sufficient material for future analysis.

DNA exonuclease—*See* Endonuclease and DNA-dependent DNA polymerase.

DNA gyrase—An enzyme that introduces negative superhelical twists into relaxed, closed circular DNA. Several DNA phages (e.g., T4) require its use during replication.

DNA ligase—An enzyme that catalyzes phosphodiester bond formation between the 5′-phosphate of one oligonucleotide and the 3′-hydroxyl of another. It is involved in DNA synthesis and in linking DNA in genetic manipulation. The ligase-produced hyphase T4 is widely used in genetic manipulation.

DNA primase—*See* Primase.

DNA sequencing—The relative order of nucleotide pairs in a stretch of DNA, a gene, a chromosome, or an entire genome.

DNA-binding protein—A protein that binds directly to DNA; often determined by specific amino acid sequences. These proteins have a number of functions including maintaining DNA in a single-stranded form for transcription or replication.

DNA-dependent DNA polymerases (I, II, III alpha, beta, gamma)—Enzymes that synthesize DNA from a DNA template.

DNase—An enzyme that breaks down DNA by hydrolysis.

Dodder—Any plant of the genus *Cuscuta* of dicotyledonous leafless vines that are deficient in chlorophyll and parasitic on other plants. Also called goldthread, strangleweed, hellbind, and love vine.

Dodecahedron (of viruses)—One of the five Platonic solids, characterized by 12 pentagonal faces.

Dog lichen—Species of *Peltigera*, especially *P. canina*, which have been used in folk lore for bites of mad dogs.

Dog stinkhorn—*Mutinus caninus.*

Dolabrate, dolabriform—Hatchet-shaped.

Dolicho- (in Greek combinations)—Long.

Dolichospore—A long spore.

Doliiform—Barrel- or cask-shaped to broadly subglobose.

Dolipore septum—Septum in a dikaryotic basidiomycete hypha that flares out in the middle, forming a barrel-shaped structure with open ends, as shown by electron microscopy; septal pore swelling. *See* Parenthesome.

Dollar spot—A turfgrass disease presumably caused by *Sclerotinia homeocarpa* or species of *Lanzia* and *Moellerodiscus,*

Dolomitic limestone—A limestone rich in magnesium carbonate, $CaMg(CO_3)_2$.

Domains—Segments of polypeptide chains folded together and stabilized by disulfide bonds.

Dominance—A genetic principle, established by Mendel, that when two contrasting characters are brought together in a hybrid as a result of a cross, one character (dominant) may mask the other (recessive).

Dominant gene—A gene that is fully expressed in the phenotype of the heterozygote; refers to the gene (or the expression of the character it influences) that, when present in a hybrid with a contrasting gene, completely dominates in the development of the character. *See* Recessive gene.

Dormant (n. **dormancy**)—Being in a state of reduced physiological activity occurring in seeds, buds, bulbs, corms, or tubers; resting; quiescent.

Dormant period—Time during that no growth occurs.

Dormant spray—A spray applied in late fall, winter, or early spring when plants are in a dormant or nongrowing condition. The temperature at time of application should be 5–8 C° (40–45°F) or above.

Dorsal—Referring to the back or upper surface of an organism. *See* Ventral.

Dorsal esophageal gland or **dorsal gland** (of nematodes)—The only dorsally located of the three to nine salivary glands; all the rest are in the subventral position in the basal portion of the esophagus.

Dorsal esophageal gland outlet or orifice (DEGO)—The point at which the duct from the dorsal gland in a nematode empties into the lumen of the esophagus.

Dorsiventral (botany)—Having distinct dorsal and ventral sides, e.g., most leaves.

Dorsoventral (of nematodes)—The imaginary median line extending from the dorsal to the ventral side; the dorsoventral axis.

Dorylaim—A common name for a nematode member of the order Dorylaimida.

Dorylaimoid (of nematodes)—*See* Esophagus.

Dosage, dose—Quantity of a substance applied per unit of plant, soil, or other surface; also called *rate*.

Dothideaceous—Having asci in locules embedded in a stroma with no distinct perithecial walls, as in *Dothidea*.

Double antibody sandwich (DAS)—A method in enzyme-linked immunosorbent assay (ELISA) in which the reactants are added to the test plate in the order of antibody, virus, antibody-enzyme complex, and enzyme substrate. *See* ELISA and Enzyme-linked immunosorbent assay.

Double clamp connection—A system with two clamp connections.

Double fertilization—The process of sexual fertilization in flowering plants, in which one nucleus from the male gametophyte fertilizes the egg nucleus to form the zygote and a second nucleus from the male gametophyte fertilizes two polar nuclei to form endosperm tissue.

Double resistance (durability)—Describing long-lasting resistance.

Double-diffusion test—Antigen-antibody precipitation reaction in agar or a similar gel in which antigen and antibody are allowed to diffuse toward one another and react at equivalence in the agar.

Downy mildew—A plant disease in which the mycelium, sporangiophores, and spores of the fungus appear as a downy growth on the host surface; caused by fungi in the oomycete family Peronosporaceae. *See* Powdery mildew.

DR curve—A log-probit, dosage-response curve.

Drift (of pesticides)—The movement of a portion of the airborne particles of a spray or dust away from the target area during or shortly after application.

Drip irrigation—A method of watering plants so that only the soil in the plant's immediate vicinity is moistened. Water is supplied by a thin plastic tube at a low rate of flow; sometimes called trickle irrigation.

Droplet nucleus—Airborne particles containing viable microorganisms (microbes).

Drumstick (capitate)—Describing the shape of bacteria sporangia, indicating that one end (that contains the terminal spore) is larger than the other.

Drupe—A simple, fleshy, indehiscent fruit, derived from a single carpel, usually one-seeded, in which the exocarp is thin, the mesocarp fleshy, and the endocarp hard and stony. Example: peach.

Druse—A stellate cluster of large crystals in the thallus of a lichen.

Dry land farming—A practice used in areas where annual rainfall is low and soil moisture is conserved by dust mulching, weed control, and in other ways to produce a crop every other year.

Dry matter percentage—The percent of the total fresh plant material left after water is removed. The percentage is calculated by weighing a sample of fresh plant material, oven-drying the sample, then reweighing the dried sample. Dry matter percentage = dry weight divided by fresh weight × 100.

Dry rot or house rot fungus—*Serpula lacrimans*.

Dry rot—Disintegration (enzymatic) of a substrate by fungi without release of liquids. *See* Soft rot.

Dry spore—A spore separated without slime from the parent hyphal cell. *See* Slime spore.

Dryad's club—*Clavaria pistillaris*.

Dryad's saddle—*Polyporus squamosus*.

ds—Double-stranded (nucleic acid molecules) with two linear nucleotide chains.

Dual phenomenon (in Deuteromycotina)—A condition in which a fungus is composed of two culturally distinct elements or individuals.

Duplex (of the context in pileate basidiocarps or polypores)—Consisting of two distinct zones of different textures. The layer adjacent to the lamellae or tubes is harder than the one over it.

Dust (of pesticides)—A dry, finely ground formulation usually containing 4–10% active ingredient mixed with an inert material, e.g., talc, powdered nut hulls, clay, volcanic ash, and similar materials.

Dutch elm disease—Vascular wilt of elms (*Ulmus*) caused by *Ophiostoma ulmi* (syn. *Ceratocystis ulmi*) or the more aggressive *O. novo-ulmi*; anamorphs in *Pesotum*.

Duvet (of dermatophytes)—A soft, thick, finely matted layer of hyphae.

Dwarf bunt—*See* Bunt.

Dwarfing—The underdevelopment of a plant or plant organs. May be caused by any type of disease agent under certain conditions, by faulty nutrition, or other unfavorable environmental conditions.

Dysgonic (of dermatophytes)—Growing more slowly in culture, commonly with less aerial mycelium than a normal *eugonic* strain.

E

E-—Ratio of length to width (of spores, basidia, etc.).

E-, ex- (prefix)—From; out of; without; not having.

EA—Erythrocyte coated with antibody; will be lysed if complement is added.

Ear—Fruiting spike or head of a cereal or grass plant, including the kernels and protective structures.

Early blight (mostly of potato and tomato)—Disease caused by *Alternaria solani*.

Early genes—Viral genes expressed early in the replication cycle; genes usually involved in the replication of the viral nucleic acid. *See* Late genes.

Earlywood (spring wood)—The less dense part of the growth ring of woody plants composed of cells with thinner walls, a greater radial diameter, and shorter length than those formed later in the year. *See* Latewood (summer wood).

Earth-balls—Fruitbodies of the Sclerodermatales.

Earth-stars—Basidiocarps of *Geastrum*.

Earth-tongues—Ascocarps of *Geoglossum*.

Eccentric (of the stipe)—Not attached to the center of the pileus, one-sided; off center; asymmetrical in growth.

Eccrinid—A member of the Eccrinales or Amoebidiales.

Ecdysis (pl. **ecdyses**)—The act of shedding the outer cuticular layer of an insect, together with internal cuticular linings; occurs between each of its growth stages or *instars*; a molt.

ECe—Symbol for electrical conductivity of a water-saturated soil extract. It is measured in millimhos per centimeter (mmhos/cm); 1 mmhos/cm is equivalent, on the average, to 640 ppm of table salt (NaCl).

Echinate (of spores, scales, etc.)—Having sharply pointed spines.

Echinulate (n. **echinulation**)—(Of spores, conidiophores, etc.) having small, slender, pointed spines or pointed processes on their surface; (of bacterial growth) with toothed or pointed margins along the line of inoculation.

Eclipse period (or phase)—The time between when a virus penetrates into a cell and the appearance of newly synthesized infectious virus.

Eclosion (of nematodes)—The act of hatching from the egg; (of fungi) an explosive series of movements resulting in the release of a germinating inner spore from a rigid exosporium, as in *Hypoxylon fragiforme*.

Ecology (as applied to plant disease)—The study dealing with the effect of environmental factors on the occurrence, severity, and distribution of plant diseases or pathogens. In gnereal, the study of life in relation to its environment.

Economic injury level—The lowest population density of pests that will cause economic damage.

Economic poison—As defined by the U.S. Federal Insecticide, Fungicide and Rodenticide Act (FIFRA), an economic poison is "any substance or mixture of substances intended for preventing, destroying, repelling, or mitigating any insects, rodents, nematodes, fungi, or weeds, or any other form of life declared to be pests." As so defined, economic poisons are now generally known as "pesticides."

Economic threshold level—The density of pests at which control measures should be implemented to prevent an increasing pest population from reaching the economic injury level; pest or pathogen population density or damage level at or above which the value of crop losses, in the absence of management efforts, would exceed the cost of management practices (especially the use of pesticides).

Ecorticate—Lacking a cortex.

Ecosystem—A community of living things and its environment; the interacting, functional system comprised of all living organisms in an area and their nonliving environment.

Ecotype—Genetic variant within a species that is adapted to a particular environment yet remains interfertile with all other members of the species; part of a population of a species showing morphological, chemical, or physiological characteristics that appear to be genetically determined and correlated with particular ecological conditions but that are thought to have no taxonomic significance.

Ectal—Outer; outermost; superficial.

Ectal excipulum—Outer layers, including the hymenium, in a nonlichenized apothecium, sometimes multilayered.

Ectendotrophic mycorrhiza—Mycorrhiza in which there is a Hartig net but the fungus also penetrates the root.

Ecto- (prefix)—Outside; external.

Ectoascus—An outer wall of a fissitunicate ascus, as in *Lecanidion*.

Ectochroic—Having pigment on the outside of the hypha.

Ectomycorrhiza (pl. **ectomycorrhizae**)—Fungus-root association with the fungal hyphae between and external to the root cells.

Ectoparasite—A parasite living on the outside of its host. *See* Endoparasite.

Ectospore—Exogenous spore.

Ectostroma—A fungus stroma on the surface of a host plant.

Ectothecal (of ascomycetes)—Not having the hymenium covered over.

Ectothrix—Living on the surface of hair.

Ectotrophic—Describing fungal development of mycorrhizae, primarily over the root surface.

Ectotropic—Curving out.

Ectotunica—The outer wall in a bitunicate ascus.

ED—Effective dose.

ED$_{50}$—The dose that gives a 50% (usually lethal) response.

Edaphic—Of or pertaining to soil as it influences plant growth.

Edema (adj. **edematous**)—Oedema; a watery swelling of plant organs or parts; an abnormal accumulation of serous fluid in connective tissue or in a serous cavity; in plants often caused by overwatering in cloudy, humid weather when evaporation (transpiration) is reduced.

Eelworm—Nematode, nema. *See* Nematode.

Effete—Dead, no longer sporulating; overmature, exhausted; (of fruiting bodies) empty.

Efflorescent—Bursting out of.

Effuse—Stretched out; flat and spreading; a film-like growth.

Effused (of basidiocarps)—Resupinate, spread out without a regular form and adhering to the substratum; no true pileus is formed.

Effused-reflexed—(Of basidiocarps) with part of the basidiocarp resupinate and part shelving into the pileus growing away from the substratum; (of Hymenomycetes) stretched out over the substratum but with the edge turned up to form a pileus.

Egg—A female gamete; (of phalloids, *Amanita*, etc.) the young basidiomata before the volva is broken; in nematodes, the fertilized egg is the first stage of the life cycle.

Eguttulate (conidia)—Lacking guttules or droplets.

Ejaculatory duct (of nematodes)—Terminal muscular portion of the male vas deferens.

Ejaculatory glands (of nematodes)—Glands associated with the male ejaculatory duct and secreting adhesive cement functional in copulation.

Elater—Hygroscopic structures produced in the sporophytes of liverworts and the spores of horsetails that aid in spore dispersal; a free capillitium thread, e.g., in myxomycetes and *Farysia*; a structure with spiral or annular markings in the gleba of *Battarraea*.

Eldredge tube—A double-compartment culture tube for collecting carbon dioxide given off during growth of a bacterial culture.

Electroendosmosis—Movement of fluids through a gel or other solid matrix during electrophoresis.

Electrofocusing—Method for separating protein molecules by gel or density gradient electrophoresis in which a pH gradient has been established. Every protein moves to a pH approximating its isoelectric point.

Electron microscope serology—*See*, or a gel. *See* Chemotaxonomy and Chromatography.

Electrophoretic mobility—The relative rate of movement of a charged particle or virus per unit potential gradient. Mobility may be toward the cathode or the anode depending on whether the molecule or virus has a net positive or negative charge at the pH used. Immunosorbent electron microscopy (ISEM).

Electron microscopy—Use of an electron microscope, in which a focused beam of electrons produces a greatly enlarged image of minute objects (such as virus particles), in the same way as light forms an image in a compound microscope. *See* Scanning electron microscope and Transmission electron microscope.

Electrons—Negatively charged particles that form a part of all atoms.

Electropherogram—A picture (autoradiograph or photograph) showing the distribution of proteins or nucleic acids separated by gel electrophoresis.

Electrophoresis—The differential migration of charged molecules and macromolecular ions (e.g., fragments of nucleic acid or proteins) in a free solution or through a porous medium in an electric field; the porous supporting medium may be filter paper, cellulose acetate

Electroporation—A method using an electric pulse whereby nucleic acids or virus particles can be introduced into protoplasts or cells by creating transient pores in the plasma membrane.

Element—A substance that cannot be divided or reduced by any known chemical means to a simpler substance; 92 naturally occurring elements are known.

Elf-cups—Ascocarps of Pezizales; scarlet elf-cups—*Sarcoscypha coccinea.*

Elicitor—A chemical compound or physical process produced by the host (or pathogen) that induces a response by the pathogen (or host).

ELISA—An immunoassay system employing enzyme bound to antibody as the immunologic probe determining the extent of antigen-antibody reaction; a widely used method for detecting and quantifying antibodies. *See also* Enzyme-linked immunosorbent assay.

Elite seeds—Seeds selected from basic stocks of known origin, cultivar purity, and freedom from disease and protected from contamination by sanitation and isolation.

Elliptic, ellipsoid, ellipsoidal (of spores, etc.)—Having the shape of an ellipse; elliptical in optical section; ovate or ovoid (egg-shaped).

Elliptic-fusiform (of spores)—Fundamentally fusiform but somewhat elliptical; more fusiform than elliptical.

Elongation factor—Protein-forming part of the ribosomal binding complex that promotes elongation of a polypeptide chain.

Elutriator—A device used to separate nematodes from soil by washing. *See* Christie-Perry and Seinhorst extraction techniques.

Eluviation—The removal of soil material in suspension (or in solution) from a layer or layers of a soil. The loss of materials in solution is usually described as leaching.

Elytron (pl. **elytra**)—Horny, veinless, front or upper wing that serves as protection; found in the beetles (Coleoptera) and some insects in Homoptera.

Emarginate—(Of lamellae) notched near the stipe; (of apothecia) lacking an excipulum thallinum (lichens) or a raised excipulum proprium. *See* Sinuate.

Emasculate—The removal of anthers from a bud or flower before pollen is shed; a normal preliminary step in hybridization to prevent self-pollination.

Embargo—A government order prohibiting transport of plants or plant parts into a country or region to prevent spread of disease, pathogens, or vectors.

Embossed—Having a small central hump; "umbonate."

Embryo—An organism in the early stages of development; a nematode before hatching from the egg; a young sporophytic plant, before the beginning of its rapid growth; the part of the seed that has developed from the fertilized egg; a rudimentary plant.

Embryo sac—Typically an eight-nucleate female gametophyte of an angiosperm; a structure within an angiosperm ovule in which the egg is fertilized. An embryo begins its development within an embryo sac.

Emergence—Appearance of the shoot above the soil surface.

Emetic—Something (e.g., swallowing soapy or salt water) that induces vomiting to get rid of poison in the stomach.

Empty virus particles—Virions that contain no nucleic acid and can be identified by negative staining and electron microscopy. These particles have a lower buoyant density than complete virus particles.

Emulsifiable concentrate (EC)—A liquid formulation of a pesticide containing a high concentration of active ingredient dissolved in a solvent together with an emulsifier. Designed to be diluted, usually with water, to the desired strength before application. When the EC is added to water, a milky mixture is formed that is a suspension of active ingredient and emulsified solvent in water.

Emulsifying agent (emulsifier)—A surface-active agent that facilitates the suspending of minute droplets of one liquid in another to form a stable mixture.

Emulsion—A dispersion system in which minute droplets of one liquid are suspended in another liquid, both liquids being immiscible, e.g., oil droplets in water.

Enation—A small abnormal outgrowth of host tissue or eruption from a plant surface, often from veins (mostly from leaves, petioles, and flowers); usually induced by certain virus infections; literally "small leaf." *See* Hyperplasia.

Encapsidated—Enclosed as if in a capsule.

Encapsidation—The enclosure of a virus's nucleic acid genome within a protein shell.

Encapsulation—Method of disposal of pesticides and pesticide containers by sealing them in a sturdy, waterproof container so the contents cannot escape. Also, a method of formulating pesticides.

Encrusted (of cystidia, etc.)—Covered with a crust of crystals or resinous matter.

Encyst (n. **encystment**)—To form a cyst or encasement.

Endemic—Pertaining to a persistent low and steady level of normal disease occurrence; a disease native to or restricted to a certain country or geographic region; also plant species native to a particular environment or locality. *See* Enphytotic.

Endo- (prefix)—Inside; within.

Endoascospores—Sporelike cells produced within ascospores.

Endoascus—The inner, often extensible wall layers of a bitunicate ascus.

Endobasidium—Basidium developing within the basidiocarp (e.g., gastromycetes).

Endobiotic—Growing within living organisms.

Endocarp—Inner layer of the fruit wall (pericarp).

Endocarpous (of gasteromycetes, etc.)—Having the mature hymenium covered over; angiocarpous.

Endochroic—Pigment inside a cell; *cytochroic* pigment is diffused in the cytoplasm; *cystochroic* pigment is in cell valuoles; and *lipochroic* pigment is in oil drops.

Endoconidiophore—A conidiophore that produces conidia within itself.

Endoconidium (pl. **endoconidia**)—A conidium formed within a hypha (endoconidiophore) and extruded from the tip, as in *Thielaviopsis basicola*.

Endocyclic—Describing a microcyclic life cycle of a rust fungus in which aeciospore-like cells produce basidia.

Endocystidium—*See* Cystidium.

Endodermis—A single layer of cells with thick walls and no intercellular spaces that surrounds the vascular tissues of stems and roots; innermost layer of the cortex and most conspicuous in roots.

Endoectothrix—Organism growing in and on hair.

Endoenzyme—An enzyme found within the cell and not excreted into the medium; intracellular enzyme.

Endogenous—Produced, living, or undergoing development inside, e.g., a conidium entirely within a phialide or conidiogenous cell; immersed in the substratum; (basal appendage) originating inside the conidiogenous cell and therefore truly basal. *See* Exogenous.

Endohypha (intrahyphal hypha)—Vegetative or fertile element initiated by differentiation within a hypha from the innermost wall layer.

Endolithic—In or on stone.

Endomycorrhiza (pl. **endomycorrhizae**)—Fungus-root association in which fungal hyphae invade cortical cells of the root; the most common type is the vesicular-arbuscular mycorrhiza.

Endonuclease—An enzyme that cleaves a polypeptide chain internally.

Endooperculation (of sporangial dehiscence in chytrids)—A process in which the operculum is forced off and carried away by the emerging sporogenous contents.

Endooperculum (pl. **endoopercula**) (of chytrids)—The operculum that develops beneath the plug of hyaline material following deliquescence of the exit tube sporangium.

Endoparasite—A parasite that enters and infects its host. See Ectoparasite.

Endoperidermal—Within the periderm.

Endoperidium—Inner layer of the peridium.

Endophloeodic, endophloeodal, endophloeic (of the thallus of a crustaceous lichen)—Almost entirely immersed in bark.

Endophyllous—Living within leaves.

Endophyte (adj. **endophytic**)—Plant or fungus developing and living inside another plant.

Endophytic system—(Of mistletoes) that part of the parasite that grows within the host; cortical strands and sinkers.

Endoplasmic reticulum—An extensive system of cytoplasmic membranes within the protoplast, the lamellar or tubular system of the colorless cytoplasm in a cell.

Endopropagule (of medical mycology)—Propagule produced inside the body.

Endosaprophytism—Destruction of the lichen alga by the fungus.

Endosclerotium—Sclerotium of endogenous origin.

Endosperm—Nutritive tissue formed within the seeds of angiosperms and on which the embryo feeds while developing. It develops from the sexual fusion of a sperm cell with two polar nuclei of the embryo sac and is thus triploid (3n).

Endospore—A highly resistant and thick-walled asexual spore, formed within a bacterial or fungal cell and capable of surviving exposure to high temperatures or long periods of dryness; inner wall of a spore.

Endosporium—The inner wall of a double-walled spore; is usually thin.

Endostroma, entostroma—Perithecial stroma formed under the ectostroma.

Endothermic—A chemical reaction in which energy is consumed overall.

Endothrix—Living inside a hair.

Endotoxin—Toxin produced in an organism (e.g., bacterial cell) and liberated only when the organism disintegrates, as opposed to the diffusible exotoxins produced by living organisms. The term is now used almost exclusively as a synonym for the lipopolysaccharides (which see) of Gram-negative bacteria.

Endotrophic—Feeding internally, as a fungus living inside the tissues of a host. Often refers to a mycorrhiza in which the mycelium grows within the cortical cells of the root (e.g., orchids, Ericaceae).

Endotunica—*See* Endoascus.

Endozoic—Living inside an animal.

Energy (kinetic)—The capacity to do work, e.g., light, heat, chemical, electrical, or nuclear energy.

Enhancement—An increased yield of one virus in joint infections with another virus.

Enniatin A and B—Peptide antibacterial antibiotics from *Fusarium oxysporum* (syn. *F. orthoceras*); avenacein, sambucinin (or enniatin).

Enokitake—*Flammulina velutipes*, cultivated for human consumption in Japan and Taiwan; winter mushroom.

Enphytotic—Describing a plant disease that regularly causes a constant amount of damage in a restricted area each year; also called endemic. *See* Epiphytotic.

Enrichment culture—A culture that favors growth of the desired organism in a mixed culture or population.

Ensate, ensiform—Narrow and pointed; sword-shaped.

Enteroblastic conidiogenesis (conidiogenous cell)—Blastic with the inner wall or neither wall contributing toward formation of the conidium.

Enterothallic (conidiogenesis)—Thallic with only the inner wall of the conidiogenous cell contributing toward formation of the conidial wall.

Entire (edges of lamellae, etc.)—Not torn; without teeth; (of microorganism colonies) having a smooth margin.

Ento- (prefix)—Inside.

Entomo- (prefix)—Of insects.

Entomogenous—Living in or on insects, especially as pathogens.

Entomology—The science or study of insects.

Entomophilous—Spore distribution by insects.

Entomycete—Fungus growing upon insects.

Entoparasitic—Parasitic within the host

Envelope (of a virus)—A nucleocapsid enclosed in one or more layers of protein. Lipoprotein membrane derived from a host cell membrane or synthesized *de novo*, surrounding the capsid or nucleocapsid; usually a bilayer carrying virus-specified proteins or peplomers.

Enveloped virus—Plant viruses of the rhabdoviridae and bunyaviridae that have an outer lipid-protein membrane surrounding the protein shell of the virus.

Envelopment (bacteria)—Surrounded by plant cell products; not necessarily immobilized.

Environment—The soil, water, the atmosphere, plants, animals, other organisms, and human beings that surround the object or organism being discussed.

Enzyme—A complex, high molecular weight protein produced in living cells by protoplasm that catalyzes a specific biochemical reaction but does not enter into the reaction itself.

Enzyme conjugate—Usually a preparation of an antibody to which an enzyme is linked covalently; the enzyme produces a color reaction with an appropriate substrate and is used in ELISA and Western blotting (which see).

Enzyme, adaptive—*See* Adaptive enzyme.

Enzyme, constitutive—An enzyme the formation of which is not dependent upon the presence of a specific substrate.

Enzyme, intracellular—*See* Endoenzyme.

Enzyme-linked immunosorbent assay (ELISA)—An extremely sensitive serologic procedure, with many variations, often used for assay of viruses and microorganisms; a serological method in which an antibody carries with it an enzyme that releases a colored compound on reaction with a substrate. *See* also Double antibody sandwich (DAS).

Epapillate—Without papillae.

EPA—The Environmental Protection Agency in the United States. It is responsible for controlling the various aspects of air, water, and soil pollution. Included are pesticide regulations such as residue tolerances, product approval, reentry standards, protective clothing, etc.

Epi- (prefix)—Upon.

Epibasidium—Name applied to each of the four structures that are separated from the basal portion of the basidium (hypobasidium) by a septum; promycelium of the rust fungi (Uredinales).

Epibiotic—Living on the surface of another organism.

Epibody—*See* Ab 2 epsilon.

Epibryophilous—Preferring to grow on bryophytes.

Epichroic—Discoloration due to injury.

Epicormic sprouts—Sprouts arising abnormally along a trunk or limb as the result of release of dormant buds or the differentiation of buds from callus; water sprouts.

Epicortex—Thin polysaccharide-like layer over the cellular cortex of some Parmeliaceae, which may have regular pores.

Epicorticate—Having a cortex.

Epicotyl—Portion of an embryo axis or seedling above the attachment of the cotyledon(s). The growing apex of the epicotyl forms the young stem of a plant.

Epicutis (of basidiomata)—Outer layer of compound hyphae parallel to the surface.

Epidemic—A widespread and rapidly developing outbreak of an infectious disease of humans in a community; used loosely of plants and animals. *See* Epiphytotic and Epizootic.

Epidemic (or epiphytotic) rate—Amount of increase of disease per unit of time in a population.

Epidemiology—The branch of plant pathology concerned with disease in plant populations; study of the factors influencing the initiation, development, and spread of infectious causes of disease; the study of epidemics (epiphytotics).

Epidermal-strip test—An electron microscope technique by which a virus can be quickly examined in crude-sap extracts. The *quick leaf-dip* and *quick dip* techniques are similar procedures.

Epidermis (adj. **epidermal**)—The outermost layer of cells of leaves, young stems, roots, flowers, fruits, and seeds, absent from the root cap and apical meristem; (of nematodes) the outer cylinder of cells consisting of a single layer of epithelium that secretes the cuticle (hypodermis).

Epidermophytosis—Dermatophytosis.

Epiflora—Surface flora, sometimes applied to the microbial flora on seed surfaces; the epibiota.

Epigeal, epigean, epigeic—On the earth.

Epigeal germination—A type of seed germination in dicots in which the cotyledons rise above the soil surface, e.g., beans (*Phaseolus*).

Epigenous—Growing on the surface; borne above.

Epigynous—Having the antheridium formed above the oogonium on the same hypha.

Epigyny—A condition in which the ovary is embedded in the receptacle, so that other floral parts appear to arise from the top of the ovary.

Epihymenium—Thin layer of interwoven hyphae on the surface of the hymenium; epithecium.

Epilithic—Living on the surface of stones.

Epinasty—Abnormal downward twisting, curling, or bending of a leaf blade or stem.

Epinecral layer—*See* Necral layer.

Epiparasite—An organism parasitic on another that parasitizes a third.

Epiphloeodal—Living on bark.

Epiphragm—Membrane over the young fruitbody in the Nidulariaceae.

Epiphyllous—Borne on the upper surface of a leaf; foliicolous.

Epiphyte—An organism (e.g., bacterium) growing on the surface of a plant, from which it gains physical and nutritional support, without causing disease.

Epiphytic—Living on the surface of plants, but not as a parasite; same as corticolous.

Epiphytology—The science or study of epiphytotics (epidemics in plants).

Epiphytotic—The sudden, widespread, and destructive development of disease on many plants, usually over large areas. Corresponds to an epidemic of a human disease.

Epiplasm (of an ascus)—Cytoplasm of the ascus not used up in "free cell formation" of ascospores.

Episome—Dispensible piece of extrachromosomal genetic material (probably almost always DNA), endowed with the capacity of independent replication; able to multiply independently of the chromosome but capable of reversible integration into the chromosome. Most viruses are episomal, as are plastids.

Epispore, episorium—Thick fundamental layer of the outer spore wall that determines the shape and rigidity of the spore; often gives rise to ornamentation.

Epistomeus—Spigot-shaped.

Epistroma—A normally conidial stroma formed in the periderm and commonly breaking through the bark. *See* Stroma.

Epithecium—Tissue by or at the surface of an apothecium formed by the branching of the tips of paraphysis above the asci in some Discomycetes.

Epithelium—(Of nematodes) a membrane-like layer of cells lining the internal or external surface of an organ, tissue, or cavity; (of basidiomata) a cuticular layer consisting of isodiametric cells of the pileus and stipe.

Epithet—The second or specific part of a Latin binomial of a plant; also the third or fourth (varietal, etc.) term.

Epitope—An antigenic determinant of defined structure, e.g., an identified oligosaccharide, or a chemical hapten that elicits the formation of antibodies, e.g., a grouping of amino acid sequences on a protein or between adjacent protein subunits.

Epitunica—*See* Exosporium.

Epitype—A family of restricted epitopes; a set of epitopes.

Epixylic, epixylous—Living on wood; lignicolous.

Epizoic—Living on animals.

Epizootic—An epidemic among animals.

Epruinose—Lacking a pruina.

Equal—(Of a stipe) having the same diameter throughout; (of gills) alike in length.

Equilibrium density gradient centrifugation—Isopycnic gradient.

Equilibrium dialysis—A technique used to measure the affinity of antibody-antigen binding, based on diffusion of unbound antigen across a dialysis membrane.

Equivalence—The ratio of antibody to antigen that gives the maximum precipitation in the quantitative precipitation reaction.

Eradicant—A chemical or physical agent that kills established pathogens. *See* Chemotherapy.

Eradicant fungicide—Fungicide that kills a pathogen after infection has been established.

Eradication (v. **eradicate**)—Control of plant disease by destroying or removing the pathogen or pest after it is already established in a given area or by eliminating the plants carrying the pathogen; may differ from mere sanitation in that legislative action may be required.

Eramose—Unbranched.

Erasure phenomenon (in *Dictyostelium*)—Loss by amoebae of the ability to recapitulate when developing cultures are disaggregated and placed on a growth medium.

Erect—Upright; straight up and not curved up.

Ergastic substances—The nonprotoplasmic components of the protoplast, for example, crystals and oil droplets.

Ergoline alkaloids—Alkaloids from sclerotia of *Claviceps*, including both lysergic acid derivatives and clavine alkaloids. *See* Ergot.

Ergometrine (D-lysergic acid propanolamide)—An ergot alkaloid from sclerotia of *Claviceps purpurea*; used in medicine against migrane. *See* Ergotamine.

Ergosterol—Commonest sterol of fungi; first isolated from *Claviceps purpurea* sclerotia; yeast ergosterol is converted to vitamin D_2 by ultraviolet radiation.

Ergot—Disease of inflorescence of cereals and grasses caused especially by *Claviceps purpurea* and *C. paspali*; also the dark sclerotium or ergot body developing in place of a healthy grain in a diseased inflorescence. The sclerotium contains a range of toxic alkaloids, which have been used in medicine as drugs; an ergot fungus.

Ergotamine—Cyclic tripeptide derivative of lysergic acid from sclerotia of *Claviceps purpurea*; used in medicine against migrane. *See* Ergometrine.

Ergotism—Disease of humans and animals from eating rye bread or inflorescences of Poaceae contaminated by ergot sclerotia; may be gangrenous (St. Anthony's Fire of the Middle Ages) and spasmodic; causes staggers in cattle, sheep, and horses and agalactia in sows, resulting in the death of baby piglets; ergot poisoning.

Ergotoxine—Cyclic tripeptide derivatives of lysergic acid from sclerotia of *Claviceps purpurea*; a mixture of ergocornine, ergocristine, and ergokryptine.

Ericaceous—Of the plant family Ericaceae.

Erinaceous, erinous—Prickly with sharp points, like a hedgehog.

Erinium—Form of a gall arising from a leaf hair.

Eriophorous—Densely cottony or woolly; tomentose.

Ermineous—Ermine colored; white, with a yellow tinge in places.

Erogen—A substance controlling the induction and differentiation of sex organs.

Erose (of a lamella, colony, etc.)—Having the margin irregularly notched as if gnawed; delicate toothlike projections from the edge.

Erosion—The wearing away of the surface soil by wind, moving water, or other means.

Erostrate—Without a beak.

Erotactin—A sperm attractant.

Erotropin—A substance inducing a chemotropic response in sex organs. *See* Hormone.

Erratic (of lichen thalli)—Not attached to the substratum and often blowing around; epigaeic; vagrant; wandering lichens.

Erubescent—Bluish red; blushing.

Eructate—Thrown up.

Erumpent—Breaking through to the host surface or substratum in the course of development; bursting forth.

Erysiphoid—Like *Erysiphe*; mildewy; cobwebby.

Erythrocyte—Red blood corpuscle.

Escape—A condition in which a susceptible plant avoids infectious disease through some character of the plant or its location.

-esce, -escent (suffix)—Denoting beginning; slightly.

Esculent—Of use as a human food; edible.

Eseptate—Same as aseptate.

Esophageal bulb (of nematodes)—Any of the several enlargements of the esophageal wall, that are either muscular, glandular, or both.

Esophageal gland lobe (of nematodes)—The basal, glandular portion (usually three glands) overlapping the intestine in many plant-parasitic nematodes.

Esophageal glands (of nematodes)—Elongated glands (salivary) of simple or branched tubules located in the esophagus, one gland dorsal and two to eight subventral; secretions are apparently of an enzymatic nature.

Esophago-intestinal or cardiac valve—The valve in a nematode between the esophagus and intestine.

Esophagus or pharynx (of nematodes)—That portion of the digestive or alimentary tract between the stoma (or buccal capsule) and the anterior portion of the intestine; may be differentiated into as many as four distinct parts: procorpus, metacorpus, isthmus, and basal region. The types of esophagi are as follows:

1. *Aphelenchoid*—Four parts, metacorpus rhomboid and exceeding two-thirds of the body width; dorsal gland outlet just anterior of the valve of the metacorpus instead of near the stylet base, basal region as in tylenchoid. Example—*Aphelenchoides*.

2. *Bulboid*—Two parts, long, slender anterior portion with the base a distinct, muscular bulb. Example—*Achromadora*. (*Plectoid*—Similar but basal bulb has a true valve. Example—*Plectus*).

3. *Criconematoid*—Three parts, procorpus and metacorpus fused into a large corpus, metacorpus with large valve, dorsal gland duct outlet near the base of the stylet, basal region a reduced bulb. Example—*Criconemella*.

4. *Cylindrical*—Esophagus is approximately the same diameter throughout its length. Example—*Mononchus*.

5. *Diplogasteroid*—Four parts, no valves, basal region a glandular bulb. Example—*Diplogaster*.

6. *Dorylaimoid*—Long, slender, anterior portion and thicker, long, bulbar portion that is muscular and glandular. Example—*Dorylaimus*.

7. *Neotylenchoid*—Three parts, metacorpus not present or only slightly developed and lacking a valve, dorsal gland duct outlet near the base of the stylet, basal region as in tylenchoid. Example—*Neotylenchus*.
8. *Rhabditoid*—Three parts, procorpus and metacorpus fused into a large corpus, basal region a muscular bulb with a "butterfly" valve. Example—*Rhabditis*.
9. *Tylenchoid*—Four parts, metacorpus round and less than two-thirds of the body width, dorsal gland outlet near the base of the stylet, basal region a distinct bulb. Example—*Tylenchorhynchus*. Or with highly glandular lobes overlapping the intestine. Example—*Pratylenchus*.

Esophagus lumen (of nematodes)—The canal of the esophagus.

Esorediate, esorediose—Lacking soredia.

Essential oils—Widely distributed organic compounds in organisms; have an oily texture and pronounced odors; often thought to be metabolic waste.

Esterase—One of a group of enzymes that hydrolyze esters.

Estriate—Without lines or markings.

Ethnomycology—Mycology as a branch of ethnology.

Ethylene—Unsaturated, colorless, flammable gas ($CH_2=CH_2$) with plant growth regulator properties commonly produced by some living plants, injured plants, fruits, and during incomplete combustion of fuels; used as an agent to color citrus fruits. Ethylene is involved in various aspects of vegetative growth, fruit ripening, abscission of plant parts, and the senescence of flowers.

Ethylenediaminetetra-acetic acid (EDTA)—A chelating agent with a high affinity for Mg^{2+} ions. It is used in solutions to remove divalent ions, inhibit DNase and to act as a bacteriostat.

Ethyleneglycolbis(aminoethylether)tetra-acetic acid (EGTA)—A chelating agent with a high affinity for Ca^{2+} ions at its optimum pH of 6.5–7.0. It has less affinity for Mg^{2+} ions and is used to remove Ca^{2+} in the presence of Mg^{2+}.

Etiolation (adj. **etiolated**)—Blanching or yellowing of tissue (absence of chlorophyll), elongating of stems (spindliness), and failure of normal leaf development caused by reduced light or complete darkness.

Etiologic agent—The parasite causing a disease in a plant or animal.

Etiology, aetiology—The science of the causes or origins of disease, together with the relations of the causal factor(s) to the host; the study of the causal factor, its nature, and relations with the host.

Eu- (prefix)—True.

Eubacteriales—An order of the class Schizomycetes; true bacteria.

Eucarpic—Having only part of a thallus used to form the sporocarp.

Euchroic—Having true pigmentation, not due to injury. *See* Epichroic.

Eucortex (of lichens)—Cortex composed of well-differentiated tissue.

Eugonic—*See* Dysgonic.

Euhymenium—A hymenium in which the basidia and their sterile homologs are the first elements to be formed as a palisade; in a *static* (nonthickening) euhymenium, the exhausted basidia are replaced at the same level by intercalation; in a *thickening euhymenium*, the tramal hyphae grow between the exhausted basidia, forming a new euhymenium above the old.

Eukaryote (adj. **eukaryotic**)—An organism having membrane-limited organelles and nuclei that divide by mitosis; all organisms except bacteria and blue-green algae. *See* Prokaryote.

Eukaryotic cell—Type of cell that possesses a well-defined nuclear membrane.

Eukaryotic initiation factors—Proteins essential for the translation of mRNAs. There are nine or more of these factors whose functions include the recognition of the 5′-terminal cap structure and binding of the 40s ribosome subunit to the mRNA.

Eumorphic—Well-formed.

Eumycetoma, Eumycetes—*See* Mycetoma.

Eumycota—The division that includes the true fungi.

Eurithermophile—An organism capable of growth at 60°C and also at temperatures near 30°C.

European pear rust—*See* Trellis rust.

Euseptate—Of a conidium having true multilayed septa, lumina not reduced; septa in which the diaphragm merges with the outside wall. *See* Distoseptate.

Eustroma—A mass of fungal cells or closely interwoven hyphae comprising a distinctly structured tissue, but *not* applied to either simple pycnidia or acervuli; may be unilocular (simple) or multilocular (complex). If multilocular, it can be *effuse*, with distinctly separate locules, or *pulvinate*, with a superficial more or less cushion-shaped structure, either convoluted or with distinctly separate regularly or irregularly arranged locules.

Euthecium—An ascoma (apothecium, cleistothecium, perithecium) of an euascomycete. *See* Pseudothecium.

Euthyplectenchyma—Hyphal tissue without "cellular" structure, i.e.,, not composed of conglutinate cells. *See* Prosoplectenchyma.

Eutrophic—Being in a well-nourished state. *See* Oligotrophic and Dystrophic.

Evaginate—To protrude or turn inside out; without a sheath.

Evanescent—Fleeting, ephemeral; soon disappearing; fugacious.

Evapotranspiration—Total loss of moisture through the combined processes of evaporation from the soil surface and transpiration from plants, for a given area and during a specified period of time.

Evelate—Without a veil.

Even—Describing a hymenial or other surface (e.g., spores) lacking warts, teeth, gills, tubes, pits, ridges, furrows, depressions, folds, striations, or unevenness of any kind.

Evergreen—Trees or shrubs that are never entirely leafless and retain their functional leaves throughout the year.

Evert (adj. **everted**)—To turn inside out.

Evolution—The development of a species, genus, or other larger group of plants, animals, or other organisms over a long period of time.

Ex- (prefix)—From, out of, without; not having.

Exalbid—Whitish.

Exannulate—Without an annulus or ring.

Exanthema—Copper deficiency in fruits.

Exappendiculate—Not appendaged.

Exarid—Dried out.

Exasperate—Roughened with hard projecting points.

Excentric, eccentric—Off center, displaced to one side, one-sided. *See* Centric.

Excentrically stipitate (of basidiocarps)—With a stipe not attached in the center of the pileus.

Exchange capacity—The total ionic charge of the adsorption complex active in the adsorption of ions; also called anion exchange capacity, cation exchange capacity, and base exchange capacity.

Excipulum proprium—Non-lichenized excipular tissue forming the margins of the apothecium of a lichenized fungus.

Excipulum, exciple—Outer wall, tissue, or tissues containing the hymemium of an apothecium or forming the walls of a perithecium; *ectal-*, forms outermost layers, including the margin, and *medullary-*, the zone enclosed by the ectal excipulum and the hypothecium.

Excipulum thallinum—Lichenized excipular tissue of a lecanorine apothecium external to an excipulum proprium.

Excise—To cut out.

Exclusion—Control of disease by excluding the pathogen or infected plant material from crop production areas (e.g., by quarantines and embargoes) and allowing them into disease-free, crop production areas only.

Excrete—To cast off products from a cell or organism.

Excretory canal or tube (of nematodes)—A tubelike structure located in the lateral chord or pseudocoelom that collects excretory products and conveys them to the excretory duct and pore.

Excretory duct or terminal duct (of nematodes)—A tube or canal, lined with cuticle, that leads to the excretory pore.

Excretory pore (of nematodes)—The exterior opening of the excretory system; located on the ventral side of the body generally near the base of the esophagus.

Excurrent—A type of erect shoot growth in which there is a single main stem, usually with many lateral branches, as in a pine; alternate, narrowly elongated.

Exergonic—Energy-yielding, as in a chemical reaction.

Exigynous—Having the antheridial stalk arising directly from the oogonial cell above the basal septum.

Exo- (prefix)—Outside.

Exobasidial—Having uncovered basidia; separated by a wall from the basidium.

Exocarp—The outermost layer of the fruit wall (pericarp).

Exocellular—Produced, moved to, or active outside the cell.

Exochthonous—*See* Autochthonous.

Exoenzyme—Enzyme secreted by a microorganism into the environment; extracellular enzyme.

Exogenous—(Of plants) produced outside of a generating structure; (of fungi) produced or originating from the surface of a protoplasmic mass (e.g., a generating structure); protruding from the end of a phialide; (basal appendage) originating outside the conidiogenous cell and therefore suprabasally inserted; opposite of *endogenous*.

Exolete (of perithecia, pycnidia, etc.)—Long over-mature; empty.

Exomycology—Mycology of outer space.

Exons—The protein-coding DNA sequences of a gene in eukaryotes. The original transcript from such genes is processed by splicing to remove the introns (noncoding sequences) and to produce the mRNA.

Exonuclease—An enzyme requiring a free end in order to digest an RNA or DNA molecule. These are both 5'- or 3'- exonucleases.

Exooperulation (of sporangial dehiscence in chytrids)—The operculum is hinged to the rim of the pore; "true operculation." *See* Endooperculation.

Exoperidium—The outer layer of the peridium.

Exopropagule (of medical mycology)—Propagule formed outside the body.

Exospore, exosporium—The outer layer of a double-walled spore wall, commonly responsible for its ornamentation; tunic.

Exothermic—A chemical reaction that gives off energy.

Exotic—An organism introduced from another country or area; not indigenous.

Exotoxin—Diffusible toxic substance produced and then excreted by a microorganism into the surrounding medium.

Expallant (of a pileus)—Turning pale on drying.

Expanded (of the pileus)—Opened out when matured or in age.

Expansa—Expanded.

Expansin—*See* Patulin.

Expersate (of oospores in Saprolegniaceae)—Having a large refractive body surrounded by a homogeneous cytoplasm.

Explanate—Spread out flat.

Explant—To take living cells or tissue from a plant and place it in an artificial medium for tissue culture.

Explosive (of asci)—*See* Ascus.

Exponential growth phase—That phase of a bacterial or viral replication in which the number of cells or particles increases at a constant rate.

Exserted—Sticking out; protruding beyond.

Exsiccatus (pl. **exsiccata, exsiccatae,** or **exsiccati** n. **exsiccatum**)—Dried or dry specimen(s) of fungi. *See* Schedula.

Exsilient—Escaping.

External basidium—Basidium that matures outside of a teliospore, i.e., the usual kind.

Extinction coefficient—Specific absorbance.

Extracellular—Outside the cell wall.

Extraction buffer—Buffer used in grinding infected leaves during the initial stages of virus purification.

Extractive—Any chemical that may be extracted by means of solvents.

Extramatrical—Living on or near the surface of the matrix or substratum.

Extrorse—Toward the edge.

Extrude (n. **extrusion**, adj. **extruded**)—To push out, emit to the outside.

Exudate—Substance, often a liquid ooze or slime, discharged or secreted from diseased, injured, or healthy plant tissue; metabolic byproducts discharged from roots into surrounding soil. The presence of an exudate often aids in diagnosis, e.g., droplets of fire blight bacterial ooze, slime flux of elm (*Ulmus*).

Eye spot—An elliptical to circular spot with a prominent border and paler center that resembles an eye; (in algae) a small, light-sensitive, pigmented structure present in certain species.

F

°F—Fahrenheit; a unit of temperature. *See* Fahrenheit and °C.

F(ab')$_2$—Fragment obtained by pepsin hydrolysis of an immunoglobulin molecule. The F(ab')$_2$ fragment consists of two Fab fragments joined by disulfide bonds.

f.sp.—*See Forma specialis.*

F$_1$—The first generation progeny of a cross between any two parents.

F$_2$, F$_3$, etc.—The second, third, etc. generations following a cross.

Fab fragment—Fragment obtained by papain hydrolysis of immunoglobulin molecules. The Fab fragment (about M_r 45,000) consists of a light chain linked to the N-terminal half of the contiguous heavy chain. Two Fab fragments are obtained from each 7S antibody molecule. Each contains one antigen-combining site and can combine with antigen as univalent antibody but cannot form precipitates.

Facial eczema—*See* Sporidesmin.

Facultative—Sometimes; not necessarily; referring to an organism being able to live under a variety of conditions; not obligate.

Facultative aerobe—An organism normally living without atmospheric oxygen, although capable of utilizing it when available.

Facultative anaerobe—An organism normally living in the presence of atmospheric oxygen, but capable of living anaerobically.

Facultative parasite—An organism normally saprophytic but that can live as a parasite under certain conditions and be cultured on laboratory media. *See* Parasite and Obligate parasite.

Facultative saprophyte or saprobe—An organism that is normally parasitic but under proper conditions can live saprophytically during part of its life cycle.

Fahrenheit (F)—A thermometer scale in which the freezing point of water (ice point) is set at 32 and the boiling point (steam point) is 212. *To convert from Fahrenheit to Celsius*: Subtract 32 from the Fahrenheit reading, multiply by 5, and divide the product by 9. Example: $132°F - 32 = 99 \times 5 = 495; 495 \div 9 = 55°C$. *See also* Celsius.

Fairy butter—*Tremella albida.*

Fairy ring—A ring of mushrooms or puffballs on the ground representing the periphery of mycelial growth of a basidiomycete; the distinct zone of grass greener than the rest inside or outside the ring of mushrooms.

Fairy ring champignon or fungus—The edible *Marasmius oreades.*

Falcarindiol—An antifungal compound produced by carrot roots.

Falcate, falciform—Curved like the blade of a scythe or sickle.

Falciphore—*See* Falx.

Fallow—Pertains to cropland not cultivated or not planted for one or more seasons; maintained without plant growth.

False membrane (of a smut)—Tissue of sterile fungal cells limiting the sorus, as in *Sphacelotheca.*

False mildew—Same as downy mildew (which see).

False morel—*See* Lorchel.

False tinder fungus—*Phellinus igniarius.*

False truffle—*Hymenogaster* spp. *See* Truffle.

Falx (pl. **falces**)—A "fertile hypha" or conidiophore of *Zygosporium* shaped like a bill-hook. Falces may be sessile or on special hyphae (*falciphores*).

Family—A taxonomic grouping of plant or animal genera according to resemblance and similarity; denotes natural relationships. A category above a genus and below an order.

Farctate (of a stipe)—Having the center softer than the outer layer; stuffed.

Farinaceous, farinose (of texture)—Mealy, with a loose, powdery appearance.

Fasciate, fasciated—Formed by abnormal growth of several buds that fuse laterally in their formative stage; massed or joined side by side.

Fasciation—A distortion of a plant caused by an injury, infection, or genetic factor that results in thin, flattened, and sometimes curved or curled shoots; the plant may look as if several of its stems were fused; the condition of being bound or bundled together.

Fascicle, fasciculate—Small group, bundle, or cluster.

Fascicled (of rust fungi)—Applied to pedicels composed of several adherent hyphae.

Fascicular cambium—Cambium situated within a vascular bundle.

Fasciculate, fascicled—Having growth in fascicles or bundles.

Fastidious—Nutritionally exacting.

Fastidious organism—An organism that is difficult to isolate or cultivate on ordinary culture media.

Fastigiate—Having leaves or branches more or less parallel and pointed upward; massed upright branches; bunched, clustered.

Fastigiate cortex (of lichens)—Made up of parallel hyphae at right angles to the axis of the thallus. *See* Fibrous cortex.

Fat—A kind of food insoluble in water and composed of carbon, hydrogen, and oxygen, with proportionately less oxygen than in carbohydrates.

Fatiscent—Cracked or falling apart.

Fatty acid—Organic compound containing carbon, hydrogen, and oxygen that combines with glycerol to make a fat.

Faveolate, favose, favoid—Honeycombed or like a honeycomb; alveolate; pitted.

Favus—Skin disease of humans caused by *Trichophyton schoenleinii*.

Favus mouse—*Trichophyton quinckeanum*.

Fc—"Crystallizable fragment." The non-antigen-binding fragment of an immunoglobulin molecule produced after digestion with papain comprising two heavy chain fragments. When first recognized using rabbit IgG, this fragment formed crystals, but the Fc of most species does not crystallize.

Fc′—Fragment obtained by pepsin digestion of an immunoglobulin molecule. The Fc′ consists of the non-antigen-binding portion of IgG, comprising two heavy chain fragments.

Fd fragment—Part of the heavy chain of immunoglobulin that lies on the N-terminal side of the site of papain hydrolysis.

Fecund, (n. fecundity)—Fruitful in offspring or vegetation; prolific.

Feedback inhibition—Inhibition of the activity of an enzyme specifically caused by a small molecule that is usually the end product of a biosynthetic pathway.

Feeder root—Fine root (rootlet) with a large absorbing area.

Fellent—Bitter like gall.

Felt (of citrus)—Superficial saprobic fungi such as *Septobasidium pseudopedicellatum*.

Felty (of context tissue)—With a compact but soft, interwoven texture like felt.

Female, white or purging agaric (agaraicum)—*Fomitopsis (Fomes) officinalis*.

Fenestra (pl. **fenestrae**)—A windowlike or transparent opening through or into a tissue or organ. Used to describe the opening in the nematode cyst through which larvae, which hatch from eggs in the cyst, escape.

Fenestrate—Having windows or openings; (of spores) muriform.

Fermentation—Oxidation (decomposition) of certain organic substances, especially carbohydrates, in the absence of molecular oxygen, by the enzyme action of microorganisms; chemical changes in organic substrates caused by enzymes, usually those of living yeasts; gaseous oxygen is not involved in this energy-yielding process. *See* Anaerobic respiration.

Ferruginous, ferrugineous—Rusty reddish brown or rust-colored.

Fertile hypha—*See* Conidiophore.

Fertilization—The fusion of two sex nuclei (gametes), resulting in a doubling of chromosome numbers to form a zygote; or the act (process) of rendering soil fertile, especially by use of needed plant nutrients, such as nitrogen, phosphoric acid, potash, and others.

Fertilization tube (of phycomycetes)—A hypha originating from the male gametangium and penetrating the wall of the female gametangium, and through which the male (gametes) nuclei pass.

Fertilizer—Any organic or inorganic material of natural or synthetic origin containing nutrients (elements) essential to the growth of plants. To be labeled and sold as such it must be state-licensed with the analysis printed on the package label.

Feulgen stain (Schiff's reagent)—A histochemical stain composed of basic fuchsin and sulphurous acid. It stains chromatin that contains thymidine.

Fiber—Elongated, tapering plant cell with thick walls (in xylem and phloem) that is dead at maturity.

Fiber layer (of nematodes)—The inner layer of the cuticle.

Fibril—A minute, thin, threadlike fiber; (in *Usnea*) short, simple branches perpendicular to the main branches.

Fibrillose, fibrillar, fibrillate—Covered by or composed of fine, silklike fibers or elongated, generally thick-walled cells; (of basidiocarp surfaces) having macroscopic, appressed, and radiating hyphal strands.

Fibrinolysin, fibrolysin—An enzyme that dissolves fibrin; also called *streptokinase* as it is produced by hemolytic streptococci that can liquefy clotted blood plasma.

Fibroblast—The stellate connective tissue cell type found in fibrous tissue; an important type of cell used in cell culture.

Fibrosin bodies—Rodlike structures found in the conidia of some powdery mildew fungi.

Fibrous—Containing, consisting of, or resembling fibers (toughish stringlike tissue).

Fibrous cortex (of lichens)—Composed of loosely woven distinct hyphae parallel with the long axis of the thallus. *See* Fastigiate cortex.

Fibrovascular—Having or consisting of fibers and conducting cells (as vessels).

Field capacity (field moisture capacity)—The maximum amount of water the soil can hold against the force of gravity.

Field immune (of plants)—Not becoming infected by a pathogen in the field, although susceptible under experimental conditions.

Field resistance—Resistance observed under natural field conditions but usually not detected under experimental conditions. *See* Tolerance.

"Fig smut"—Disease caused by *Aspergillus niger*.

Filament (adj. **filamentous**)—Stalk of a stamen bearing the anther at its tip; thin, flexible, threadlike, or filiform structure (row of cells); (of bacterial growth) composed of long, irregularly placed or interwoven threadlike forms.

Filamentous phages—*See* Inovirus.

Filiform—Thread- or hairlike; long and slender.

Filoplasmodium—Netlike pseudoplasmodium of the Labyrinthulales. *See* Plasmodium.

Filopodium (pl. **filododia**)—Slender, unbranched, hairlike appendage (pseudopodium) from a plasmodium. *See* Rhizopodium.

Filter, bacterial—A special type of filter through which bacterial cells cannot pass.

Filterable virus—Capable of passing through the pores of a bacterial filter; applies to most viruses.

Filtrate—Liquid that has passed through a filter.

Fimbra (pl. **fimbroae**)—Hairlike process on the surface of certain Gram-negative bacteria, probably composed of protein subunits and about 7 nm in diameter. Fimbroae are shorter and thinner than flagella and do not contribute to motility; also called *pili*.

Fimbriate—Fringed, or delicately toothed; edged; (conidioma) with free hyphae at the apex; (margin of resupinate basidiocarps or dissepiment edges) with radiating hyphae that give a minutely fringed or torn appearance.

Fimbrillate—Having a very small fringe; minutely fringed.

Fimicolous—Living on animal droppings. *See* Coprophilous fungi.

Fine structure—*See* Ultrastructure.

Finger printing—A procedure for characterizing DNA, RNA, or proteins using electrophoretic or chromatographic analysis of specific fragments.

Finger-and-toe-disease—*See* Clubroot.

Fire blight—A common disease of pome fruits and many ornamental shrubs in the rose or apple family (Rosaceae) caused by a bacterium (*Erwinia amylovora*). Infected portions turn brown or black and appear to be scorched by fire.

Fireplace fungi—Fungi characteristic of burned ground. *See* Pyrophilous fungi.

Firing—Drying and dying of leaves, especially of grasses and corn (maize).

First-year needles—Of conifers; the age class of needles formed most recently; needles of the current year.

Fish molds—Saprolegniales.

Fiss-, fissi-, fid- (prefix)—Cleft, divided, split.

Fissile—Cleft; ruptured; splitting readily.

Fission—Transverse splitting in two of bacterial cells (binary fission); asexual form of cell division; becoming two by division of the complete organism; (of conidial liberation) secession by the separation of a double septum. *See* Budding.

Fissitunicate ascus—Same as bitunicate ascus (which see).

Fissure (of nematodes)—A cleft or groove separating adjacent tissues, parts, or organs.

Fistular, fistulose—Hollow, like a pipe; tubular.

Fix—To harden and preserve specimens, e.g., nematodes and plant tissue, especially for permanent mounting and microscopic study.

Fixation—Preservation of biological structures for microscopic examination by killing in a suitable chemical or physical condition so as to avoid changes in structure; (of soil) the process by which certain chemical elements essential for plant growth convert from a soluble or exchangeable form to a much less soluble or to a nonexchangeable form. *See* Nitrogen fixation.

Fixative—Generally a mixture of chemicals that harden and preserve specimens.

Fixed—Not easily detached.

fl oz—Fluid ounce (U.S.) = 2 tablespoons = 0.125 cup = 0.0625 pt = 0.03125 qt = 0.00781 U.S. gal = 29.573 ml = 0.029573 liter.

Flabellate, flabelliform—Fan-shaped, tapering to a narrowed lateral base.

Flaccid—Flabby; lacking firmness; limp; soft and limber.

Flag, flagging—A branch with drooping, yellowing, or dead leaves on an otherwise healthy-appearing shrub or tree.

Flag leaf—Leaf originating from the first culm node below the rachis; uppermost leaf on cereals and grasses.

Flag smut of wheat, barley, and grasses—Disease caused by *Urocystis agropyri*.

Flagellate—Having one or more flagella or whiplike processes.

Flagellum (pl. **flagella**, adj. **flagellar**)—A long, delicate, flexible, hair-, whip-, or tinsel-like filament projecting from a cell of certain bacteria and zoospores of the lower fungi that enable them to swim through a liquid; a flagellum is functionally similar to a cilium but is longer. *See* Axoneme, Blepharoplast, and Cilium.

Flammulin—An antitumor antibiotic from *Flammulina velutipes*.

Flange (of nematodes)—A rib, rim, or expansion developed for strength, guiding, or attachment to another part or organ; the basal expansions of the odontostylet.

Flaring (of volva or annulus)—Spreading away from the stipe at the free margin.

Flask fungi—Ascomycotina with perithecioid ascocarps (Pyrenomycetes).

Flav-, flavi-, flavo- (prefix)—Yellow.

Flavescent—Becoming yellowish.

Flavous—Golden-yellow.

Flavovirent—Yellowish green or greenish yellow.

Fleck—A small or minute (0.2–2 mm), white to tan lesion, often translucent and visible through a leaf; a hypersensitive reaction by the suscept to attempted infection by obligately parasitic fungi; a characteristic response of tobacco and numerous other plants to toxic amounts of ozone in the atmosphere.

Flesh (of basidiocarps)—Inner sterile substance (trama) of pileus or stipe exclusive of gills, spines, tubes, etc., i.e., context.

Fleshy—(Of stem and cap of basidiocarps) of rather soft consistency, tending to disintegrate or decay fairly rapidly, as opposed to leathery, corky, woody, membranous, etc.

Fleshy fruit—Any fruit formed from an ovary that has fleshy or pulpy walls at maturity; any fruit that includes fleshy parts of the perianth, floral tube, or receptacle.

Flexuous, flexuose—Having turns, bends, or windings alternately in opposite directions; capable of bending.

Flexuous (receptive) hyphae (of rust fungi, Uredinales)—Uunbranched or branched, haploid hyphal projections from a spermogonium (pycnium) that may be dikaryotized by a spermatium (pycniospore) of the opposite mating type; same as trichogyne. *See* Receptive hypha.

Flexure—A turn or fold.

Flocci—Like groups or tufts of cotton.

Floccose—Cottony soft, downy; woolly or with soft and very small cottony or woolly tufts; like cotton flannel; byssoid.

Flocculate (n. **flocculation**)—To aggregate (clump together) into a loose fluffy mass; (of serology) precipitation reaction of antibody and antigen in which the precipitate appears as flakes of insoluble protein; usually seen with horse antibody-toxin reactions.

Flocculent (of a liquid culture)—Having small adherent masses of microorganisms or other materials throughout or as a deposit.

Flocculose—With small tufts of soft woolly hair; delicately cottony.

Flooding—A method of irrigation by which water is released from field ditches and allowed to spread over the land.

Flor effect (of yeasts)—Formation of a pellicle. *See* Yeasts.

Flora—Plants of a given geographical area, habit, or region; a description, catalog, or list of all or some groups of plants in a particular region; formerly used for bacteria, fungi, lichens, and other organisms. *See* Microflora.

Floral incompatability—A genetic condition in which certain normal male gametes are incapable of functioning in certain pistils.

Floret—Small flower, usually part of a dense cluster.

Florigen—Name given to a hypothetical substance believed to be associated with flowering.

Flowable (pesticide)—A finely ground, wettable powder formulation sold as a thick suspension in a liquid. Flowables require only moderate agitation and seldom clog spray nozzles.

Flower—The characteristic reproductive structure of angiosperms, interpreted morphologically as a specialized branch system of floral leaves grouped together.

Flower break—Break or stripe in flower color.

Flower bud—Bud that grows into a flower.

Flower primordia—Meristematic tissues differentiable into flower parts.

Flowers of tan—The myxomycete *Fuligo septica.*

Fluorescein—A yellow dye used to label antibody for fluorescence technique.

Fluorescein isothiocyanate (FITC)—A fluorescent compound used for labeling proteins or nucleic acids. It is excited by wavelengths of light in the range of 450–490 nm and emits in the range of 520–560 nm.

Fluorescence (adj. **fluorescent**)—Emission of radiation (light) when a substance or organism is placed in ultraviolet (UV) or other radiation.

Fluorescence microscope—A compound light microscope arranged to emit radiation of specific wavelengths, such as UV, to the spectrum, which then fluoresces.

Fluorescence microscopy—Microscopy in which the microorganisms are stained with a fluorescent dye and observed by illumination with ultraviolet (UV) light.

Fluorescent antibody—An antibody labeled with a fluorescent dye, e.g., fluorescein isothiocyanate. The antibody can then be used in a fluorescent microscope to detect viral antigen in cells. *See* Enzyme conjugate.

Fluorography—Photography of an image produced on a fluorescent screen; now more widely used to describe a technique in which ^{3}H-labeled molecules are detected in chromatograms and polyacrylamide gels. The scintillator PPO is introduced into the chromatogram or gel, which is then exposed to photographic film.

5-Fluorouracil—*See* Base analogue.

Flush (of growth)—Sudden development.

Fly agaric, fly fungus, fly mushroom—*Amanita muscari.*

Fly fungus, house fly fungus—*Entomophthora muscae.*

Fly speck fungi—Members of the Micropeltaceae, Microthyriaceae, and conidiomata of their anamorphs.

Foaming agent—A chemical that causes a pesticide mixture to form a thick foam. It is used to prevent or reduce drift.

Focus (pl. **foci**)—Site of initial disease infection from which secondary spread may occur.

Fodder—Coarse grasses, such as corn (maize) or sorghum, harvested with the seed and leaves and cured for animal feed.

Foetid—Ill-smelling; of a nauseating odor.

Foliage—The leaves of a plant.

Foliar burn—Injury to shoot tissue caused by dehydration due to contact with high concentrations of chemicals, e.g., certain fertilizers or pesticides.

Foliar, foliose—Pertaining to leaves; leaflike.

Foliicole, foliicolous—Attacking or living on leaves.

Foliole—Small leaf-like excrescence on the surface of a foliose lichen.

Foliose—Leaflike; (of lichens) having a stratose thallus, usually with a lower cortex, and attached to the substratum either by rhizines or at the base, but not by the whole lower surface.

Follicle—A dry, dehiscent, many-seeded pod originating from a single, simple pistil and opening along a single suture.

Fomites—Inanimate objects that carry viable pathogenic organisms.

Food—An organic substance that furnishes energy for vital processes or is transformed into living protoplasm and cell walls.

Food chain—The path along which food energy is transferred within a natural plant and animal community.

Food poisoning—A general term applied to all stomach or intestinal disturbances due to food contaminated with certain microorganisms or their toxins.

Foot—Basal area of a culm; (in mosses, liverworts, ferns, and other plant groups) a portion of a young sporophyte that attaches the sporophyte to the gametophyte and often absorbs food from the latter.

Foot candle (ft-c)—A standard measure of light (English system). The illumination of a standard light source of one candle, one foot away, falling at all points on a square foot surface; 1 ft-c = 10.76 lx. *See* Lux.

Foot cell—Basal cell of a spore (e.g., *Fusarium*) or conidiophore (e.g., *Aspergillus*), obliquely attenuated at the base.

Foot rot—Rot involving the lower part of the stem-root axis.

Forage—Green or dehydrated vegetation used as food by livestock, e.g., hay, pasture, and silage.

Foramen—An opening; aperture.

Forate (of gasteromycete basidiocarps)—Invagination of the primordial tissue, resulting in a series of pits; opposite of the type of development usually known as "coralloid."

Forcing crops—Inducing crops to reach maturity outside of their natural season or indoors.

Forest—A formation (or biome) in which the dominant plants are trees.

Forficate, forficulate—Spoon-shaped.

Form (Latin forma)—A subdivision of a species below the rank of variety.

-form (suffix)—Shape.

Form genus (or **species**)—A genus or species based on morphology and not on evolutionary relationships, e.g., in Deuteromycotina (Fungi Imperfecti) and certain rusts (Uredinales).

Form genus—A genus unassignable to a family but possibly referable to a taxon of higher rank.

Forma specialis (pl. *formae speciales*; abbr. *f.sp.* or *ff.sp.*)—Special form; a biotype (or group of biotypes) of a species of pathogen that differs from others in the ability to infect selected genera or species of susceptible plants; an infraspecific taxon characterized from a physiological standpoint (especially host adaptation) but scarcely or not at all from a morphological standpoint. *See* Physiologic race.

Formalin—A 37–40% aqueous solution of formaldehyde.

Formation—A large natural assemblage of plants, such as tundra or coniferous forest.

Formulation (in relation to pesticides)—A specific mixture containing active ingredients, the carrier, and other additions (adjuvants) required to make it ready for sale (as an emulsifiable concentrate [EC], dust [D], flowable [F], granule [G], wettable powder [WP], oil solution, etc.).

Formvar—A trade name for polyvinyl formol used for making films for grids in electron microscopy.

Fornicate—Arched; vaulted; having the fibrous and fleshy layers of the fruitbody (e.g., *Geastrum*) becoming arched over the cuplike mycelial layer.

Forssman antigens—Heterophile antigens widely distributed in nature.

Fossil—The imprint or remains of an organism (plant or animal) of past geologic eras preserved in the earth's crust.

Fossores (of nematodes)—Modified structures of head and dorsal and ventral lips, fitted for "digging." Example: *Diplogaster coronata*.

Foundation planting stock—A stock of very high quality, grown separately for genetic purity and very carefully rogued to produce pathogen-free daughter plants for sale.

Foundation seed—Seed stocks increased from breeder seed, and handled to closely maintain the genetic identity and purity of a cultivar.

Foveate—Having small holes or cavities; pitted.

Foveolate—Having small dimples or delicately pitted.

Fracture (of conidial liberation)—Secession that involves rupture of the wall of an adjacent vegetative or degenerate cell at a point removed from the septum. *See* Fission, Llysis, and Rhexolytic.

Fragmentation—The segmentation of a multicellular organism (e.g., fungal thallus) into a number of fragments, each of which is capable of growing into a new individual; a method of asexual reproduction.

Fragmentation spores—Conidia produced by hyphae breaking up into individual cells.

Frame shift—A mutation caused by the insertion or deletion of one or more nucleotides that changes the reading frame of a codon, giving a changed amino acid sequence starting at the mutated codon; also, a change from the reading frame of one cistron to that of an overlapping cistron.

Framework—*See* Cephalic framework.

Framework region (of serology)—Sequences of amino acids in variable regions of an H or L immunoglobulin chain other than hypervariable sequences.

Frass—The wet or dry sawdust-like material excreted by boring insects; usually evident at their exit holes.

Free (of gills [lamellae] or tubes)—Not attached to the stem or stipe. *See* Sinuate and Seceding.

Free cell formation—The process by which the eight nuclei, each with some adjacent cytoplasm, are separated by walls in an immature ascus to become ascospores; the new wall material is synthesized within the cell, not in contact with previously existing wall material.

Free-living—Of a microorganism that lives freely, unattached; a nonparasitic nematode or a pathogen living in the soil, outside its host.

Freeze-dried—*See* Lyophilization.

Freeze-drying or lyophilization—Preservation of living microorganisms (or concentration of macromolecules with little or no loss of activity), etc., by removing water under a high vacuum while tissue remains in a frozen state.

Freund's adjuvant—A substance containing an emulsifier (lanolin) and mineral oil that is mixed with an antigen before it is injected into the muscles of an animal to produce antiserum. The adjuvant allows slow release of the antigen following injection and enhances immune responses when emulsified with antigen for immunization. *Freund's complete adjuvant* includes killed mycobacteria; *Freund's incomplete adjuvant* does not contain mycobacteria.

Friable—Easily crumbled into small aggregates or powder when handled, as with soil or context tissue in fungal fruitbodies.

Frog cheese—Young puffballs.

Frogeye—A leaf spot caused by a fungus with the central portion lighter than the colored margin.

Frond—The leaf of a palm or fern; also applied to other very large leaves.

Frost crack—Linear separation of a woody stem along radial lines due to unequal shrinkage of the sapwood, compared to the heartwood, during periods of very low temperatures.

Frost resistance—Genetic resistance (or hardiness).

Frost rib—Structure of callus tissue that forms over a frost crack.

Frost tolerance—Ability of plants to withstand a lower temperature than they originally could, by being conditioned through repeated exposures to temperatures just above the freezing point.

Fruct-, fructi- (prefix)—Fruit.

Fructicole, fructicolous—Living on fruit.

Fructification—Any structure in or on which spores are formed. *See* Fruitbody.

Fruit—A matured ovary, or cluster of matured ovaries.

Fruit scar—Scar left on a twig after separation of the fruit.

Fruitbody, fruiting body—A complex, multicellular fungal structure (fructification) that contains or bears spores, e.g., acervulus, apothecium, ascocarp, ascoma, basidiocarp, coremium, pycnidium, perithecium, sporocarp, sporodochium, etc.

Frustose (of the surface of a pileus)—Cracked or fissured into more or less polygonal bits.

Fruticole, fruticolous—Living on shrubs.

Fruticose—Shrublike; (of lichens) having an upright thallus of radiate structure.

Fruticulose (of lichens)—Having a minutely shrubby habit.

Ft—Foot = 12 in. = 0.333 yard = 30.8 cm = 0.3048 m.

Fucicole, fucicolous—Living on lichens.

Fucoxanthin—Brown xanthophyll pigment found in brown algae.

Fugacious—Fleeting, disappearing early (evanescent).

Fuliginous, fuligeneous—Sooty or soot-colored; obscure; having a dark or dusky brown color.

Full coverage (FC)—The total amount or volume of pesticide spray to thoroughly cover the target plant to the point of runoff or drip. *See* Low volume, Ultra low volume.

Fulvous—Of a dull yellowish brown color; tawny.

Fumigant—Vapor-active (volatile) pesticide or material used to kill disease-causing organisms, insects, nematodes, weeds, and other pests; a gaseous or readily volatilizable disinfectant or disinfestant used to destroy organisms by vapor action in an enclosed area or under plastic laid on the soil. Fumigants may be volatile liquids and solids as well as substances already gaseous.

Fumigatin—A benzoquinone antibacterial antibiotic from *Aspergillus fumigatus*.

Fumigation—The application of a fumigant, applied in an enclosed area (e.g., a greenhouse) or usually under plastic laid on the soil, for disinfesting an area.

Fumitremorgin—A tremorgenic metabolite of *Aspergillus fumigatus*; an indole derivative.

Fungaemia, fungemia—Fungi in the blood.

Fungal—Relating to fungi.

Fungi Imperfecti (Deuteromycotina)—A form class erected to contain asexually reproducing fungi with affinities to the Ascomycotina and Basidiomycotina. Two classes are recognized:

1. *Hymenomycetes*—sterile mycelial forms (agonomycetes) or conidia, borne on separate hyphae or aggregations of hyphae (as synnemata or sporodochia), but not within discrete conidiomata.

2. *Coeleomycetes*—forms that produce conidia in acervuli, pycnidia, pycnothyria, and cupulate or stromatic conidiomata.

Fungi—A kingdom of organisms.

Fungi, higher—Ascomycotina, Basidiomycotina, Deuteromycotina (mitosporic fungi).

Fungi, lower—Myxomycota, Chytridiomycota, Hyphochytriomycota, Oomycota, Plasmodiophoromycota, and Zygomycota. *See also* Phycomycetes.

Fungicide (adj. **fungicidal**)—A chemical or physical agent that kills fungi, sometimes including agents that only inhibit the growth of fungi; a type of pesticide. May be used as disinfectants or eradicants to kill fungi in soil, seed, or plants. Often applied as protectants covering susceptible plant parts before the pathogen can infect.

Fungicole, fungicolous—Describing fungi that grow on other fungi as parasites (*mycoparasites*) or saprobes.

Fungiform—Shaped like a mushroom.

Fungistat (n. **fungistasis,** adj. **fungistatic**)—A chemical or physical agent that inhibits fungal growth, sporulation, or spore germination but does not cause death. *See* Genestat.

Fungitoxicant—Compound that exhibits either fungistatic or fungicidal properties.

Fungivore (adj. **fungivorous**)—An organism using fungi as food.

Fungizone—Trade name for amphotericin B (which see).

Fungoid—Funguslike.

Fungous—Of, or pertaining to fungi; fungal.

Funguria—The presence of fungi, especially yeasts, in urine.

Fungus (pl. **fungi**)—A single- to many-celled eukaryotic organism (thallophyte) that lacks chlorophyll, usually with a chitinous wall, reproducing by sexual or asexual spores, and commonly producing mycelium; a nonchlorophyllous organism whose vegetative body (thallus) consists of threadlike filaments (hyphae) usually aggregated into branched systems (mycelia).

Fungus root—Mycorrhiza (which see).

Fungus, ray—An actinomycete.

Funicular—Cordlike

Funicular cord, funiculus—The cord of hyphae by which the peridioles in Nidulariaceae are at first fixed to the inner wall of the peridium. See Splash cup.

Funiculose (of hyphae)—Occurring in ropelike strands or bundles.

Funiculus (adj. **funicular**)—Stalk of an ovule; cordlike.

Funoid—Composed of ropelike strands or fibers.

Furcate—Forked; branching like a fork.

Furfuraceous—Powdery, scurfy.

Furrow—Small V-shaped ditch made for planting seed or for irrigating.

Furrow irrigation—A method of irrigation by which water is applied to row crops in ditches.

Furrowed—Upper surface of the reflexed portion of a basidiocarp with gently rounded, concentric troughs or furrows.

Furvous—Black and lusterless.

Fusaric acid—A pyridine-carboxylic acid from Fusarium oxysporum f. sp. lycopersici, F. oxysporum f. sp. vasinfectum, and other Hypocreaceae able to induce wilt symptoms in tomato (Lycopersicon exculentum). See Lycomarasmin.

Fuscate—Darkened.

Fuscous, fuscid—Brownish gray, smoky, dusky or somber color.

Fuseau—Fusoid macroconidium of a dermatophyte, e.g., *Microsporum.*

Fusicoccin—A tricarboxylic terpene from *Fusicoccum amygdali* that induces stomatal opening and promotes spore germination and cell elongation.

Fusidic acid—An antibacterial antibiotic from *Fusidium coccineum*, same as ramycin from *Mucor ramanniana.*

Fusiform—Spindle-shaped; slender and tapered to each end.

Fusiform rust (of pines)—Disease caused by *Cronartium quercuum.*

Fusiform-elliptical (of spores)—Fundamentally elliptical, but somewhat fusiform; more elliptical than fusiform.

Fusoid—Somewhat fusiform; tapering toward each end.

Fuzzball—*See* Puffball.

G

g—Gram = 1,000 mg = 0.035274 oz = 0.0222046 lb = 0.001 kg (1 lb = 453.59 g).

G and C content—The total guanine (G) and cytosine (C) content of a nucleic acid, usually refers to double-stranded DNA. The G and C content is a good measure of such physical properties as melting temperature; it also affects the banding density in isopycnic gradients.

Galeate, galeiform—Hat- or helmet-shaped; hooded.

Galeriform—Cup-shaped.

Gal—Gallon (U.S.) = 4 qt = 8 pt = 16 cups = 128 fl oz = 0.1337 ft^3 = 0.83268 British or Imperial gal = 3785.4 ml or cm^3 = 231 in.^{3} = 8.337 lb water = 3.782 kg.

Gall—An abnormal swelling or localized outgrowth (tumor) on leaves, stems, roots, and other plant parts, often more or less spherical, produced by host plant cells as the result of attack by a fungus, bacterium, nematode, insect, mite, or other agent; composed of unorganized cells; a cecidium; a cecidium.

Gallery—Insect tunnel in bark and wood.

Gallic acid agar—Malt extract agar medium with 5 g gallic acid per liter; darkens in the presence of polyphenol oxidases.

Gamet-, gameto-, gamo-, gamous-, -gamy—Combining forms meaning union, marriage, copulation.

Gametangial copulation, gametangiogamy—Fusion of two sexual organs (gametangia).

Gametangium (pl. gametangia)—Any cell or differentiated structure containing gametes or nuclei that act as gametes; a gamete mother-cell.

Gamete—A differentiated sex cell or a haploid (1n) nucleus capable of fusion with a gamete of the opposite mating type in sexual reproduction to form a zygote; especially a cell formed in a gametangium that fuses with another sex cell in sexual reproduction—usually an egg with a sperm.

Gametogenesis—The development of gametes.

Gametophyte—A phase in the life history of a plant that arises from a haploid spore, resulting from meiosis in a diploid sporophyte; plants have haploid nuclei during the gametophyte phase; a haploid or sexual plant; haplont or haplophase. *See* Sporophyte.

Gametothallus—Thallus that produces gametes. *See* Sporothallus.

Gamma globulin—A fraction of serum globulin rich in antibodies; an obsolete term for immunoglobulin G (IgG). *See* IgG.

Gamma particle—A DNA-containing cytoplasmic organelle in the zoospore of *Blastocladiella emersonii*.

Gangliform—Knotted; having knots.

Gangligenous—Bearing knots or swellings.

Ganglion (pl. **ganglia**) (of nematodes)—A group of nerve cell bodies (dorsal, lateral, amphidial, retrovasicular, etc.) with their conspicuous nuclei forming a nerve center.

Gangliospore—An aleuriospore in the sense of "holoblastic conidium."

Gas chromatography—*See* Chromatography.

Gasterer-, gastero-, -gaster—Combining forms meaning stomach.

Gasteroconidium, gasterospore—Thick-walled, globose, chlamydospore of *Ganoderma*; probably apomictic.

Gasteromycetes—A class of Basidiomycotina with spores borne in cavities within the fruitbody.

Gasteromycetous—Of, pertaining to, or having the characteristics of Gasteromycetes.

Gel—A jellylike colloidal mass; coagulated colloid.

Gel chromatography—A molecular sieving procedure, by which viruses or proteins are separated from molecules of different sizes when passed through the pores of gel beads such as agarose. Used for purification of viruses, proteins, nucleic acids, etc.

Gel diffusion—*See* Immunodiffusion.

Gel double-diffusion—A serological test in which the antibody and antigen reactants diffuse toward each other in gel and react to form a visible precipitation line. *See also* Radial diffusion and Immunodiffusion.

Gel electrodiffusion—Electrophoresis of macromolecules in a matrix of agarose, polyacrylamide, or similar gel.

Gel filtration—A type of column chromatography that separates molecules on the basis of size.

Gel tissue—Mixture of gel and hyphae found in members of the Leotiales and Tremellales. The gel may arise by direct secretion or disintegration of hyphae. *See* Gliatope.

Gelatin—A protein obtained from bones, hair, skin, tendons, etc., used in culture media for the determination of a specific proteolytic activity of microorganisms (especially bacteria) or for the preparation of a peptone; obtained by boiling collagen; soluble in water above about 40°C.

Gelatinase—An exoenzyme that degrades gelatin.

Gelatinous—Resembling gelatin or jelly.

Geminate—Paired; twinned.

Gemma (pl. **gemmae**)—An irregular, thick-walled cell borne singly or in chains (young bud or conidium-like structure) similar to a chlamydospore, as in certain Phycomycetes; an asexual or vegetative outgrowth of a parent body, capable of growing into a new individual directly, as in liverworts. *See* Stilboid.

Gemmation—Budding.

Gemmifer—A gemma bearer; in *Mycena citricolor* (syn. *Omphalia flavida*), the true airborne, leaf-infecting bodies.

Gene—The base triplets of the DNA molecule in a chromosome that determines or conditions one or more hereditary characters. The smallest functioning unit of genetic material on a chromosome; bearer of an inherited factor; the fundamental unit of inheritance; an ordered sequence of nucleotides that specify the manufacture of a single type of protein (or for some genes, certain RNAs). See Allele.

Major gene—Many genes that have large observable effects upon the phenotype.

Minor gene—One gene that has a specific effect upon the phenotype.

Gene cloning—The isolation and multiplication of an individual gene sequence by its insertion, usually into a bacterium, where it can multiply.

Gene expression—The expression of the genetic material of an organism as specific traits; the transcription of mRNA from the DNA sequence of a gene and the subsequent translation of the mRNA to give the protein gene product.

Gene pool—All the genes of an interbreeding program.

Gene therapy—Insertion of more desirable DNA into cells to correct a given genetic defect.

Gene-for-gene hypothesis—The concept that corresponding genes for resistance and virulence exist in the host and pathogen, respectively. *See* Vertical resistance.

Genera—*See* Genus.

Generation time—The length of time necessary for an organism to complete its life cycle.

Generative hyphae—The essential hyphal component of any basidiocarp; typically thin-walled, simple-septate or with clamp connections; profusely branched hyphae capable of indeterminate growth, giving rise to basidia and in some species the skeletal and binding hyphae.

Genestasis—Inhibition of sporulation.

Genetic—Relating to heredity; describing heritable characteristics as influenced by germ plasm.

Genetic code—The sequence of nitrogen bases in a DNA molecule that codes for an amino acid or protein. In a broader sense, for example, the full sequence of events from the translation of chromosomal DNA to the final stage in the synthesis of an enzyme. The arrangement of three nucleotides (a codon), each of which specifies a single amino acid. The code is degenerate, as 64 codons specify 20 amino acids, and thus there are numerous amino acids that are determined by more than one triplet. *See* Start codon and Stop codon.

Genetic complementation—*See* Complementation.

Genetic diversity—*See* Genetic variability.

Genetic engineering or genetic manipulation—Alteration (insertion or removal) of inherited genetic material from an organism by various tissue culture procedures (transformation, protoplast fusion, etc.) so that the cell can produce more or different chemicals or perform completely new functions. *See* Protoplast fusion, Somaclones and Transformation (DNA).

Genetic map—A graphic representation of the linear arrangement of genes on a chromosome or genome.

Genetic marker—A mutation in a gene that allows its phenotypic identification.

Genetic reassortment—*See* Reassortment.

Genetic transmission—*See* Vertical transmission.

Genetic variability—The property or ability of an organism to change its inherited characteristics from one generation to succeeding ones.

Genetics—The science or study of patterns of inheritance of specific traits.

Geniculate—Bent like a knee.

Geniculation—A kneelike bend or joint.

Geniscar—Abscission conidial scars on the "knees" of conidiophores.

Genistat—Substance that prevents or reduces sporulation in fungi without materially affecting vegetative growth; "anti-sporulator." *See* Fungistatic.

Genital papillae or setae (of nematodes)—Sensory organs of a tactile nature located posteriad in the male; may be preanal, postanal, or caudal in position.

Genital tract (of nematodes)—The reproductive system, the principal components of which are the ovary, oviduct, spermatheca, uterus, vagina, and vulva in the female and the testis, vas efferens, vas deferens, cloaca, and anus in the male.

Genocentric—*See* Reproductocentric.

Genom—A haploid set of chromosomes.

Genome—The complete genetic complement of an organism or virus; size is usually denoted in gene pairs; (of viruses) the nucleic acid component, either DNA or RNA, which may consist of a single (*monopartite*), two (*bipartite*), three (*tripartite*), or more (*multipartite*) molecular species of RNA. *See* Multicomponent virus.

Genomic library—A collection of clones made up of a set of overlapping DNA fragments representing the entire genome.

Genomic masking—*See* Phenotypic mixing.

Genotype—The aggregate of genes in an organism that influences the phenotype; class or group of individuals sharing a specific genetic makeup; the total genetic constitution, expressed and latent, of an organism.

-genous (suffix)—Producing, yielding, produced by, arising in.

Genus (pl. **genera,** adj. **generic**)—A taxonomic category that includes a group of closely related (structurally or phylogenetically) species; the genus or generic name is the first name in a Latin binomial; (more generally) a class of concepts or objects.

-geny (suffix)—Denotes generation, production, origin, development of.

Geo-, ge- (prefix)—Earth, ground, soil.

Geofungi—Soil fungi.

Geophilous—Earth loving, e.g., fungi developing underground fruit bodies. *See* Terrestrial, Terricolous.

Geotrichosis—Disease of humans or animals caused by *Geotrichum*.

Geotropism—A growth curvature in plants induced by gravitational stimulus.

Germ—A historically popular word for a pathogen or disease-causing microorganism (microbe); also refers to the living portion of a seed.

Germ plasm—Material capable of transmitting heritable characteristics sexually or asexually; the total genetic variability available to an organism represented by the pool of germ cells or seed.

Germ pore—A small, round, thin area or hollow within a spore wall (especially in rusts and Ascomycotina) through which a germ tube may emerge.

Germ slit—A thin area in a spore wall that extends the length of the spore.

Germ sporangium—A sporangium borne on the germ tube of a zygospore in the Mucorales.

Germ tube—Hyphal thread resulting from an outgrowth of the spore wall and/or cytoplasm during germination of a fungus spore; also put forth by bulbils and sclerotia. The hyphal threads elongate, branch, and become a new fungus body (*mycelium*).

Germfree—Not having any known germs.

Germicide—A substance capable of killing germs (usually pathogenic microorganisms). *See* Disinfectant.

Germinal zone of ovary or testis (of nematodes)—A region in which rapid division of the germ cells occurs.

Germinate (n. **germination**)—To begin growth of a seed, spore, sclerotium, zygote, or other reproductive body starting with inbibition of water.

Germination by repetition—Producing secondary spores in place of germ tubes as in Heterobasidiomycetes and *Sporobolomyces*.

Ghost—A tailed phage particle in which the contents of the head have been lost. Ghost particles are used in virus adsorption studies and to determine the effects on the host other than those due to virus replication.

Ghost fungus—*Pleurotus nidiformis*, an Australian luminous agaric.

Giant cell—A large, multinucleate mass of protoplasm formed by abnormal cell enlargement with nuclear division but without cell division. Found in plants infected by certain nematodes (e.g., root knot). *See* Syncytium and Nurse cell.

Giant puffball—*Langermannia gigantea*. *See* Puffball.

Gibber- (prefix)—*See* Gibbous.

Gibberellic acid—Plant hormone with similar physiological action to gibberellin A_1. First recognized as the cause of Bakanae disease of rice (*Oryza*).

Gibberellins (A_1, A_2, A_3, and other fractions)—Group of natural plant hormones (growth-regulating compounds) important in many physiologic processes, including cell elongation, parthenocarpy, seed germination, and flowering; first identified as a metabolic product in *Gibberella fujikuroi*, anamorph *Fusarium moniliforme*; first recognized as the cause of Bakanae disease of rice.

Gibbous (of a pileus)—Having a swelling, wide umbo, or having a convex top and a flat underside; gibber, gibbose.

Gill, gills—Platelike hymen (lamella) on the underside of the cap of higher basidiomycete fungi (e.g., mushrooms), often radiating from a point, covered by the hymenium of basidia. *See* Lamella.

Gill fungi—Members of the Agaricales.

Gill trama—Gill tissue between the two hymenial layers.

Gilvous—Pale yellow.

Ginger-beer plant—A mixture of a yeast (*Saccharomyces pyriformis*) and a bacterium (*Bacterium vermiforme*) used for fermenting a sugar solution to make a drink.

Girdle—To encircle and cut through the bark of a woody plant, disrupting the phloem, resulting in death from root starvation; to encircle.

Glabrate—Nearly glabrous or becoming glabrous with age.

Glabrous—Smooth, without hairs or scales.

Glacial till—A product of glacial weathering in which rock particles varying in size are deposited by the glacier on the land surface as it melts and recedes.

Glaireous—Slimy.

Glancing (of pore surfaces)—Showing a change in appearance from dull to lustrous when the orientation of the pore surface in regard to incident light is changed.

Glandular—Having sticky drops or glands.

Glaucescent—Turning bluish green or sea-green.

Glaucous—Covered with a bluish gray waxy "bloom" or powdery coating that gives a frosty appearance; sea-green in color.

Gleba (pl. **glebae**)—The inner fertile or sporulating tissue in a fungal fruit body, especially the gasteromycetes and hypogeous Pezizales.

Glebal mass—The "projectile" of *Sphaerobolus*.

Glebula—A rounded extension of a lichen thallus.

Gliatope—Site of heavy gel production. *See* Gel tissue.

Gliotoxin—An antibacterial and antifungal antibiotic from *Aspergillus fumigatus, Gleocladium virens, Penicillium cinerascens*, and probably other fungi. *See* Viridin.

Globoid, globate, globose, globular, globulose—Globe-shaped; spherical or nearly so.

Globule—Tiny globe or ball; usually lipid.

Globulin—A protein soluble in dilute solutions of neutral salts but insoluble in water; serum globulin. Antibodies are globulins.

Glochidiate—Covered with barbed bristles.

Gloeocystidium (pl. **gloeocystidia**)—Sterile, hymenial or imbedded, thin-walled element (includes vesicles and sulfocystidia) with distinctly oily globular, resinous, granular, or refractive contents; in some species becoming blackish in sulfuric benzaldehyde. *See* Cystidium.

Gloeopleurous hyphae—Hyphae with strongly refractive contents, usually staining brightly with phloxine or Melzer's reagent.

Glomerate—Collected into heads; agglomerate.

Glomerule, glomerulus—A clump or cluster; commonly used for clusters of phycobiont cells in lichens; headlike compacted clusters.

Glottoid apparatus (of nematodes)—A sclerotized, valvular structure at the base of the stoma.

Glucose—A simple, six-carbon sugar (monosaccharide, $C_6H_{12}O_6$), a basic building block of polysaccharides (e.g., starch, cellulose) and other plant chemicals. Used as an energy source by many microorganisms. *See* Dextrose.

Glucoside—A substance that, on decomposition, yields glucose and certain other compounds.

Glume—One of the two sterile chaffy bracts at the base of a grass spikelet.

Gluten—A viscose and toughish substance on the surface of some agarics and other fungi that is very sticky when wet; also used for a spore mass in *Phallus*.

Glutinous, glutinose—Sticky or gluey; covered or made up of gluten.

Glyceollin—A phytoalexin from soybean (*Glycine max*).

Glycogen—A starchlike carbohydrate, a polysaccharide commonly found in animals and in some fungi and algae. Glycogen yields glucose on hydrolysis.

Glycolysis—Anaerobic process whereby glucose is broken down to form pyruvic acid or lactic acid with release of energy; anaerobic respiration.

Glycoprotein (adj. **glycosylated**)—A conjugated protein in which the nonprotein group is at least one carbohydrate covalently attached to an amino acid.

Glycosidase (glycosylase)—A general and imprecise term for an enzyme that degrades the linkage between a sugar and another molecule. The enzymes may distinguish between alpha- and beta-links but are not very substrate-specific.

Glyoxal—A chemical (OHC-CHO) used to maintain nucleic acids in a denatured state.

Glypholecine—Having an especially labyrinth-like lirella, as in *Glypholechia*.

GMP—Guanosine 5'-phosphate.

Gnotobiotic culture—A culture in which all the living components are known. *See* Axenic culture.

Gon-, gono-, gony-, gonium-—Combining forms meaning a reproductive unit or structure.

Gongylidius (pl. **gongylidia**)—A bulbous structure developed by fungi cultivated by termites.

Gonidia—Minute, asexual reproductive structures produced by some bacteria.

Goniocyst, goniocystula—Group of algal cells surrounded by hyphae and forming a rounded structure that is not a soralium.

Goniocystangium—Cuplike structure bearing goniocysts on foliicolous species of *Catillaria* and *Opegrapha*.

Goniospore—An angular spore.

Goniosporous—Having angled spores.

Gonoplasm (of higher Peronosporales)—Protoplasm at the center of an antheridium that later undergoes fusion with the oosphere.

Gonopore (of nematodes)—The exterior opening of the reproductive system; the vulva in the female, the anus or cloacal opening in the male.

Gonotocont—Organ in which meiosis occurs.

Grade—To establish elevations and contours prior to planting.

Gradient of infection—*See* Disease gradient.

Graft, grafting—Transfer of aerial parts of one plant (e.g., buds or twigs—the scion) into close cambial contact with the root or trunk (the rootstock) of a different plant; a method of plant propagation. Also, the joining of cut surfaces of two plants so as to form a living union.

Graft indexing—A procedure used to determine the presence or absence of a virus, mycoplasma, spiroplasma, etc., in a plant. The plant is grafted to another plant that is known to show symptoms if affected by the disease in question. The method is used to detect the presence of disease agents that are not readily transmitted mechanically.

Graftage—Method of inserting buds, twigs, or shoots in other stems or roots for fusion of tissues.

Grain—The simple, indehiscent fruit of a grass, chiefly a cereal or grain crop, whose ovary walls are fused to the seed. *See* Caryopsis.

Gram's stain—A differential stain by which most species of bacteria are classified into two broad groups, Gram-positive or Gram-negative, depending on whether they retain or lose the primary stain (*crystal violet*) when subjected to treatment with iodine, decolorizing in 95% ethanol, followed by a safranin-O counter stain.

Gramineae—Former taxonomic family name for grasses; now the Poaceae.

Gramineous—Of or relating to a grass; grassy.

Graminicole, graminicolous—Growing on grass.

Gram-positive or Gram-negative—Bacteria staining violet or pink-red, respectively, following treatment with Gram's stain.

Grana—Small, chlorophyll-bearing bodies found in chloroplasts, usually seen as stacks of parallel lamellae with the aid of an electron microscope.

Graniform—Shaped like a corn kernel.

Granular, granulate, granulose—Describing a surface covered with very small particles (granules).

Granulation—*See* Capping.

Granule, metachromatic—A deeply staining body in the protoplasm of some bacterial and yeast cells.

Granules (of pesticides)—A ready-to-use formulation in which the pesticide is attached to particles of an inert carrier such as clay or ground corncobs. Granules are generally in the size range of 15–40 mesh. They can be metered out in dry form more easily and accurately than dusts or wettable powders. Some degree of volatility is necessary for granular formulations to be effective.

Granulose—With a grainy or granular outer layer.

Graphium (pl. **graphia**)—Synnema of *Graphium*.

Grassland—A formation (or biome) in which the dominant plants are grasses.

Gravid (of nematodes)—Containing an egg or eggs; capable of depositing eggs.

Gravitational water—Water that moves downward through soil due to the earth's gravitational force upon it.

Gray mold—*Botrytis cinerea*.

Green chop—Forage chopped in the field while succulent and green and fed directly to livestock, made into silage, or dehydrated.

Green islands (in virology)—Nonchlorotic regions in a leaf showing mosaic or ringspot symptoms.

Green manure—A crop plowed under while still green and growing to improve the soil.

Green mold—A species of *Penicillium*.

Gregarious—In groups or companies scattered in a restricted area; closely associated but not joined together or tufted.

Griseofulvin (trade names: Fulvicin, Grifulin, Grisactin, etc.)—A chlorine-containing, antifungal antibiotic from *Penicillium griseofulvum* and *P. janczewskii (P. nigricans)*; described and named as "curling factor."

Griseous—Grayish or dirty-white.

Grisette—The edible *Amanita vaginata*. *See also* Tawny grisette.

Ground cover—Any plant used to cover the ground, hold soil, and give foliage texture (e.g., ivy, *Pachysandra*, *Vinca*), usually as a substitute for grass.

Ground meristem—Meristematic tissue that develops into pith and cortex.

Groundwater—Water that fills all unblocked pores of underlying material below the water table, which is the upper limit of saturation.

Group—Any taxonomic unit; taxon.

Group species—A cluster of populations that appear similar but are biologically distinct, all known by the same name, that have not been differentiated by traditional taxonomic criteria.

Group-specific antigen—An antigen specific to a group of viruses. *See* Type-specific antigen.

Growing medium (soil mix)—Soil or soil substitute prepared by combining such materials as peat moss, vermiculite, perlite, sand, or composted organic material and used for growing potted plants, cuttings, and germinating seeds.

Growing season—The period between commencing of growth in the spring and cessation of growth in the fall.

Growth—An irreversible increase in cell size and/or cell number; an increase in dry weight, regardless of cause.

Growth curve—Graphic representation of the growth (population changes) of bacteria or other microorganisms in phases in a culture medium.

Growth inhibitor—A natural substance that inhibits the growth of a plant.

Growth layer—Layer of secondary xylem or phloem produced (ordinarily) in a single growing season.

Growth or plant regulator—A synthetic or natural substance that in low concentrations regulates the enlargement, division, or activation of plant cells.

Growth retardant—A chemical that selectively interferes with normal hormonal promotion of plant growth, but without appreciable toxic effects.

Growth ring—A growth layer seen in cross section, e.g., the annual ring of xylem in a woody plant.

Growth zone of ovary or testis (of nematodes)—A region of gradual increase in size of the oogocyte or spermatozoa.

Grub—The larva of a beetle.

Guaic (gum guaic)—Spot test reagent; resinous substance derived from members of the neotropical genus *Guaiacum* that turns blue when applied in alcoholic solution to an active fungal mycelium or basidiocarp secreting phenoloxidase(s).

Guanine—One of the purine bases found in DNA and RNA

Guanosine—A nucleoside of guanine and ribose. *See* Nucleoside and Nucleotide.

Guard cells—Paired, specialized, epidermal cells that contain chloroplasts and surround a stoma.

Gubernaculum (of nematodes)—Spicule guide; sclerotized accessory piece, posterior to the spicules.

Guiding ring—A sleevelike structure that surrounds and guides the stylet of nematodes in certain genera of the Dorylaimoidea; position varies among the genera from apex to posterior portion of stylet.

Gully erosion—A process in which water accumulates in narrow channels and, over relatively short time periods, removes the soil from these narrow areas to considerable depth.

Gum—Gelatinous, sugary aggregate that is synthesized and exuded by plant tissues;

complex mucilaginous, poysaccharidal substances formed by cells in reaction to wounding or infection.

Gum guaiac solution—A 5% solution of gum guaiac in 95% ethyl alcohol; used to test for polyphenol oxidases in culture, rot, or basidiocarps. A blue color indicates a positive reaction.

Gummose—Gummy.

Gummosis (pl. **gummoses**)—Pathologic condition characterized by excessive secretion of sap, "gum", latex, or resin by or in a plant tissue; the products of cell degeneration. May be due to a parasite working within the plant, to unfavorable growing conditions, or to other environmental factors.

Gummosis of cucurbits (especially cucumber)—Disease caused by *Cladosporim cucumerinum*.

Guttate—Having tearlike droplets; (of a pileus) spotted as if by drops of liquid.

Guttation—Exudation of water and solutes from the hydathodes of plants, especially along the leaf margin and at the tip.

Guttiferous—Bearing droplets.

Guttiform—Having the shape of a drop.

Guttulate (of spores)—Having one or more oil droplets (guttulae) within.

Guttule (pl. **guttulae**)—An oillike droplet of refractive material, commonly seen in spores and other structures.

Guttulose—Covered or full of droplets.

Gymno- (prefix)—Uncovered or naked.

Gymnosperm—Vascular plant that produces seeds not enclosed within carpel tissues (naked seed).

Gyn-, gyneco-, gyno-, -gyne, -gynous—Combining forms meaning female.

Gynoecium—The female part of a flower or pistil formed by one or more carpels and composed of the stigma, style, and ovary.

Gynophore (of Pyronemataceae)—Multinucleate female structure undergoing development.

Gyrate, gyrose—Curved backward and frontward in turn; folded and wavy; convoluted like a brain; (of lichen apothecia) concentrically folded.

Gyre—A circular or spiral motion or form.

H

H antigens—Antigens of motile, Gram-negative enterobacteria (e.g., the *Salmonella* group) found in the flagella. H antigens are more labile than O antigens (which see), and are destroyed by 70°C for 20 min.

H bodies—Sporidia of *Tilletia* fused in pairs while still attached to the promycelium.

H chain, heavy chain—One pair of identical polypeptide chains, each $M_r \pm 50,000$, of four-chain immunoglobulin molecules. The other pair of chains is the L (light) chains.

ha—*See* Hectare.

Habitat—The natural environment of an organism.

Hadromycosis—A plant disease restricted to the xylem. *See* Tracheomycosis and Wilt.

Haemolymph (of nematodes)—The watery lymphlike nutritive fluid of the pseudocoel.

Haerangium—Organ of spore formation and dispersal of certain Ascomycotina, including *Ceratostomella, Endoxyla, Fugascus,* and *Ophiostoma*, in which the eight ascospores (developed from the *octophore*) are contained by a membrane and surrounded by a circle of hairs (the *tentacle*) around the ostiole of the perithecium.

Hair (in Agaricales)—One of the hair-shaped epicuticular elements forming a pilose covering (or down under a lens) and not homologous with a cystidium, pseudoparaphysis, or seta.

Hairy root—The development of large numbers of small roots on a limited area of a root caused by *Agrobacterium rhizogenes*.

Hallucinogenic fungi—Basidiocarps of *Psilocybe* (especially *P. mexicana*), *Lycoperdon cruciatum, L. mixtecorum, Paneolus,* and *Stropharia*, eaten by Mexican Indians during magical ceremonies. The active principle is the crystalline psilocybin.

Halmophagous (of ectotrophic mycorrhizae)—Having a mantle and Hartig net.

Halo—Symptom normally characterized by an area of diseased tissue(s), often discolored or watersoaked, that surrounds a lesion; an area of chlorotic tissue that surrounds a necrotic area (spot) or a sign (e.g., uredium).

Halonate—Describing a leaf spot with concentric rings; a "frog-eye" type; (of a spore) covered with a transparent coat.

Halophile (adj. **halophilic**)—Organism that is salt-tolerant; living in salt water; preferring salt.

Hamanatto—An edible Oriental product obtained by fermenting soybeans (*Glycine max*) with *Aspergillus oryzae*; called *tao-cho* in Malaysia and *tao-si* in the Philippines.

Hamate, hamose, hamous—Hooked at the apex; unicate. *See* Hamulate.

Hamathecium—A neutral term for all kinds of hyphae or other tissues between asci or projecting into the locule or ostiole of ascomata; usually of carpocentral origin. Seven categories have been recognized: 1) *interascal pseudoparenchyma*, carpocentral tissues unchanged or compressed between developing asci; 2) *paraphyses*, hyphae originating from the base of the cavity, usually unbranched and not anastomosing; 3) *paraphysoids* (trabecular pseudoparaphyses; tinophyses); interascal or preascal tissue stretching and coming to resemble pseudoparaphyses; often only remotely septate, anastomosing and very narrow; 4) *pseudoparaphyses* (cellular pseudoparaphyses; cataphyses), hyphae originating above the level of the asci and growing downward between the developing asci, finally becoming attached to the base of the cavity and often also then free in the upper part; often regularly septate, branched, anastomosing and broader; 5) *periphysoids*, short hyphae originating above the level of the developing asci but not reaching the base of the cavity; 6) *periphyses*, hyphae confined to the ostiolar canal; unbranched and not anastomosing; can occur with paraphyses, pseudoparaphyses or periphysoids; and 7) *hamathecial tissue absent.*

Hamulate, hamulose—Having small hooks.

Hanging drop technique—A technique for observing microorganisms suspended in a drop of liquid. *See* van Tieghem cell.

Hanging drop—*See* van Tieghem cell.

Haplo- (prefix)—Single; one only.

Haplobiontic (of an organism)—Completing the life cycle as one kind of plant or thallus, the sexual one, homothallic or heterothallic, haploid or diploid, and with or without spores. *See* Diplobiontic.

Haploconidium (pl. **haploconidia**)—A uninucleate conidium formed on the mycelium of Tremellales.

Haplodioecious, haploheteroecious—Heterothallic.

Haplohyphidium—*See* Hyphidium.

Haploid—Having a single complete set of unpaired chromosomes (1n) in each cell nucleus; characteristic of the gametophyte generation; (of a mycelium) composed of haploid cells.

Haplomonoecious—Homothallic.

Haplont, haplophyte—Thallus in the haplophase; the gametophyte.

Haplophase—The part of the life history of an organism in which the cells are haploid (1n).

Haplostromatic—Describing a fungus with an ecto- or endostroma. *See* Stroma.

Haplosynoecious—Homothallic.

Hapten—A partial antigen; substance that can combine with antibody but cannot initiate an immune response unless bound to a carrier before introduction into the body. Most haptens are small molecules (M_r <1000) and carry only one or two antigenic determinants, but some macromolecules, e.g., pseumococcal polysaccharides, are haptenic. Haptenic groups can be conjugated to carriers *in vitro* and may then be regarded as isolated determinants capable of reacting with antibody but requiring the carrier molecule to become immunogenic *in vivo*.

Hapteron—Mass of adhesive hyphae at the base of the funicular cord in the Nidulariaceae; (of lichens) an aerial organ of attachment of some fruticose lichens formed by a secondary branch that becomes attached to the substratum.

Hard (of fungal fruit bodies)—May be somewhat corky when fresh, but drying hard and brittle.

Hard pine—A pine with two or three needles per cluster.

Hardening, hardening-off—Adapting plants to unfavorable outdoor conditions by withholding water and lowering the temperature or nutrient supply to hasten maturing of tissues for increasing hardiness.

Hardiness—The ability of plants to resist injury from unfavorable temperatures or other stresses.

Hardpan—An impervious layer of soil or rock in the lower A horizon or upper B horizon that prevents downward drainage of water and root growth. A chemical "hard pan" is caused by the presence of toxic ions and often is related to extremes in soil pH.

Hardwood—The wood of angiosperms.

Hardy plants—Plants adapted to cold temperatures or other adverse climatic conditions in an area. *Half-hardy* describes some plants that may be able to take local conditions with some protection.

Hart's truffle—*Elaphomyces.* See Truffle.

Hartig net—The intercellular net of hyphae formed by an ectomycorrhizal fungus on the surface of a root.

Hastate—Shaped like a spear or arrowhead.

HAT medium—A cell culture medium (containing hypoxanthine, aminopterin, and thymidine) used in monoclonal antibody production to select hybridomas from unfused myeloma cells.

Hatching factor—A material produced by the roots of certain plants that increases the hatching of eggs of certain nematodes such as *Globodera* and *Heterodera* spp.

Haustorial cap—An electron-dense, caplike mass at the end of a lobe of the haustorial apparatus of *Exobasidium camelliae.*

Haustorium (pl. haustoria)—Specialized, simple or branched structure of a fungal parasite, especially with a projection within a living host cell, for absorption of food; often associated with obligate parasites (rusts, downy and powdery mildews, parasitic flowering plants), but also produced by some facultative saprophytes and between mycobionts and phycobionts in lichens. There are three main types: *wall-to-wall apposition* without penetration; *intracellular haustoria*, in which the fungus penetrates into the host cell and may or may not form a special sheath, neckband, or collar; and *intraparietal haustoria*, in which penetration is restricted to the wall layers.

Haustrulum (of nematodes)—The sphaeroidal cavity of the esophageal valvular apparatus that functions as a pump.

Hay—Herbage of forage plants, including seed of grasses and legumes, that is harvested and dried for animal feed.

Hazard—The probability that injury (to applicator, bystanders, consumers, pets, livestock, wild life, or crops, etc.) will result from the use of a substance in the quantity and manner proposed.

Head—A type of inflorescence, typical of the composite family (Asteraceae), in which numerous small flowers are densely crowded upon a usually disc-shaped common receptacle; as in sunflowers (*Helianthus*).

Head cap (of nematodes)—A distinctively off-set head region.

Head region (of nematodes)—That portion anterior to the base of the stoma (buccal aperture) or to the anterior of the esophagus; (of tailed phages) may be isometric or elongated and contain the DNA genome.

Heading back—Pruning off the terminal parts of a twig, branch, or tree.

Headland—Roadway or field margin.

Healing over—The process whereby a wound is closed or protected by a new growth (callus) without replacing the lost parts.

Heart rot—Decay of the nonliving heartwood of living trees by fungi.

Heartwood—The central cylinder of nonconducting xylem tissue in a woody stem or trunk that contains no living cells; heartwood is usually darker than sapwood.

Heaving—The partial lifting of plants out of the ground, frequently breaking their roots, due to alternate freezing and thawing of the surface soil during the winter.

Heavy soil—Soil with a high content of fine separates, especially clay, or one with a high tractor power requirement and hence difficult to cultivate.

Hectare (ha)—A land area in the metric system equal to 2.471 acres = 395.367 rods = 10,000 m^2 = 0.01 km^2 = 0.0039 mi^2.

Heeling in—Temporary planting for bare-rooted nursery stock. Roots are placed in a shallow trench, covered with soil or sawdust, and watered.

Helenin, helenine—An antiviral antibiotic from *Penicillium funiculosum*; considered to be an RNA of viral origin.

Helical, helicoid—In a coil; in the form of a spiral; coil-like.

Helical symmetry (in virology)—A form of capsid structure in many RNA viruses in which the protein subunits interact with the nucleic acid to form a helix. The only axis of symmetry is the length axis of the particle. All rod-shaped plant viruses have coat protein subunits arranged in helical symmetry.

Helicospore—A cylindrical, coiled, or spiral-shaped spore, usually septate. *See* Amerospore and Scolecospore.

Helicosporous—Having spores that are cylindric, spiral, or convolute and usually septate.

Heliophilous—Preferring direct sunlight.

Heliotropism—Turning toward or following the sun during the day.

Heliozooid—Amoeba-like, but having well-marked raylike pseudopodia.

Helix (adj. **helical**)—Coiled or spiral in shape; often used in reference to the double spiral of the DNA molecule.

Helminth—A worm, especially a parasitic worm, e.g., nematode.

Helminthoid—Worm-shaped; vermiform.

Helminthology—The branch of zoology dealing with helminths, especially parasitic worms.

Helminthosporal—A terpenoid mycotoxin from *Bipolaris (Drechslera) sorokiniana*, teleomorph *Cochliobolus sativus*, toxic to wheat (*Triticum*) and barley (*Hordeum*).

Helminthosporoside—Host-specific toxin produced by *Bipolaris sacchari* in sugarcane (*Saccharum*).

Helotism—The physiologic relation of an alga to a fungus in a lichen.

Helper virus—A virus required for replication of a defective or satellite virus or a satellite RNA. *See* Dependent transmission.

Helveous, helvelous—Pale yellow or light ochraceous yellow.

Hemagglutination—Agglutination (clumping) of red blood cells.

Hemagglutinin—An antibody or substance that causes agglutination of red blood cells.

Hemi- (prefix)—Half; in part. *See* Semi-.

Hemiangiocarpous (of a sporocarp)—Opening before quite mature.

Hemiascospore—Ascospore of a hemiascus.

Hemiascus—The atypical multispored ascus of the Hemiasci.

Hemicompatible—Homokaryon compatible with one of the two components of the dikaryon. *See* Compatible.

Hemiform (of the Uredinales)—Having only uredial and telial stages.

Hemiparasite—A partial parasite; same as facultative parasite (which see).

Hemisphaeroid, hemispherical—Having the shape of half a globe or sphere; semiglobose.

Hemispore (especially of dermatophytes)—A cell at the end of a filament that later divides to form several deuteroconidia; protoconidium; one of the two cells produced by a primary transeptum in an ascospore. *See* Septum.

Hemizonid (of nematodes)—A lenslike structure or nerve commissure situated between the cuticle and hypodermal layer on the ventral side of the body just anterior to the excretory pore; generally believed to be associated with the nervous system.

Hemizoniom (of nematodes)—Companion structure to hemizonid (smaller and posterior to the hemizonid).

Hemoglobin—The constituent of red blood cells that gives them their color and carries oxygen.

Hemolysin—A substance that lyses red blood cells and liberates hemoglobin.

Hemolysis—The process of dissolving (lysis) red blood cells.

Hemorrhagic—Showing evidence of hemorrhage (bleeding). Tissue becomes reddened by the accumulation of blood that has escaped from capillaries into the tissue.

Herb—Botanically, a plant with fleshy, not woody, stems. Commonly a plant used for flavoring, fragrance, or medicinal purposes.

Herbaceous—Describing a higher plant with mostly soft, succulent, nonwoody stems (e.g., annuals, biennials, and perennials) that normally die back to the ground in the winter.

Herbals—Medieval and Renaissance books picturing and describing plants.

Herbarium—A collection of dried and pressed plant or fungal specimens, cataloged for easy reference (arranged systematically); the place where such a collection is stored.

Herbarium beetle—*Stegobium paniceum*; may eat spores of certain fungi.

Herbicide—Any chemical or physical agent used for killing or inhibiting the growth of plants where they are not wanted; a weed or grass killer; a type of pesticide.

Herbicole (adj. **herbicolous**)—An organism growing on herbs.

Herbivore, herbivorous—Describing an animal that subsists principally or entirely on plants or plant products.

Heredity—The tendency of an organism to resemble its parents; the transmission of morphological and physiological characters from parents to their offspring.

Hermaphrodite flower—A flower having both stamens (male) and pistils (female).

Hermaphroditism (n. **hermaphrodite**, adj. **hermaphroditic**)—The condition, in species of animals, including nematodes, in which recognizable and functional male and female reproductive organs are produced by the same individual.

Hetero- (prefix)—Other; not normal; different.

Heterobasidium—A basidium of the Heterobasidiomycetes, usually a phragobasidium.

Heterocaryosis (heterokaryosis)—*See* Heterokaryosis.

Heterocyst—A large cystlike cell in an algal filament that separates from other cells, causing the filament to separate into hormogonia (which see).

Heteroduplex analysis—The study of structures produced by the hybridization of two single-stranded DNA molecules derived from different sources. The study uses the Kleinschmidt procedure, usually by electron microscopy.

Heteroecious (n. **heteroecism**)—Requiring two unrelated host plants for completing the life cycle, as in many rust fungi (Uredinales). *See* Autoecious.

Heterofermentation (adj. **heterofermentative**)—A type of fermentation used by leuconostoes and some lactobacilli in which the end products include CO_2 or acetic acid in addition to lactic acid. *See* Homofermentation.

Heterogametes—Gametes that differ in size, structure, and behavior. *See* Isogamy.

Heterogamic copulation—Fusion of gametes unlike morphologically.

Heterogamy (adj. **heterogamous**)—The union (copulation) or conjugation of gametes differing in size, structure, and behavior; sexual reproduction. See *Isogamy*.

Heterogeneous—Differing in structure, tissues, qualities or nature. *See* Homogeneous.

Heterogenic incompatibility—The limiting of interfertility to pairings that bring like factors together.

Heterogonic or indirect cycle (of nematodes)—A situation in which eggs of parasitic parents develop into free-living males and females, the offspring of which then proceed to the parasitic phase. *See* Homogonic or direct cycle.

Heterokaryon—Cell or mycelium with two or more nuclei of different constitution (n+n); also used to denote a fungal strain with this characteristic; a hybrid cell formed by the fusion of two cells from different species. *See* Homokaryon.

Heterokaryosis (adj. **heterokaryotic**)—The condition in which a mycelium contains two (or more) slightly genetically different nuclei per cell; may result from anastomosis (hyphal fusion) between similar but genetically different fungi; heteroplasmic. *See* Dikaryon.

Heterokaryotise (of rusts and pyrenomycetes)—Fusion of haploid structures of opposite sex that does not give a conjugate arrangement of the nuclei. *See* Diploidization.

Heterokont, heterokontous—Having flagella in pairs of different lengths; condition in which a flagellum has two rows of tubular hairs.

Heterologous—Derived from a different type or species.

Heterologous reaction—Different although apparently similar; e.g., a serological reaction between an antiserum and an antigen closely resembling but not identical to the antigen causing the production of antibody. *See* Homologous reaction.

Heteromerous—(Of trama in Russulaceae) having sphaerocyst nests among filamentous hyphae; (of a lichen thallus) having the mycobiont and photobiont in well-marked layers, usually between the medulla and the upper cortex. *See* Homoiomerous.

Heteromixis (adj. **heteromictic**)—*See* Heterothallism.

Heteromorphic, heteromorphous—Having variation from normal structure; having organs of different length; (of agaric lamella edge) sterile due to the pressure of cystidia; made up of different elements. *See* Homomorphous.

Heterophile antibody—An antibody that reacts with totally unrelated species of cells, or microorganisms from unrelated species of animals.

Heterophile antigen—An antigen found in different, apparently unrelated, organisms and/or tissues, usually carbohydrate, that reacts with antibodies stimulated by unrelated species.

Heterophyllous (of agarics)—Having lamellae (gills) that are dissimilar, of a different form or length.

Heterophytic (and **homophytic**)—The equivalents in the sporophyte generation of dioecious and monoecious, respectively, in the gametophyte generation.

Heteroplanogametic (n. **heteroplanogamete**)—Having motile gametes that are dissimilar.

Heteroplastic—Heterokaryotic.

Heteroploid—Describing a cell, tissue, or organ that contains more or fewer chromosomes per nucleus than the normal 1n or 2n for that organism.

Heterosis—Hybrid vigor; a type of increased vigor, growth, size, yield, or function often occurring in an organism that is the offspring of parents of different inbred lines, varieties, or species.

Heterospory, heterosporous—Production of spores of two or more types (differing in the mating type, + or –).

Heterostrophic—Coiled in a direction opposite to the usual one.

Heterothallism (adj. **heterothallic**)—Self-sterility; a sexual condition in which an individual produces only one kind of gamete; used chiefly in reference to fungi and algae. See Homothallism.

Heterotroph (adj. **heterotrophic**)—An organism living only on organic food substances (produced by other organisms) as primary sources of energy and is thus a parasite or saprophyte; a microorganism unable to use carbon dioxide as its sole source of carbon. See Autotroph.

Heteroxenous—Having more than one host.

Heterozygote—A zygote that possesses two different alleles in a single gene pair, e.g., Bb.

Heterozygous—Having mixed hereditary factors; not a pure line; having heterokaryosis resulting from the fusion of unlike gametes; having different alleles at various loci. See Homozygous.

Hexa- (prefix)—Meaning six.

Hexagonal—Six-angled or six-sided.

Hexamer—A group of six protein subunits on the triangular faces of virus capsids with icosahedral symmetry.

Hexaploid—Having six sets of chromosomes (6n).

Hexasporous—Six-spored.

Hexon—Capsomer with six subunits surrounded by six identical capsomers, said to be "hexavalent."

Hiascent—Becoming wide open; gaping.

Hibernate—To lie dormant.

Hidden determinant, cryptotype—Antigenic determinant so positioned on a molecule or cell that it is not accessible for recognition by lymphocytes or antibodies and is neither capable of binding specifically to them nor of stimulating an immune response, unless some stereochemical change causes the hidden determinant to be revealed.

High-pressure (performance) liquid chromatography (HPLC)—A method for separating peptides, oligonucleotides, etc., with high resolution. See Chromatography.

High voltage electrophoresis—Electrophoresis at potential differences of more than 1,000 volts; used in nucleic acid sequencing and in paper electrophoresis of nucleotides.

Higher fungi—Ascomycota, Basidiomycota, and Deuteromycota (mitosporic fungi).

Hilar appendage (of basidiospores)—The small, cone- or wartlike projection (diverticulum) that connects the spore with the sterigma; sterigmatal appendage; apicule. *See* Apophysis.

Hilar depression (of basidiospores)—The depression near the hilum; a dorsibasal depression.

Hill—Raising the soil in a slight mound for planting, or setting plants some distance apart.

Hilum—A dot, mark, or scar, especially on a spore at the point of attachment to a conidiophore, conidiogenous cell, or sterigma; scar on a dicot seed coat marking the place of attachment of the seed stalk to the seed; a small depression of a motile cell in which the flagellum is inserted.

Himantioid (of mycelium)—In spreading fan-shaped cords, as in *Himantia*.

Hinge region—Amino acid segment between the first and second constant regions of the immunoglobulin heavy chain of IgG that permits bending of the molecule.

Hip—Rose fruit formed by a group of achenes surrounded by a receptacle and hypanthium.

Hippocrepiform—Horseshoe-shaped.

Hircinol—*See* Orchinol.

Hirsute—Covered with coarse, stiff, long fibers or hairs; (of nematodes) rough with hairs or bristles.

Hirt supernatant—A method for separating viral from cellular DNA by lysing cells with sodium dodecyl sulfate (SDS) in the presence of $1M$ NaCl. The cellular DNA is precipitated, leaving the viral DNA in the supernatant.

Hirtose, hirtous—Hairy; hirsute.

Hispid—Covered with coarse, stiff, erect hairs or bristles visible to the naked eye.

Hispidulous—Somewhat or minutely hispid.

Hist-, histi-, histo-—Combining forms meaning web, hence tissue.

Histochemical—Pertaining to chemical reactions in tissues.

Histogenic—Developing or originating from tissues.

Histogenous—Produced from tissue; (of spores) produced from hyphae or cells, without conidiophores (conidiogenous cells).

Histology—The science or study that deals with the microscopic structures of plant or animal tissues.

Histolysis (adj. **histolytic**)—The dissolving of walls or tissues.

Histones—The five main classes of basic water-soluble proteins in cell nuclei in close association with DNA forming chromatin. They contain tyrosine but little tryptophan.

Histopathology—Microscopic study of diseased tissues.

Histoplasmin—An antigen prepared from *Histoplasma capsulatum*, especially for skin testing.

Histoplasmosis—A widespread human and animal disease caused by *Histoplasma capsulatum*, characterized by emancipation, anemia, and other symptoms.

Histotype (in immunology)—A reaction between different types of cells.

Hoary (especially of a pileus or stipe)—Densely covered with silklike hairs; canescent.

Hol-, holo- (prefix)—Whole, entire, all.

Holdfast—A special structure by means of which fungi become attached to their hosts (e.g., appressorium, hyphopodium, stigmatopodium, and stomatopodium). *See* Rhizoid, Haustorium, Hapteron.

Holobasidium—A basidium in which the metabasidium is not divided by primary septa but may be adventitiously septate. It may be cylindrical with longitudinal nuclear spindles at different levels (*stichobasidium*) or clavate with transverse nuclear spindles at the same level (*chiastobasidium*).

Holoblastic conidium ontogeny—Development in which both outer and inner walls of the blastic conidiogenous cell contribute to the formation of a conidium.

Holocarpic, holocarpous—Describing an organism whose thallus is entirely converted into one or more reproductive structures (e.g., sporangium).

Holocoenocytic—*See* Astatocoenocytic.

Hologamy—The condition in which the entire thallus becomes a gametangium, i.e., there is copulation between two mature, indistinguishable individuals, as in *Polyphagus*.

Hologonic—The proliferation of germ cells along the entire length of a nematode's gonad.

Holomorph—The whole fungus in all its states (sexual and asexual) in the life cycle.

Holomyarian—An arrangement of the somatic musculature in nematodes with only two chords or no chords, in which the musculature is divided into two fields or is continuous.

Holophyte (adj. **holophytic**)—A physiologically independent, chlorophyllous plant.

Holosaprophyte—A true saprophyte.

Holosericeous—Covered with fine and silky pubescence.

Holosporous (of conidial maturation)—Describing a conidium that is almost its full size and shape before delimiting cells and maturing as a whole.

Holotype—The single specimen (strain, isolate) or other element designated or indicated as the "type" by the original author at the time of the publication of the original description.

Holozoic—Ingesting food as solid particles in the manner of amoebae.

Holy Fire—One of the names given to the great ergot plagues of the Middle Ages; also called St. Anthony's Fire and St. Michael's Fire.

Homo- (prefix)—One and the same; common; like.

Homobasidium—A nonseptate basidium of the Homobasidiomycetes, usually a holobasidium. See Holobasidium.

Homobium (pl. **homobia**)—The independent association of a fungus and an alga, as in lichens.

Homobody—See Ab 2 beta.

Homofermentation (adj. **homofermentative**)—Type of fermentation carried out by streptococci and some lactobacilli, in which carbohydrate is converted entirely to lactic acid. See Heterofermentation.

Homogeneous—Similar in certain characteristics, such as in chemical nature or physical properties; (of context tissue) uniform in consistency and color. See Heterogeneous.

Homogenic incapatibility—The limiting of interfertility to pairings that bring together different factors.

Homogonic or direct cycle (of nematodes)—A situation in which eggs of parasitic parents hatch into free-living larvae that develop directly into parasitic forms. See Heterogonic or indirect cycle.

Homoheteromixis—See Heterothallism.

Homoiology—Similarity due to parallelism.

Homoiomerous—(Of a lichen thallus) having the same mycobiont and phycobiont evenly intermixed throughout the thallus; (of trama in agarics) composed only of hyphal tissue. See Heteromerous.

Homokaryon (adj. **homokaryotic**)—An individual (cell or mycelium) whose nuclei are genetically alike (of one genotype); hybrid cell formed by the fusion of two cells of the same species. See Heterokaryon.

Homologous, homology—(In serology) the specific relationship between an antigen and the antibody induced by its interaction with animal immune systems; likeness in structure; similarity of origin of two or more parts or organs; the same with respect to type or species; (of fungi) showing a resemblance in form or structure, but not necessarily function, and is considered to be evidence of evolution relatedness. See Analogous and Alternation of generations.

Homologous antiserum—A serum containing antibodies raised against a specific antigen and that will react with that antigen.

Homologous chromosomes—The two members of a chromosome pair.

Homologous reaction—A serological reaction in which an antiserum is reacted against the antigen used for its preparation.

Homomixis—Haplomonoecism or monoecism.

Homomorphic, homomorphous (of an agaric lamella edge or polypore)—Edges of the gills or tubes of the same composition as the hymenium; all parts alike; of one form. See Heteromorphic.

Homonym—A name of a taxon that (under the Code) must not be used because of an earlier name used in a different sense; i.e., the names are the same but the types are different.

Homophytic—*See* Heterophytic.

Homoplasmic—Karyotic.

Homosporous—Producing spores of a single type.

Homothallism (adj. **homothallic**)—Self-fertile; the condition in which sexual reproduction can occur without the interaction of two different thalli; compatible male and female gametes on the same mycelium; a dikaryotic mycelium develops from a single basidiospore and can eventually give rise to basidia. Three types have been recognized: *primary homothallism*, as occurs in a homokaryotic individual; *secondary homothallism*, also called pseudo- or facultative heterothallism; and *amphithallism*, as in a thallus derived from a heterokaryotic spore containing nuclei of compatible mating types. *See* Heterothallism.

Homozygous—Having paired identical genes present in the same cell or organism; without allelic differences; a pure breed or line. *See* Heterozygous.

Honey agaric—*Armillaria mellea*; shoestring fungus; honey mushroom.

Honeydew—A sweet, sticky secretion given off by aphids, whiteflies, mealybugs, scales, and other sucking insects in which sooty mold fungi grow; a secretion, attractive to insects, associated with the *Sphacelia* phase of *Claviceps* and the spermagonial state of some rust fungi.

Hopperburn—Marginal yellowing, scorching, and curling of leaves (e.g., alfalfa, dahlia, potato) due to the feeding of certain leafhopper species.

Horizon, soil—A layer of soil, approximately parallel to the soil surface, with distinct characteristics produced by soil-forming processes. *See* A, B, and C horizons.

Horizontal resistance—Partial resistance that is effective against most or all races (genetic variants) of a pathogen.

Hormocyst—A propagule or diaspore produced in a special *hormocystangium* made up of a few algal cells and fungal hyphae; produced by a few gelatinous lichens.

Hormocystangium—*See* Hormocyst.

Hormogonia—Reproductive segments of algal filaments.

Hormone—A chemical substance produced in one part of a plant and used in minute quantities to induce a growth response in another part; frequently referring particularly to auxins. A naturally occurring or synthetic, organic compound that stimulates plants in a specific manner. (For sex hormones of fungi *see* Antheridiol, Progamone(s), Sirenin, Trisporic acid(s), and Erogen, etc.).

Horn of plenty—The edible *Craterellus cornucopioides*.

Horny (of textures)—Hard and brittle, homogeneous in texture and difficult to section.

Horse mushroom—The basidiocarp of the edible *Agaricus arvensis*.

Horse-hair blight fungi—The rhizomorphic mycelia of tropical species of *Marasmius*. *See* Thread blight.

Horseradish peroxidase—An enzyme derived from horseradish (*Armoracia rusticana*) used in ELISA tests to give the color reaction with its substrate.

Horticulture—The art and science dealing with fruits, vegetables, flowers, ornamentals, shrubs, and amenity trees.

Host—A living organism (e.g., a plant) harboring or invaded by a parasite and from which the parasite obtains part or all of its nourishment; (in virology) an organism or cell culture in which a given virus can replicate. *See* Suscept.

Host indexing—1) A procedure to determine whether a given plant is a carrier of a pathogenic virus, mycoplasma-like organism, or spiroplasma; 2) a procedure in which material is taken from one plant and transferred to another plant that will develop characteristic symptoms if affected by the pathogen in question.

Host range—The complete range of plants that may be attacked by a given pathogen.

Host-specific—Describing pathogens that attack only certain species of hosts.

Hot caps—Waxpaper cones, paper sacks, or cardboard boxes with bottoms removed that are placed over individual plants in spring for frost and wind protection.

Hot spot (in genetics)—A region usually within a gene in which mutations occur at an unusually high frequency.

Hot water insoluble nitrogen (HWIN)—Fertilizer nitrogen not soluble in hot water at 100°C. Used to determine the activity index of ureaforms. *See* Nitrogen activity index.

Hotbed—A plant bed similar to a cold frame but provided with a source of heat (electric cables, fermenting organic matter, steam pipes, hot air flues) to supplement sunlight. *See* Cold frame.

House or dry rot fungus—*Serpula lacrimans.*

Hue (of color)—Tint, shade.

Hugh and Leifson test—A method of determining whether carbohydrates are broken down by anaerobic fermentation or by oxidation.

Hülle cells—Terminal or intercalary thick-walled cells that occur in large numbers in association with ascocarps (cleistothecia) of *Emericella nidulans*, anamorph *Aspergillus nidulellus.*

Humectate, humectous—Moist, wet.

Humicole, humicolous—Growing in or on soil, decaying organic matter.

Humidity, relative—The weight of water vapor in a given quantity of air as compared to the total weight of water vapor that the air is capable of holding at a given temperature and expressed as a percentage.

Humus—The more or less stable, decomposing organic matter from any source, remaining after microbial decomposition, that may become fine, rich, black soil. May come from vegetable refuse, leaf mold, manure, peat, or animal matter.

Hyaline—Transparent or nearly so; translucent; clear or colorless.

Hyalo- (prefix) (of spores)—Hyaline or brightly colored (not dark colored), especially for groups of Deuteromycotina (Mitosporic fungi).

Hybrid (v. **hybridize**)—The offspring of two individuals of different genetic character; the crossing of two individuals differing in one or more heritable characteristics; also used for the offspring resulting from a cross between two species that result in viable progeny.

Hybrid arrested translation—A method used to identify proteins encoded by a cloned DNA sequence. *See* Hybrid released translation.

Hybrid released (selected) translation—A method used to identify proteins encoded by a cloned DNA. A preparation of mRNA is hybridized to the cloned DNA immobilized on a solid matrix, e.g., nitrocellulose. The mRNA, homologous to the DNA, is retained on the filter and can then be removed by melting the RNA:DNA duplex. The purified RNA is then translated *in vitro* and the protein product(s) identified, generally by gel electrophoresis.

Hybrid vigor—*See* Heterosis.

Hybridization—The cross of two individuals that differ in one or more heritable characteristics. The joining of two complementary single strands of DNA, or of RNA and DNA, usually from different sources, to form a double-stranded molecule.

Hybridoma—A hybrid animal cell or cell line produced from the fusion of a spleen cell (lymphocyte) and a tumor (myceloma) cell and able to multiply and to produce monoclonal antibodies.

Hydathode—Special epidermal gland or pore structure in leaves and stems at the ends of vascular tissue through which water of guttation is exuded.

Hydnaceous (of the hymenophore)—Bearing the hymenium on spinelike protuberances.

Hydnoid—A toothed hymenial surface.

Hydration—Absorption of water by a substance with attendant swelling; incorporation of water into a complex molecule (e.g., proteins, nucleic acids, and virus particles).

Hydraulic sprayer—A machine that applies pesticides by using water at high pressure and volume to deliver the pesticide to the target.

Hydrofungi—Aquatic fungi.

Hydrogen bonding—A relatively weak noncovalent bond formed between the hydrogen atom and an -OH or -NH group and an oxygen or nitrogen atom. Hydrogen bonds are crucial in maintaining the secondary structure of nucleic acids and proteins; they maintain the double-helical structure of DNA.

Hydrogen ion concentration (pH)—A measure of the acidity of a chemical in solution. It is expressed in terms of the pH of the solution. *See* pH.

Hydrologic cycle—The movement of water from the atmosphere to the earth and its return to the atmosphere.

Hydrolysis (v. **hydrolyze**)—Enzymatic disintegration of a large molecule into smaller component ones; to undergo or cause to undergo a chemical reaction with ions of water; digestion is a process of hydrolysis.

Hydrophyte—Plant that lives in water or exceedingly wet soils.

Hydroponics—Growing plants in aerated water containing all the essential mineral elements rather than in soil. *See* Solution (or soilless) culture.

Hydroseed—To seed in a water mixture by pumping through a nozzle that sprays the mixture onto a seedbed. The water mixture may also contain amendments such as fertilizer and certain mulches.

Hydrosis—*See* Watersoaked.

Hydrotropism—A growth curvature due to the stimulus of water.

Hygiene—*See* Sanitation.

Hygrophanous—Having a water-soaked appearance when wet, with the moisture disappearing rapidly and becoming paler and nontransparent when dry.

Hygrophilous—Preferring a moist habitat.

Hygroscopic—Capable of readily absorbing water and expanding; becoming soft in wet air, hard in dry; (of a fruitbody) opening and discharging spores in dry air.

Hymenial algae (or gonidia)—Algal cells in the hymenium of a lichenized ascomycete.

Hymenial cystidium—*See* Cystidium.

Hymenial veil—*See* Annulus.

Hymeniderm (of basidiomata)—An outer layer (derm) composed of an unstratified layer of single cells or hyphal tips arranged like a palisade, tightly crowded and of the same height.

Hymenium (pl. hymenia)—Spore-bearing (or sporogenous) layer of a fungus fruitbody consisting of asci or basidia (and cystidia, paraphyses, and/or hyphidia, if present) in a continuous layer (palisade). *See* Catahymenium, Euhymenium.

Hymenomycetes—Class of fungi that produce basidia in hymenia.

Hymenophore, hymenophorum—That portion of the sporocarp, especially of a basidiomycete, on which the hymenium develops. *See* Sporophore.

Hymenophyte—Fungus with a hymenium.

Hymenopode, hymenopodium—The layer of tissue under the hymenium; subhymenium or hypothecium.

Hypanthium—Floral tube formed by the fusion of the basal portions of the sepals.

Hyper-, hypero- (prefix)—Above, beyond, over.

Hyperchromicity—The increase in absorbance of light of 260 nm wavelength by nucleic acid at its melting temperature, an indication of the amount of base-pairing in the nucleic acid.

Hyperimmune serum—Serum from an animal that has received two or more injections of a foreign antigen for the purpose of producing a reagent for use in serology.

Hyperimmunization—A condition resulting from immunization, usually with extensive booster doses, designed to stimulate the production of high affinity antibody (relatively late in the immune response).

Hyperparasite—An organism parasitic on another parasite.

Hyperplasia (adj. hyperplastic)—A plant overgrowth (gall, enation, tumor, witches' broom) due to *increased cell division*; excessive, abnormal, usually pathological multiplication of the cells of a tissue or organ. *See also* Hypoplasia.

Hypersaprophyte—Saprophyte found only on substrates invaded by other saprophytes.

Hypersensitive (in virology)—An extreme reaction to a virus, e.g., the production of local lesions or the necrotic response of a leaf to a plant virus.

Hypersensitivity—Increased sensitivity, as in the rapid death of localized plant cell and tissue death at the point of attack by a pathogen, so infection does not spread; especially the reaction of a rust fungus or other obligate parasite; intolerance; state of extreme sensitivity to foreign proteins (e.g., allergens); (in virology) inflammatory immune response in mammals on second exposure to an antigen.

Hypertonic (of culture media)—Having a higher osmotic pressure than the organism being cultured. *See* Hypotonic.

Hypertrophy (adj. **hypertrophic**)—A plant overgrowth (gall or tumor) due to *abnormal cell enlargement*; excessive, abnormal, usually pathological enlargement of individual cells in a tissue or organ. *See also* Hyperplasia.

Hypertrophyte—A parasitic fungus that causes hyperplasia.

Hypervariable region—Portions of either the heavy or light chains of an immunoglobulin that exhibit extreme amino acid sequence variability and that make up part of the immunoglobulin's (antibody's) antigen-combining site.

Hypha (pl. **hyphae**, adj. **hyphal**)—The basic vegetative unit of structure and function of most fungi; a largely microscopic tubular filament that increases in length by growth at its tip. New hyphae arise as lateral branches. Hyphae constitute the body (mycelium) of a fungus. They may be divided into cells by cell walls (*septa*) or be one long cell. Some hyphae are specialized for producing spores, penetrating host tissues, absorbing food, overwintering, or trapping nematodes, etc. *See* Ampoule hypha, Arboriform hypha, Ascogenous hypha, Binding or ligitive hyphae, Conducting hyphae, Endo hypha, Inflated hypha, Oleiferous hypha, and Skeletal hyphae.

Hyphal analysis—Procedure to investigate the development and structure of the sporocarps of Ascomycotina and Basidiomycotina.

Hyphal fusion—Joining of fungal hyphae, usually with some exchange of cell contents; anastomosis.

Hyphal net ("**Hyphenfilz**")—Organ of attachment in some squamulose or placodioid lichens in which a delicately branched reticulate net penetrates the substrata; rhizinose strand.

Hyphal peg—Ridges or fascicles of sterile, somewhat interwoven hyphae that appear as peglike projections in cross sections of tubes of polypores; projection from a hypha for fusion, peg hypha.

Hyphal rhizoid—A hypha acting as a rhizoid by penetrating the substratum. *See* Rhizoid.

Hyphal sheath—Mantle composed of fungal hyphae, covering a root.

Hyphidium (pl. **hyphidia**)—A slender, sterile hymenial or imbedded element often characterized by distinctive branching; a little or strongly modified terminal hypha in the hymenium of hymenophytes. Several types have been distinguished: *Haplo-* or *simple hyphidium*—unmodified, unbranched or a little branched;

dendrohyphidium (dendrophysis)—irregularly and strongly branched; *dichohyphidium* (dichophysis)—repeatedly dichotomously branched; *acanthohyphidium* (acanthophysis)—bottle-brush paraphysis, with pinlike outgrowths near the apex. In Corticiaceae, the hyphidium may be botryose, clavate, coralloid, or cylindrical. *Paraphysis, pseudoparaphysis, paraphysoid, dikaryoparaphysis,* and *pseudophysis* are synonyms or near synonyms. *See* Cystidium.

Hyphocystidium—*See* Cystidium.

Hyphoid—Hyphalike; cobwebby; (of cystidia) not clearly differentiated from vegetative hyphae; (of aecia of *Dasyspora*) having ascospores on tips of branching hyphal projections from stomata.

Hyphomycetes—A class of the Deuteromycotina (Mitosporic fungi) in which conidia are borne on conidiophores not organized into a fruit body.

Hyphophore—Erect, stalked, peltate, asexual sporophore in the Asterothyriaceae.

Hyphopode, or **hyphopodium** (pl. **hyphopodia**)—Short, lateral, hyphal branch or appressorium (one or two cells) of an epiphytic fungus, usually flattened or lobed, specialized for host attachment or penetration; characteristic of the black mildews (Meliolales). The end is rounded in a *capitate hyphopodium* while a *mucronate hyphopodium* is a phialide or conidiogenous cell. A pale spot in the center often indicates the point of origin of a narrow filament that penetrates the cuticle. *See* Stigmatopodium.

Hypnocyst—An *Alternaria*-like group of cells.

Hypnospore—A thick-walled resting spore.

Hypo- (prefix)—Under; lower.

Hypobasidium—*See* Basidium.

Hypochnoid (of texture)—With the hyphae loosely compacted into a soft, cottony to feltlike mat; fruiting layer supported on a loose weft of hyphae.

Hypochromicity—Reduction in absorbance of light at 260 nm wavelength by complementary, hybridizing strands of nucleic acid or nucleic acid that is increasing its secondary structure.

Hypocotyl—That portion of a developing embryo or seedling just below the point at which the cotyledons are attached; the root meristem is situated at the tip of the hypocotyl, the radicle.

Hypocreaceous—Fleshy and brightly colored like *Hypocrea.*

Hypocystidium—*See* Cystidium.

Hypodermal glands (of nematodes)—Glands formed by the hypodermis (caudal glands, lateral hypodermal glands, amphidial and phasmidial glands).

Hypodermataceous (of asci)—*See* Ascus.

Hypodermis (epidermis in some nematodes)—A thin cellular layer of the body wall beneath the cuticle that thickens to form the dorsal, lateral, and ventral chords that extend the length of the body (also secretes the cuticle); tissue lying under the epidermis.

Hypogaen, hypogeal, hypogeic, hypogeous—In the earth; underground; subterranean; developing and becoming mature in the soil.

Hypogeal germination—In dicotyledons, a type of seed germination in which the cotyledons remain below the soil surface.

Hypogenous—Arising beneath; produced farther down.

Hypogeous fungi—Fungi with subterranean sporocarps (e.g., truffles).

Hypogyny (adj. **hypogynous**)—(Of fungi) having the antheridium situated below the base of the oogonium and on the same hypha; (of flowers) condition in which the ovary of a flower surmounts a receptacle, and in which the sepals, petals, and stamens are attached below the ovary.

Hypolithic—Below the surface of rocks. *See* Endolithic.

Hyponecral—*See* Necral layer.

Hyponym—The name of a taxon that cannot be recognized because of its vagueness or the inadequacy of the published description(s).

Hypophloeodal, hypophloedic—Under the periderm or bark; subcortical; within the bark; endophloeodal.

Hypophyllous—Borne on the lower surface of a leaf.

Hypoplasia (adj. **hypoplastic**)—Underdevelopment (malformation) of plant tissue due to decreased cell division. *See* Hyperplasia, Hypertrophy, and Hypotrophy.

Hypopodium (pl. **hypodia**)—A stalk or support.

Hypostroma (pl. **hypostromata**)—The pseudoparenchyma at the base of superficial conidiomata by which they are attached to the substrate; the fungal tissue immersed in the substrate in pycnothyrial genera.

Hypothallus (of lichens)—Prothallus or protothallus. The first hyphae of a thallus to grow, usually of a crustaceous lichen without algal cells or cortex.

Hypothecium (pl. **hypothecia**)—The layer of hyphal tissue under the subhymenium in an apothecium; medullary excipulum.

Hypothesis—A proposition or supposition provisionally adopted to explain certain facts. Once proven by ultimate scientific investigation, it becomes a theory or a law.

Hypotonic (of culture media)—Having an osmotic pressure lower than that of the organism being cultured. *See* Hypertonic.

Hypotrophy—Underdevelopment of plant tissue due to reduced cell enlargement.

Hypovirulence—Reduced virulence of a pathogenic strain due to the presence of transmissible double-stranded RNA (dsRNA).

Hysteriaceous, hysterioid, hysteriiform—Elongated, with a median cleft, like the hysterothecium (fruit body) of the Hysteriaceae; lirellate.

Hysterochroic—Slowly discoloring when old, from base to apex.

Hysterothecium (pl. **hysterothecia**)—A specialized fruit body (ascocarp) of needle cast and other fungi that produces ascospores; is usually elongated, boat-, or elliptical-shaped, covered, and opens at maturity by a longitudinal slit (cleft); characteristic of the Hysteriaceae.

I

I—See Iodine.

IAP—*See* Index of atmospheric purity.

IAP$_{50}$ or median inoculation access period—An inoculation access period during which 50% of the inoculative insects will transmit. *See* Inoculative insects and Inoculation access period.

Ibotenic acid—A metabolite of *Amanita muscaria* that is toxic to humans and flies (*Musca* spp.); amanita factor C.

Iceland moss—A lichen, *Cetraria islandica*, that is habitually eaten by reindeer and caribou. It has been used as a substitute for flour during hard times in Scandinavia and Iceland and is rich in digestible carbohydrates.

Icones—Printed illustrations; pictures; figures; plates.

Icosadeltahedron—A deltahedron (a polyhedron with $20 \times T$ equilateral triangles on its surface) with icosahedral symmetry (T = triangulation number).

Icosahedral (n. **icosahedron**)—Having the form of an icosahedron (a regular polyhedron with 20 equilateral triangular faces or sides); the symmetry forms the basis for the arrangement of the protein subunits of isometric virus particles. One of the five Platonic solids; has 20 identical triangular faces, 30 edges, 12 apices, and two-, three-, and fivefold axes of symmetry.

ID—Infective dose. The number of pathogens required to infect a host.

ID$_{50}$—Median infective dose.

-idae (suffix)—Used in virology and zoology added to the stem (of a type genus name) to form a family name.

Identification—The study of the characters of an organism to determine its name.

Ideolectotype—A lectotype by the author of the species.

IDI (private idiotope)—An idiotope exhibited by only those antibodies derived from a common B lymphocyte.

Idiotope—An antigenically distinct region associated with an immunoglobulin's paratope or hypervariable region.

Idiotype network—A series of idiotype-anti-idiotype reactions postulated to control production of idiotype-bearing antibody molecules or cells.

Idiotype—A specimen identified by the author as typical of his/her species; (in serology) the total antigenic signature associated with an immunoglobulin's paratope or hypervariable region.

Id-reaction—*See* Dermatophytid.

IDX (public idiotype)—*See* Cross-reactive idiotope.

Ig—Immunoglobulin.

IgA—Secretory immunoglobulin class, dimeric in secretions, monomeric in blood.

IgD—Immunoglobulin class on the surface of B cells.

IgE—Immunoglobulin class that fixes to most cells and is responsible for anaphylactic and allergic sensitivity.

IgG—The major immunoglobulin class in the serum of humans. Similar immunoglobulins are found in most species from amphibians upwards but are not present in the serum of fish. Human IgG has been intensively studied, and its structure is known in some detail, e.g., molecular weight is 150,000, $S_{20,w} = 7S$, fixes complement, and crosses the human placenta.

IgM—Pentameric immunoglobulin; first class of antibody produced to most antigens during an immune response.

Igneous rock—Rock formed from the cooling and solidification of magma that has not been changed appreciably since its formation.

Illegitimate—Contrary to the International Code of Botanical Nomenclature; impriorable. *See* Legitimate.

Illuminance—The luminous flux (brightness of light) per unit area on an intercepting surface at any given point.

Imbibition—Process by which solid (chiefly colloidal) particles absorb liquids or vapors into the ultramicroscopic spaces, in materials like cellulose, and swell.

Imbricate (n. imbrication) (of pilei, scales, squamules, etc.)—Partly covering one another like shingles on a roof; developing in overlapping groups of pilei.

Immaculate—Without spots.

Immarginate—Without a margin or a well-defined edge.

Immersed (mycelium, conidioma, ascocarp, etc.)—Enclosed, embedded, or formed in either fungal or host tissue; below the surface in the substratum.

Immobilization (of bacteria)—Prevention of growth and movement following envelopment by plant cell products; (of fungi) controlled, intentional attachment of fungal cells in fermentation technology.

Immobilized DNA or RNA—Nucleic acid linked to nitrocellulose or activated paper. *See* Northern blotting, Southern blotting, Diazobenzyloxymethyl paper.

Immortalization—Continued growth of cells in culture after it would normally be expected to cease; also the cloning of cDNA in a bacterial plasmid or the production of monoclonal antibodies in a hybridoma.

Immune (n. immunity)—Not affected by or responsive to disease; exempt from infection due to its inherent properties (e.g., tough outer wall, hairiness, nature of natural openings, waxy coating, thick cuticle, etc.). *See* Acquired immunity, Natural immunity and Resistance.

Immune adherence—Binding of antibody-antigen complexes or antibody-coated particles to primate erythrocytes, rabbit platelets, or white blood cells due to activation of C3 and formation of C3b.

Immune response—The ability of an animal to produce antibodies as a result of foreign antigens, such as proteins or carbohydrate, entering its body either by infection with a pathogenic agent or by artificial injection. The ability of the antigen to induce this response is referred to as *immunogenicity*. The substances capable of inducing the response are called *immunogens*.

Immune serum—The liquid portion of blood containing one or more specific protein antibodies.

Immunity—The natural or acquired state of being immune, exempt from infection; or in an animal, having developed antibodies against a foreign substance (usually a protein).

Immunization—The process of increasing the resistance or of giving resistance to a living organism; (in serology) administration of antigen to an animal so as to produce an immune response to that antigen. Administration either of antigen to produce active immunity, or of antibody to produce passive immunity, so as to confer protection against the harmful effects of antigenic substances or organisms.

Immunoblot (Western blot)—Reaction of labeled antibodies with proteins adsorbed to nitrocellulose paper.

Immunochemistry—The identification of the sites of antigens in cells using antibodies to which a reporter molecule, e.g., ferritin, gold, or a fluorescent dye, is attached.

Immunocytological methods—Procedures used to study cytopathological disorders in ultrathin sections of virus-diseased tissues, using labeled antibodies to diagnose the virus.

Immunodiffusion—A serological procedure in which the antigen-antibody reaction occurs by allowing the reactants to diffuse toward each other in a gel matrix. *See* Gel double-diffusion and Radial diffusion.

Immunoelectrophoresis—A technique involving separation of proteins in a gel, using an electric field, followed by a precipitation reaction in the gel with antibodies to the separated proteins.

Immunogen—A substance that elicits a specific immune response when introduced into the tissues of an animal. This may take the form of antibody production together with the development of cell-mediated immunity, or of specific immunological tolerance. To stimulate a response, immunogens must normally be foreign to the animal to which they are administered, of a molecular weight greater than 1,000, and a protein or polysaccharide. However, the definition of an antigen is arbitrary, since specific responsiveness is a property of the host tissues, not of the injected substance. *See* Antigen.

Immunogenic—Producing immunity, usually describing a protein (antigen) capable of causing antibody formation when injected into an animal.

Immunoglobulin classes—Subfamily of immunoglobulins based on large differences in H-chain amino acid sequence: Isotypes IgA, IgD, IgE, IgG, and IgM.

Immunoglobulin G (IgG)—Serum globular glycoprotein with least mobility to the positive electrode during electrophoresis, constituting a distinct class of antibodies. *See* IgG.

Immunoglobulin subclasses—Subpopulations of an Ig class based on more subtle structural or antigenic differences in the H chains than are class differences (e.g., IgG1, IgG2, IgG3, IgG4).

Immunology—The study of acquired immunity in animals and humans against infectious disease.

Immunosorbent electron microscopy (ISEM) (syn. **electron microscope serology** and **immunoelectron microscopy**)—Techniques involving the visualization of the antibody-antigen reaction in the electron microscope (EM). These include:

Trapping—The EM grid is first coated with antiserum (referred to as an antibody-coated grid (ACG), which then attracts virus particles from a virus preparation placed on it.

Decoration—Virus particles are attached to the EM grid and antiserum is then added. Homologous antibodies react with the particles to coat or "decorate" them.

Impalpable—Extremely fine and minute.

Imperfect flower—A flower lacking either stamens or pistils; unisexual.

Imperfect fungus—An asexually reproducing fungus that may or may not produce sexual spores (the teleomorph state). *See* Fungi Imperfecti.

Imperfect state (or, improperly, *stage*)—Asexual state (anamorph) of the life cycle of a fungus; asexual spores (such as conidia) or no spores are produced; there may or may not be a perfect or teleomorph state.

Imperfect yeasts—Yeasts that do not form ascospores.

Imperforate—Without an opening.

Impervious—Resistant to penetration by fluids or by roots.

Impingement—The entrapment of aerosol particles, spores, etc., on a solid surface in a sampling device.

Impriorable—Illegitimate.

in.—Inch = 25.4 mm = 2.54 cm = 0.0254 m = 0.085333 ft = 0.027778 yard = 0.00505 rod.

in situ—In its original place or environment.

in vitro—Biological reactions in glass or in an artificial environment; in culture; outside the living host, as opposed to in vivo; Latin for "in glass."

in vivo—Biological reactions taking place in the host; applied to laboratory testing of agents within living organisms; used in contrast to in vitro; Latin for "in living."

Inactivate—To destroy the activity of a substance, e.g., to heat blood serum to 56°C for 30 min to destroy complement.

Inactivation—The loss of the ability of a virus to initiate infection.

Inaequi-hymeniiferous—Having basidia that mature and shed their spores in zones; coprinus type. *See* Acqui-hymeniiferous.

Inamyloid—Not turning blue or grayish in Melzer's reagent.

Inane—Empty, void.

Inapparent infection—An infection that does not give obvious symptoms.

Inarch (adj.)—A type of plant graft often used to detect plant viruses, MLOs, and viroids with an indicator plant(s). Cuts are made upward on the diseased stem (scion) and downward on the healthy stem (stock). The healthy plants should be at least partially defoliated. One third of the stem of each of similar diameter are cut diagonally, and both are bound at this point usually with grafting tape.

Inbred line—A pure line usually originating by self-pollination and selection.

Inbreeding—The formation of heterokaryons from plasmogamy between compatible homokaryons from the same paternal source; breeding of closely related individuals or races; used in plant breeding to obtain an homozygous condition.

Inbreeding depression—Loss of yield and vigor caused by inbreeding, which may be particularly significant if species normally cross-pollinated are self-pollinated.

Incanescent—Becoming gray or "canescent."

Incanous—Hoary; quite gray.

Incarcerate—Hidden.

Incarnate, incarnadine—Rosy flesh-color.

incertae sedis—Of uncertain taxonomic position.

Incidence, of disease—Number of plants affected within a population; disease incidence should be distinguished from disease severity.

Incipient—Early in the development of a disease or condition.

Incised—As if cut into, especially of the margin of a pileus or lobes of a lichen; deeply and sharply notched.

Incisures (of nematodes)—The longitudinal markings within the lateral fields (sometimes called involutions or lines).

Inclusion body—Subcellular structures (any matrix or array of virus particles, abnormal proteinaceous bodies, or areas of abnormal staining) in the cytoplasm or nucleus of virus-infected plant cells.

1. Virus-coded intracellular body containing protein and viruses.

2. Any array of viruses or assembly of abnormal protein detectable by light microscopy.

Incompatability (floral)—Failure to obtain fertilization and seed formation after pollination, usually because of slow pollen tube growth in the stylar tissue.

Incompatability (graft)—Failure of two graft components (scion and stock) to unite and develop into a successfully growing plant.

Incompatible—Pertains to different kinds or varieties of plants that do not successfully cross-pollinate or intergraft; not cross-fertile; not able to be cross-mated; (of homokaryons) lacking the ability or potential to undergo plasmogamy and establish a heterokaryon; failure of a pathogen to infect a host plant; an interaction between a plant and a pathogen *not* resulting in disease, usually a hypersensitive reaction occurs. *See* Compatible.

Incomplete antibody—An antibody that binds antigen but does not precipitate or agglutinate the antigen.

Incomplete dominance—A condition occurring when a dominant gene is only partially expressed in the phenotype of the heterozygote; a condition in heritance in which neither member of a pair of contrasting characters masks the other.

Incomplete flower—A flower that lacks one or more of the four kinds of flower organs (sepals, petals, stamens, or pistils). *See* Complete flower.

Incomplete virus—*See* Defective virus.

Incorporate—To work or blend a pesticide, fertilizer, or other chemical completely into the soil.

Incrassate—Thickened.

Incrusted—Lightly to densely covered with crystalline material; (of hyphae) having matter excreted on the walls.

Incubate—Hold in a steady, environmental state to await development; to allow a microorganism to grow undisturbed under a given set of conditions.

Incubation period (or *stage*)—The elapsed time between penetration of a host by a pathogen and the first appearance of disease symptoms and/or signs on the host; the time period during which microorganisms inoculated into a medium are allowed to grow; a period of rest of a pathogen prior to infection; maintaining inoculated plants or pathogens in an environment favorable for disease development.

Indefinite—Not precise or sharply limited.

Indehiscent (of sporocarps, sporangia, etc.)—Not opening, or without a special method of opening; also, a fruit that does not split or open naturally at maturity, along regular seams.

Indentate—Having an irregular margin. *See* Serrate.

Independent assortment—A genetic principle, established by Mendel, that the factors (genes) representing two or more contrasting pairs of characteristics are segregated to gametes independently of each other in meiosis. Linkage of genes may prevent independent assortment.

Indeterminate inflorescence—A condition in which the central stem of the inflorescence continues to grow, forming new bracts and flowers in succession on lateral branches, with the oldest flowers at the base of the inflorescence.

Indeterminate—Having the edge not well defined, especially of fruitbodies and leaf spots, (of conidiophores) continuing indefinite growth.

Index Bergeyana—A compilation of the thousands of names that have been used in bacteriology.

Index of atmospheric purity (IAP)—A numerical estimate of air purity based on the lichens present on trees. *See also* Air pollution.

Indexing—Any procedure for demonstrating the presence of a mycoplasma, virus, viroid, or other pathogen in susceptible plants. *See* Graft indexing and Host indexing.

Indian bread—*See* Tuckahoe.

Indian paint fungus—Fruitbody of *Echinodontium tinctorium*.

Indicator—A plant that reacts to certain pathogens or environmental factors with the production of specific symptoms and is used for detection and identification of these factors; a substance that changes color as conditions change, e.g., pH indicators reflect changes in acidity or alkalinity. *See* Differential hosts.

Indicator cell—A cell that reacts in a characteristic manner to infection with a specific virus.

Indicator plant or host—A plant that responds specifically to certain viruses, other pathogens, or environmental factors with specific characteristic symptoms; used for identification of the specific pathogen(s) or environmental factor(s); also used in diagnosis of pathogens.

Indigenous—Native; not foreign; produced or living naturally in a specific environment of a country or region.

Indirect (of fruitbody development)—Describing cell enlargement occurring chiefly after cell division. *See* Direct.

Indoleacetic acid (IAA)—A natural or synthetic organic compound belonging to the auxin class of growth-regulating substances.

Induced enzyme—*See* Adaptive enzyme.

Induced resistance—*See* Acquired resistance.

Inducible enzyme—An enzyme that can be produced in substantial amounts only when the inducer is present in the medium.

Induction—The inactivation of a latent virus infection; also an increase in the rate of synthesis of an enzyme, specifically caused by a small molecule, which is generally the substrate or a compound closely related to it.

Indumentum—A covering, such as hairs, etc.

Indurate, indurated—Hardened; made hard.

Indusium—Cover; (of fern sorus) membranous cover; (of certain phalloids) a netlike structure hanging from the apex of the stipe beneath the pileus.

Inequilateral—Having unequal sides.

Inequipolar—Having unequal poles.

Inerm, inermous—Lacking spines or prickles.

Inert ingredient (of pesticides)—Any ingredient in a formulation that does not contribute to the activity of the active ingredient (e.g., water, sugar, dust, wetting or spreading agent, emulsifier, carrier, diluent, conditioning agent, propellant, etc.).

Infarcate—Turgid; solid.

Infected (of an organism)—Successfully attacked by a pathogen.

Infection (v. **infect**)—Process or act of a pathogen entering (invasion, penetration) and establishing a parasitic (often pathogenic) relationship with a host plant; to enter and persist in a carrier. Once infection has been effected, colonization may begin. *See* Colonization.

Infection court—Site in or on a host plant in which invasion of a host can occur (e.g., root, shoot, leaf, flower, fruit).

Infection cushion—An aggregation of hyphae, amassed on a surface of a host, that serves as a support for one or more infection pegs.

Infection peg—Small hyphal protrusion that penetrates the host cell wall.

Infection propagules—Infectious units of inoculum.

Infection thread—The specialized hypha of a pathogenic fungus that invades tissue of the susceptible plant; the tube formed during early infection of a root hair through which *Rhizobium* strains penetrate the root cortex of susceptible legumes.

Infection type—Gross appearance or size of lesions, as from infection by a rust fungus.

Infectious—Capable of infection and spreading disease (a pathogen) from plant to plant.

Infectious disease—A disease caused by a pathogen that can multiply and spread from a diseased to a healthy plant.

Infectious particle—A virion containing the complete viral genome and thus capable of infecting a susceptible cell.

Infectious unit—The smallest number of virus particles that, in theory, can cause infection.

Infective (n. **infectivity**)—(Of a pathogen) able to infect a living organism; (of a vector, medium, etc.) having the ability to affect the transmission of a pathogen.

Infectivity assay—A bioassay using mechanical sap transmission to quantitatively determine the number of infectious virus particles.

Inferior (of an annulus)—Attached to the middle or lower portion of the stipe (stem of a mushroom); on the under surface.

Inferior ovary—An ovary embedded in the receptacle, or one whose base lies below the point of attachment of the perianth.

Infested (v. **infest,** n. **infestation**)—Containing or covered with large numbers of pests, especially insects, mites, nematodes, weeds, or rodents as applied to an area or field; to "contaminate" as with microorganisms (e.g., fungi and bacteria) without infection or parasitism; to be present in numbers. Do not confuse with *infected* by microorganisms, which applies only to living diseased plants or animals.

Infiltration rate—The maximum velocity at which water can enter the soil under specified conditions, including the presence of an excess of water.

Inflammation—Reaction in a tissue resulting from an irritation by a foreign material and causing a migration of leucocytes and increased flow of blood to the area, producing swelling, reddening, heat, pain, and tenderness.

Inflated (of cystidia)—Swollen like a bladder.

Inflated hypha—Hypha in which cells behind the growing apex enlarge, causing the apparent rapid growth of most agaric and gasteromycete basidiocarps.

Inflect, inflected, inflexed—Curved, bent or directed abruptly inward, downward, or toward the body axis.

Inflorescence—An axis bearing a flower or cluster of flowers, e.g., an umbel, spike, panicle.

Infossate—Sunken.

Infundibuliform—Funnel-shaped.

Infuscate—Of a brownish tint.

Ingredient statement—The part of the label on a pesticide container that gives the name and amount of each active ingredient and the amount of inactive material(s) in the mixture.

Ingress—The act, by a plant pathogen, of gaining entrance into the tissues of a susceptible plant.

Inheritance—The acquisition of characters or qualities by transmission from parent(s) to offspring.

Inhibition—Prevention of growth or multiplication of microorganisms.

Inhibitory substances—Substances that retard or inhibit the growth of *other* organisms. *See* Staling substances.

Initiation codon—*See* Start codon.

Initiation—The start of synthesis of a polypeptide or nucleic acid chain.

Injury—Momentary (transitory) damage by a causal agent, e.g., insect feeding, action of a chemical, physical or electrical agent, or an adverse environmental factor. *See* Disease.

Ink caps—Basidiocarps of *Coprinus*.

Innate—Immersed in substratum; embedded in; not superficial.

Inner bark—Interior, living portion of the bark; secondary phloem.

Inner veil—The hyphal membrane covering the gills (lamellae) of some young agarics (mushrooms).

Innervation process (of nematodes)—A neuromuscular process of the muscle cell that extends to the motor nerve.

Inoculate (n. **inoculation**)—1) To artificially introduce a pathogen at the site of infection of a host (the infection court) to induce disease or into a culture medium. Loosely, and incorrectly, used to describe the transfer of living cells of a microorganism to any place in which they will grow and develop; 2) to treat seeds of leguminous plants with bacteria to induce nitrogen-fixation in the roots.

Inoculation access period—The time that vectors are caged on test plants.

Inoculation feeding period (syn. **test feeding period**)—The length of time a vector feeds on a test plant during pathogen transmission experiments.

Inoculation threshold period—The minimum feeding period an inoculative vector needs on a susceptible test plant to transmit a disease agent.

Inoculative insects—Infective insects that will transmit during a given test access period.

Inoculum (pl. **inocula**)—The pathogen or its parts (e.g., fungus spores, mycelium, bacterial cells, nematodes, virus particles, etc.) used for inoculating to produce disease.

Inoculum potential—The number of independent infections and amount of tissue invaded per infection that may occur in a population of susceptible hosts at any time or place; the energy of a microorganism available for growth and colonization of a substratum at the surface of the substratum to be colonized.

Inoperculate (of an ascus or sporangium)—Without a lid; without an opening through a pore or opening by an irregular apical split to discharge the spores. *See* Operculate.

Inordinate—Without order, pattern, or arrangement.

Inorganic compound—A chemical compound that is generally not derived from life processes; compounds that lack carbon.

Inoviridae—A family of rod-shaped phages containing two genera; *inovirus* (filamentous phages) and *plectrovirus* (mycoplasma virus 1 phages). The virions are enveloped rods, 84–1,940 nm in length. The viral genome is circular, single-stranded DNA.

Insculptate—Hollowed in.

Insect—Member of the class Hexapoda (phylum Arthropoda). In the adult stage, true insects have six walking legs, wings, and three body divisions.

Insect vector—An insect that transmits a disease-inducing organism or agent.

Insecticide—A physical or chemical agent used to kill or protect against insects; a type of pesticide.

Insectivorous—Insect-eating or insect-digesting.

Insert—Foreign DNA cloned into a bacterial plasmid or other gene vector.

Inspissate—Thickened.

Instar—Growth phase of an insect or nematode between successive molts.

Integrase—Virus-induced enzyme that mediates the integration of the viral DNA into the host genome.

Integrated (of conidiogenous cells)—Incorporated in the main axis or branches of a conidiophore, where it is either terminal or intercalary. *See* Discrete.

Integrated pest management (IPM), integrated control—An attempt to use all available methods (biological, chemical, cultural, genetic, legal, and physical) to control the diseases and pests of a crop plant for best control results but with the least cost and the least damage to the environment.

Integration (in virology)—The process of insertion of viral DNA into the host genome, usually involving a virus-coded enzyme, the integrase.

Integuments—The layers of tissue (usually two) enclosing a seed; later they become the seed coats of the mature ovule; (of nematodes) cuticle, covering, investment.

Inter- (prefix)—Between; among; during.

Interbiotic—Living as a parasite on or near one or more living organisms as in certain rhizoidal chytrids.

Intercalary (of growth)—Between the apex and the base; stem elongation typical of grasses. Elongation progresses from the lower internodes to the upper internodes through the differentiation of meristematic tissue at the base of each internode; (of cells, spores, etc.) between two cells.

Intercalary cell—A small, sterile cell between two others (spores) within hyphal segments as opposed to hyphal tips.

Intercellular—Between or among (host) cells.

Intercostal—Between veins, interveinal.

Interfascicular cambium—Cambium situated between vascular bundles.

Interference (in virology)—Prevention of the replication of one virus by another.

Interferon(s)—An antiviral substance produced by animal tissue; animal cell proteins formed in response to virus infection that cause other cells to resist infection and also induced by other agents, including double-stranded RNA and synthetic ribonucleotide.

Intermolecular recombination—Recombination due to the reassortment of species of nucleic acid between viruses whose genomes are segmented.

Internal basidium—A basidium developed by septation of the protoplast of a "teliospore."

Internal image antibody—*See* Ab 2 beta.

Internode—The region between two adjacent nodes on a stem.

Interocalation—The insertion of planar molecules between the adjacent basic pairs of double-stranded DNA or double-stranded RNA or single-stranded nucleic acid with secondary structure.

Interphase—Condition of a nucleus undergoing mitosis or meiosis; the phase between divisions.

Intersex (of nematodes)—An individual more or less intermediate in phenotype between male and female (shows secondary male or female characters); may function as one or the other or neither sex, but never as both.

Interspace (of a pileus)—Space between the lamellae (gills).

Interspecific cross—A natural or intentional cross between two species.

Interspersed—Scattered.

Interthecial—Between asci.

Interveinal—Tissue or area between the veins.

Intervenose (of a pileus)—Veined in the interspaces or surface of the lamellae.

Interwoven (of the trama)—Having intertwined hyphae, not parallel, convergent or divergent.

Intestine (of nematodes)—A simple tube, composed of a single layer of epithelial cells in which digestion of food occurs; gut.

Intestinorectal valve (of nematodes)—The sphincter muscle separating the intestine from the rectum or cloaca.

Intra- (prefix)—Inside; within.

Intracellular—Inside or through a cell or cells.

Intrahypha—*See* Endohypha.

Intramatrical—Located within the matrix or substratum.

Intramolecular recombination—Recombination leading to the exchange of sequences between molecules of nucleic acid.

Intraperitoneal— Describing a route commonly used for injection (e.g., antigens and viruses) into an animal via the peritoneum.

Intravenous—Describing a route sometimes used for the injection of antigens into the venous blood system of an animal.

Intricate—Intertwined; interwoven, but not coalescent.

Intricate cortex (of lichens)—Composed of hyphae twisted together. *See* Textura intricata.

Introns—The DNA sequences that interrupt the protein-coding sequences of a gene; introns are transcribed into mRNA but are eliminated from the message before it is translated into protein; the region of DNA transcribed into RNA initially but then removed by splicing, in the formation of functional RNA.

Introrse—Inward; toward the direction of the central axis.

Intruded, intruse—Pushed or projecting forward.

Intumescence—A knoblike blister or pustule formed by outgrowths of elongated cells on leaves, stems, or other plant parts that have burst from sudden water excess following dry periods; (of fungi) a swelling.

Inulin—White complex carbohydrate very similar to starch.

Invaginated—Enclosed in a sheath.

Invasion (v. **invade**)—Growth or spread of a pathogen into a plant and its establishment in it.

Invasion court—The place on the host where the pathogen enters.

Invertase—An enzyme that hydrolyzes sucrose into glucose and fructose.

Inverted—Attached by a dorsal, stemlike prolongation of the pileus.

Involucre—A circle of bracts surrounding a flower cluster or single flower.

Involucrellum—Tissue forming the upper part of the pseudothecia and surrounding the parathecium.

Involute—Rolled or curled inward.

Iodine (I,J)—Used as Lugol's solution of Melzer's reagent giving blue, red, lavender, or violet colors and seen best after pretreatment with potassium hydroxide (KOH) in spores, asci, hymenial tissues, etc. *See* Amyloid and Dextrinoid.

Ion accumulation—Absorption of ions against a concentration gradient as a result of expenditure of energy by living cells. *See* Active transport.

Ion exchange chromatography—A chromatographic procedure in which the stationary phase consists of an ion exchange resin, which may be acidic or basic.

Ionomidotic reaction—Release of a dark pigment into aqueous potassium hydroxide (KOH) mounts; an important taxonomic criterion for apothecia of some dark-colored Leotiales.

Ions—Atoms, groups of atoms, or molecules bearing an electrical charge; formed by the dissociation of molecules (loss of electrons [cations] or the gain of electrons [anions]) of certain substances called electrolytes.

IPM—*See* Integrated pest management.

Ipomearone—A phytoalexin from sweet potato (*Ipomoea batatas*).

Irpiciform (of the hymenophore)—In the early stages of development, the splitting of dissepiments to form an hydnaceous hymenophore.

Irpicoid—Having flattened teeth or becoming toothed, as in *Irpex*.

Irradiation—Exposure of pollen, seed or other plant parts to radiant energy of various types (X-rays or other short wavelength [gamma] radiations) to increase mutation rates.

Irregular flower—A flower in which one or more members of at least one whorl are of different form from other members of the same whorl; zygomorphic.

Irrigation—Applying water to soil, other than by natural rainfall.

Irritability—Ability of living protoplasm to receive stimuli and to react to stimuli.

Isabelline—A dingy, dirty gray white to pale yellowish brown.

Isarioid—Synnematous, as in the genus *Isaria*.

ISEM—*See* Immunosorbent electron microscopy.

Isidiiferous, isidiate (of a lichen thallus)—bearing isidia.

Isidium (pl. **isidia**)—A phycobiont-containing protuberance of the cortex in lichens that may be warty, cylindrical, clavate, scalelike, coralloid, simple or branched; occurring directly on the thallus; may become sorediate.

Islandicin, islanditoxin—Toxins of *Penicillium islandicum*, a cause of hepatitis (yellow rice disease) in humans.

Iso- (prefix)—Equal; the same.

Isocutis (of a pileus)—A slimy cuticle.

Isodiametric—Having equal diameters.

Isoelectric focusing (electrofocusing)—A separation technique in which mixtures of proteins and/or viruses are resolved into their components by subjecting them to an electric field in a supporting gel or stabilized solution in which a pH gradient is established.

Isoelectric point—The pH at which a virus particle or protein molecule has a zero net charge; the point of highest probability of crystallizing or precipitating out of solution.

Isogametangia (sing. **isogametangium**)—Gametangia, presumably of opposite sex, that are morphologically indistinguishable.

Isogametes—Gametes, presumably of opposite sex, that are morphologically indistinguishable.

Isogamic copulation—Fusion of morphologically alike gametes; conjugation in a narrower sense.

Isogamy (adj. **isogamous**)—The condition in which gametes are morphologically similar, as in the Zygomycetes; the conjugation of isogametes; producing or possessing morphologically similar gametes; zygogamy.

Isohaplont—A haplont having genotypically similar cell nuclei. *See* Miktohaplont.

Isokont, isokontous (of cilia and flagella)—Of equal length. *See* Heterokont.

Isolate—(Verb) to separate a culturable microorganism from an infected suscept and grow it in the absence of other organisms; (noun) a sample (e.g., a virus) from a defined source.

Isolation—Prevention of crossing among plant populations due to distance or geographic barriers; the process of getting an organism into pure culture or the culture itself.

Isolichenin (I+)—The carbohydrate that forms the walls of fungal hyphae in many lichens; an isomer of lichenin. *See* Lichenin.

Isometric—Pertaining to a structural form that has three equal axes at right angles to one another; equally long, as a virus particle with all axes of equal length (essentially spherical in shape).

Isomorphic—Similar in shape or form but not in essential structure.

Isonym—Any new name and new combination based on the basionym. *See* Basionym, Synisonym and Toponym.

Isophagous (of a fungus)—Describing a fungus that attacks several allied species.

Isoplanogamete—One of two motile sex cells alike in form.

Isopycnic gradient—A separation technique in which a sample containing macromolecules is centrifuged through a gradient of increasing density. Gradients are formed by layering and diffusion or by the effects of centrifugal force and diffusion on small, heavy metal ions.

Isotomic dichotomic branching (of fungi)—A condition in which both branches are about the same size so the dichotomic pattern is visible even in older parts of the thallus.

Isotope—Any of several or more forms of a chemical element with the same number of protons in the nucleus, or the same atomic number, but with different numbers of neutrons in the nucleus, or different atomic weights.

Isotrichoderm, isotrichodermium (of a pileus)—A trichodermium composed of gelatinized hyphae.

Isotype—A duplicate or part of the type collection other than the holotype; (in immunology) part of the immunoglobulin molecule from the animal used in the characterization of sera; (in serology) the class or subclass of antibodies within a given animal species.

Isthmospore—A conidium composed of four or more thick-walled cells separated by thin-walled cells as in *Isthmospora*; a spore of two or more cells interconnected by a narrow region as in ascospores of *Vialaea*.

Isthmus—(Of nematodes) the relatively narrow portion of the esophagus between the metacorpus and the basal region; (of fungi) the thickened medial perforated septum of a polarilocular ascospore or the narrower or thinner-walled portion of an ishmospore.

Iwatake—*See* Rock tripe.

Ixo- (prefix)—Sticky.

Ixocutis (of a pileus)—A slimy cuticle.

J

J—*See* Iodine.

J chain—A polypeptide chain that joins individual four-chain immunoglobulin units to form polymeric immunoglobulins (IgA, IgM).

Jack-o'lantern—The bright orange-yellow basidiocarp of the luminous *Omphalotus olearius*, syn. *Clitocybe illudens*.

Japanese mushroom—*See* Pine mushroom.

Jelly fungi—The Tremellales.

Jew's ear—The basidiocarp of *Auricularia auricula*.

Jointing—Growth stage in grasses manifested by rapid culm elongation.

Jonquilleous—Bright yellow, as in jonquils (*Narcissus*).

Jove's beard—The basidiocarp of *Grandinia (Odontia) barba-jovis*.

Juglone—The oxidized form of hydrojuglone found in members of the genus *Juglans*, which is toxic to fungi as well as to the roots of higher plants

Juvenescence—Process of maturing at a normally immature stage of development.

Juvenile—Any immature nematode, usually applied to animal forms in which there is little dimorphism between immatures and adults.

K

Kappa light chain—One of two antigenically distinct regions associated with the constant region of an immunoglobulin's light chain.

Karyochorisis (of fungi)—Somatic nuclear division resulting from a constriction of the nuclear membrane.

Karyogamy—The fusion of two sex nuclei after cell fusion, i.e., after plasmogamy.

Karyokinesis—Nuclear division. *See* Mitosis.

Katothecium—See Catathecium.

kb—Kilobases. An abbreviation for 1,000 base pairs of DNA.

kbp—Kilobase pairs.

Keel—A structure of the legume type of flower made up of two petals loosely united along their edges.

Kefiran—A water-soluble polysaccharide from kefir grains.

Kerion—An inflammatory form of ringworm of the scalp causing loss of hair; tinea kerion.

Kernel—The whole seed of a cereal; a fruit seed; the inner softer part of a seed, fruit stone, or nut.

Kerosene fungus—*See* Creosote fungus.

Kg—Kilogram = 1,000 g = 35.273957 oz = 2.20462 lb.

Killer particles—Particulate bacteriocins that resemble bacteriophage particles.

Kilobase (pairs) (kb, kbp)—One thousand nucleotides in a polynucleotide chain; a measure of the size of a nucleic acid molecule. *Kb* refers to single-stranded nucleic acid, with *kbp* referring to double-stranded nucleic acid.

Kinase—Any enzyme that catalyzes phosphorylation reactions. *See* Polynucleotide kinase and Protein kinase.

Kinetosome—*See* Blepharoplast.

Kleinschmidt procedure—A technique in which small amounts of nucleic acid are coated with a base protein, e.g., cytochrome C, and spread on a denatured protein monolayer at an air-water interface. After being picked up on an electron microscope grid, the nucleic acid molecules are shadowed with a heavy metal.

Klendusity—A special kind of disease escape, in which a susceptible plant or variety avoids disease because of an intrinsic property of the plant or variety itself that greatly reduces the chances of its being inoculated, even though there may be an abundance of inoculum in the area.

Kleptotype—A fragment broken off from the holotype. *See* Holotype.

Km—Kilometer = 1,000 m = 3280.8 ft = 1093.6 yards = 0.62137 statute mi = 0.53961 nautical mi.

Knot—Knoblike overgrowth on roots or stems with an imperfect vascular system; a localized abnormal swelling; a gall.

Koch's postulates—Four rules, proposed by Robert Koch, to be followed to prove the pathogenicity of a microorganism. These rules are: 1) consistent association of a suspected causal agent with a disease syndrome; 2) isolation of the infectious agent and growth in pure culture, 3) reproduction of the disease syndrome after transmission of the infectious agent to healthy plants; and 4) reisolation of the infectious agent from the host (which should be identical to that found originally).
For plant viruses, the criteria have been modified to: 1) isolation of the virus from a diseased host; 2) cultivation of the virus in experimental hosts or cells; 3) filterability of the pathogen; 4) production of a comparable disease in the original or related host species; and 5) reisolation of the virus.

KOH—Potassium hydroxide, used as a 4% aqueous solution to rehydrate sections of dried fungal specimens; a 3% solution is used as a standard mounting medium for fungal tissue of basidiocarps; a 0.03% aqueous solution is used in microscopic mounts for viewing fibrous or fibrosin bodies in spores of some powdery mildew fungi (Erysiphaceae).

Koji mold—*Aspergillus oryzae* and related species used as a starter for various fermented Japanese foods (e.g., amazaké, mirin, miso, saké, shoyn) by inoculating rice.

Kojic acid—An antibiotic, metabolic substance produced by *Aspergillus oryzae*, *A. flavus*, and *A. tamarii* groups that gives a blood-red color in ferric chloride ($FeCl_3$).

Krebs cycle—A system of enzymatic reactions that convert pyruvic acid to carbon dioxide in the presence of oxygen with concomitant release of energy, which is captured in the form of ATP molecules. Also referred to as the citric acid cycle or the tricarboxylic acid (TCA) cycle.

L

l—Combining form for liter, as in ml.

"L"—Mean body length of a nematode in millimeters; usually given as a range; a deMan's value.

L chain (light chain)—A polypeptide chain of M_r 22,000 present in all immunoglobulin molecules in two forms. Each four-chain Ig molecule has one of the two light chains.

Label—A printed statement affixed to a pesticide container or wrapper by the manufacturer listing the contents, directions for use and precautions. A pesticide label must be approved and registered by a federal and state agency (e.g., U.S. Environmental Protection Agency [EPA] and the State Department of Agriculture).

Labeling—A term that usually refers to the attachment of reporter groups to a nucleic acid, a protein, or other macromolecule.

Labial disc (of nematodes)—A raised, more or less circular area in the cuticle of the most anterior annule and delimited posteriorly by the first transverse striation.

Labiate—Lobed, or having lips, hence bilabiate (two lobes), trilabiate (three lobes), etc.

Labile—Unstable.

Labium (pl. **labia**, adj. **labial**)—Lip; having or pertaining to the lips.

Laboratory strain(s)—Isolates (e.g., viruses) propagated in a laboratory for some time.

Labriform—Lip-shaped; commonly used for terminal soralia of lichens with this shape.

Labyrinthiform, labyrinthine—Having or composed of sinuous lines; like a labyrinth or maze.

Laccate—Varnished or shellacked; shining; polished.

Lacerate—As if deeply split or roughly torn.

Lacina—A delicate branch of a foliose lichen thallus with an anatomical structure typical of foliose lichens.

Lacinate (of an edge, etc.)—As if torn or cut into delicate strips.

Lacrimiform, lachrymiform, lacrimoid—Tear-shaped, tearlike.

Lactase—Enzyme that brings about the hydrolysis of lactose with the formation of two sugars, glucose and galactose.

Lacteous—Milky.

Lactescent—Becoming like milk.

Lactifer—A latex-bearing hypha or hyphal element.

Lactiferous—Having a milk-like juice.

Lactiferous hypha—A hyphal element containing a white to colored liquid that exudes from cut or broken surfaces.

Lactose—A carbohydrate (disaccharide) also known as milk sugar. On hydrolysis, it splits into glucose and galactose.

Lacuna (pl. **lacunae**)—A hole, hollow, or depression.

Laevigate, levigate—Smooth.

Lag phase—(In bacteriology) the period of physiological activity and diminished cell division when a medium is first inoculated with a culture; (in virology) sometimes used for latent period.

Lageniform—Shaped like a long-necked flask (a Florence flask); gourd-shaped with a swollen base and a narrowed top.

Lamella (pl. **lamellae**)—Gill; a platelike structure of the hymenium on which some Basidiomycotina (e.g., mushrooms) produce their basidia; thin sheet, plate, or layer of cells vertically oriented and radially aligned; also, the membrane or primary septum between any two cells.

Lamellate—Composed of thin sheets, plates, or layers (e.g., lamellae or gills).

Lamelliform—Shaped like a plate or gill.

Lamellula (pl. **lamellulae**)—A small lamella that extends from the edge of the pileus towards the stipe as in *Russula*.

Lamellule—A short lamella (gill) that does not reach the stipe.

Lamina (pl. **laminae**)—The flat, expanded part (blade) of a leaf; a layer; epithecium and hymenium plus subhymenium in an apothecium; the main part of the thallus of a foliose lichen.

Laminal—On the lamina.

Laminar air flow—Fluid flow in which streams do not intermingle.

Laminarin—Principal stored food in brown algae; a complex carbohydrate.

Laminated rot—A rot in which the wood separates along the annual rings because of more rapid decay of springwood; sometimes called ringshake.

Lampro- (prefix)—Bright.

Lamprocystidium—Sterile hymenial element with a uniformly to partially thickened wall, often incrusted with crystalline or amorphous material. *See* Cystidium.

Lanate, lanose, languinose—Covered with short hair-like processes; woolly. *See* Nematogenous.

Lanceolate—Lance- or spear-shaped; oblong and tapering to a point.

Land-grant university—A group of universities in the United States, founded in the late 19th century from the sale of public lands; system set up by the Morrill Land-Grant College Act signed by President Lincoln in 1862. Each of the 50 states and the territories has such a university. The purpose of each is to teach, conduct research, and extend information to all citizens in the state.

Land-races—Plant stocks selected by farmers on a local basis over many years, that are very well adapted for local conditions.

Languid—Hanging down; drooping; weak.

Lanuginose—Downy. *See* Lanate.

Larva (pl. **larvae**)—Juvenile; immature growth stage(s) between the embryo or egg and the adult, as in insects and nematodes.

Larvacide—An insecticide used to kill larvae of insects.

Late blight—A disease, mostly of potato (*Solanum*) and tomato (*Lycopersicon*), caused by *Phytophthora infestans*.

Late genes—Genes in a viral nucleic acid that are expressed late in the virus replication cycle; often the genes coding for capsid proteins. *See* Early genes.

Latebrose—Hidden.

Latency—Stage of an infectious disease, other than the incubation period, where no symptoms are expressed in the host.

Latent—Present but not manifested or visible.

Latent infection, latency—Infection in a plant without visual symptoms.

Latent period (phase)—Period after a vector has acquired a pathogen (e.g., virus) before it can transmit it. Often observed in the case of persistent or propagative transmission of a pathogen; (in virology) the time in the virus infection cycle between the apparent disappearance of the infecting virus and the appearance of newly synthesized virus.

Latent virus—A virus that does not induce symptom development in its host; often used to describe inapparent virus infection.

Lateral—At the side; (of plants) a shoot originating from a vegetative bud in the axil of a leaf or from the node of a stem, rhizome, or stolon.

Lateral bud—A bud that grows out from the leaf axil on the side of a stem.

Lateral field (of nematodes)—Longitudinal cuticular thickening situated on top of the lateral chords; the field may be divided by longitudinal striae (incisures) and at times by transverse markings (areolations).

Lateral guiding pieces (of nematodes)—Structure that consists of two, very small, lineate, cuticularized pieces lying lateral to the distal portion of the spicules and joining with the muscular sheath surrounding the anus (present in dorylaims and some related nematodes).

Lateral pores (of nematodes)—Minute openings to the lateral hypodermal glands.

Laterally stipitate (of basidiocarps)—With a stipe attached at the margin of the pileus or cap.

Latericeous, lateritious—A dark brick red color.

Lateritiin—*See* Enniatin A.

Laterodorsal—The position on the nematode body situated 45° from the dorsomedian line and perpendicular to the anteroposterior axis; submedian.

Lateroventral—The position on the nematode body situated 45° from the ventromedian line and perpendicular to the anteroposterior axis; submedian.

Latewood (summer wood)—The denser part of the growth ring of woody plants produced late in the season. It is made up of xylem cells with thicker walls, smaller radial diameter, and generally longer than those formed earlier in the growing season. *See* Earlywood.

Latex—A milky secretion; rubberlike, produced by various kinds of plants.

Latex agglutination—A serological test in which the antibody or antigen is adsorbed onto polystyrene latex particles that are then incubated with the other reactant. Aggregates of latex particles (positive reactions) can be easily seen with the naked eye.

Lathhouse—An open structure built of wood lath or plastic screen for protecting plants against excessive sunlight or frost.

Latin binomial—The scientific name of an organism. It is composed of two names, the first designating the genus, and the second, the species.

Latticed—Like a network; cross-barred.

Lattice-work fungus—*Clathrus* spp.

Layby application—The procedure in which a material is applied with or after the last cultivation of a crop.

Layering—A form of vegetative propagation in which an intact branch develops roots as the result of contact with the soil or another rooting medium.

Lb—Pound = 16 oz = 14.5833 troy oz = 453.5924 g = 0.4535924 kg.

LC_{50}—Median lethal concentration. The lower the LC_{50} value, the more poisonous the pesticide. It is often used as a measure of acute inhalation toxicity.

LD—Lethal dose. The number of pathogens required to cause death in a given species of plant or animal.

LD_{50}—The milligrams of toxicant per kilogram (mg/kg) of body weight of a test organism that will kill 50% of the test organism; an approximate measure of toxicity to humans; chemicals with a lower LD_{50} are more toxic. LD_{50} values are the commonly used measure for acute oral and dermal toxicity.

Leaching (v. **leach,** adj. **leached**)—Removal of soluble nutrients and other materials from the soil or plant tissue by flowing or percolating water.

Leader (of terminal shoots)—The topmost or dominant shoot of a woody plant.

Leader sequence or signal peptide—The 5' noncoding part of a mRNA.

Leaf axil—Upper angle between a leaf petiole and the stem from which it grows.

Leaf bud—Bud that develops into a leafy shoot and does not produce flowers.

Leaf curl or blister—Common disease of almond, apricot, cherries, peaches, nectarine, plums, and other *Prunus* spp. caused mainly by *Taphrina deformans*. Other species of *Taphrina* that cause leaf curl include *T. communis, T. farlowii, T. minor, T. pruni,* and *T. wiesneri.*

Leaf felt (of citrus)—*Anthina citri* and other fungi.

Leaf gap—A break in the vascular cylinder, caused by the branching of vascular tissue from the cylinder into a leaf.

Leaf mold—Partially decomposed leaves used in potting mixtures or to add organic matter to garden soil and useful for improving soil structure and fertility.

Leaf or stalk smut (of rye)—Disease caused by *Urocystis occulta.*

Leaf primordium—An outgrowth that develops from the growing point of a bud and grows into a leaf.

Leaf rust (brown rust) of barley—Disease caused by *Puccinia hordei.*

Leaf rust (of wheat, other cereals and many grasses) Disease caused by *Puccinia recondita.*

Leaf scar—Scar left on a twig following the fall of a leaf.

Leaf spot—A self-limiting necrotic lesion on a leaf.

Leaf trace—Branch of the vascular tissues of a stem, extending out into leaves.

Leafhopper—Active insects (order Homoptera, family Cicadellidae) with sucking mouth parts; often a vector of pathogens, especially mycoplasmas, viruses, and MLOs; may also cause direct plant injury during feeding.

Leaflet—One of the several blades of a compound leaf.

Leafstalk—A petiole.

Leaky mutant (in virology)—Mutant that has some residual wild-type activity under "non-permissive" conditions or reverts readily to the wild type.

Leather fungi—Members of the Thelephoraceae.

Lecanoralean (of asci)—*See* Ascus.

Lecanorine (of a lichen apothecium)—Having an excipulum thallinum; lecanoroid.

Lecideine (of a lichen apothecium)—Without an excipulum thallinum, the margin usually consisting only of excipulum proprium.

Lectins—A group of plant proteins that bind to specific carbohydrates.

Lectotype—A specimen or other element (strain, culture) chosen from the original material, but not by the author, where no holotype was named. *See* Type.

Lecythiform—Shaped like a bowling pin (ninepin) or stoppered bottle.

Leggy (of plants)—Weak-stemmed and spindly with sparse foliage caused by excess heat, shade, crowding, or overfertilization. *See* Etiolation.

Legitimate (of validly published names or epithets)—In accordance with the International Code of Botanical Nomenclature; priorable; a recognized taxon may have more than one legitimate name (*see* Illegitimate); (of mating types) compatible.

Legume—A simple, dry, dehiscent fruit with one carpel, splitting along two sutures; plant belonging to the family Fabaceae (formerly Leguminosae), including alfalfa, beans, clovers, and peas, with the characteristic capability to fix atmospheric nitrogen in nodules on the roots if inoculated with proper bacteria.

Leiodisc (of a lichen apothecium)—Having a smooth glazed disc.

Leiosporous—Having smooth spores.

Lemma—The lower and larger of the two bracts enclosing a grass flower.

Lenthionine—A smelly metabolic product of *Lentinula edodes*.

Lenticel—A small pore (natural opening) on a stem, tuber, root, or fruit, made up of loosely arranged cells in the periderm through which carbon dioxide and other gases pass as a by-product of respiration. Pathogens often enter lenticels.

Lentic—Of lakes.

Lenticular—Shaped like a biconvex lens.

Lentiginose, lentiginous—Having minute, freckle-like spots.

Lentogenic strain (of a virus)—A mild or avirulent strain.

Leochromous—Tawny, like a lion's skin.

Lepidopterous—Resembling a butterfly or moth.

Lepidose—Growing among stones.

Lepidote—Covered with small scales, scurfy.

Lepiochlorin—An antibacterial antibiotic from *Lepiota* cultivated by the gardening ant (*Cyphomyrmex costatus*).

Leprose (of lichens)—Having the entire surface of the thallus dissolved into soredia.

Lepto- (prefix)—Thin, small.

Leptocystidium (pl. **leptocystidia**)—Sterile hymenial element, much larger than the basidia, with a thin wall of variable shape and lacking incrustation or refractive contents. *See* Cystidium.

Leptoderan (of nematodes)—Caudal alae that do not meet posterior to the tail tip.

Leptodermatous (of hyphae)—Having the outer wall thinner than the lumen. *See* Mesodermatous.

Leptotichous (of tissue)—Thin-walled.

Lesion—Well-marked, localized, often sunken area of diseased or disordered tissue; a wound.

Lethal—Capable of causing death.

Leucocidin—A substance that destroys leucocytes.

Leucocyte—A type of white blood cell, characterized by a beaded, elongated nucleus.

Leucoplast—Colorless plastid involved in the formation of starch in many types of plant cells.

Leucosin—An albuminoid protein; the reserve food produced by some Chrysophyta.

Leucosporous—Having spores that appear white in mass.

Levigate—Smooth.

L-form bacteria—Bacteria that have, temporarily or permanently, lost the ability to produce a cell wall as a result of growth in the presence of antibiotics inhibiting cell wall synthesis; an unusual morphological form that lacks a rigid cell wall, arises spontaneously, and is not inheritable. L-forms may be found in natural material and can be produced in the laboratory by growing the organisms under conditions unfavorable for normal growth. The morphology of the bacterium changes, with large swollen forms appearing.

Liberty cap—A basidiocarp of the hallucinogenic *Psilocybe semilanceata*.

Library (of genetic engineering)—A collection of unordered clones whose relationships may be established by physical mapping.

Lichen—An intimate, symbiotic (self-supporting) association of a particular fungus (mycobiont) and a particular alga or cyanobacterium (photobiont).

Lichen alga—Phycobiont; photobiont.

"Lichen desert"—An area in or around a town, city, or source of air pollution in which all lichens are absent. *See* Air pollution.

Lichenicolous—Inhabiting lichens as pathogenic gall-forming parasites, parasymbionts, or saprobes.

Licheniform—Shaped like a lichen.

Lichenin, lichenen (I+)—A linear polymer of beta-D-glucose with 1,3 and 1,4 linkages in the ratio of 3:2. Isolichenin (I+) is an isomer of lichenen, lilac or lavender. The compounds (carbohydrates) are found in the walls on many lichen-forming fungi.

Licheniverous—Lichen-eating.

Lichenoid—Lichenlike.

Lichenologist—A person engaged in the pursuit of lichenology.

Lichenology—The science or study of lichens.

Life cycle, life history—Cyclical progression of stages in the growth and development of an organism (plant, animal, or pathogen) that occur between the appearance and reappearance of the same stage of the organism.

Ligase—*See* DNA ligase and RNA ligase.

Ligatin—The process of joining two linear nucleic acid molecules via a phosphodiester bond, e.g., by using ligase enzymes.

Light flux—The total light energy used by plants in the photosynthetically active region (PAR); the quantity of light (intensity × duration of light [daylength]).

Light reactions—The reactions of photosynthesis in which light energy is required; the photo (light) activation or excitement of electrons in the chlorophyll molecule, transfer of the electrons, photolysis of water, and associated reactions.

Light scattering—The process by which energy that has been removed from a beam of light is re-emitted with appreciable change in wavelength. In a suspension of virus particles, light scattering can affect the absorption spectrum of UV light.

Light soil—A coarse-textured sandy soil; hence easy to till.

Lign-, ligni-, ligno- (prefix)—Wood.

Ligneous, lignose—Woodlike.

Lignicole, lignicolous—Growing in or on wood.

Lignification, lignified—Hardening of mainly xylem tissue from wall deposition of lignins.

Lignin—Complex organic material derived from phenylpropane and distinct from carbohydrates that, in part, constitutes wood; imparts rigidity and strength, especially to woody tissues; associated with cellulose in secondary cell walls, especially xylem of many plants. Wood is composed of about 15–30% (by weight) lignified xylem cells.

Lignituber—*See* Papillae.

Lignivorous—Wood-destroying.

Ligulate, liguliform—Flat and narrow; strap-shaped; lorate.

Ligule—In grass leaves, an outgrowth from the upper and inner side of the leaf blade where it joins the sheath; a scalelike projection or row of hairs.

Limaciform—Shaped like a slug or shell-less snail.

Limbate—Bordered with another color; (of a volva) adnate to base of stipe with a narrow, free, membranous margin.

Lime (agricultural)—A material containing the carbonates, oxides, and/or hydroxides of calcium and/or magnesium. Lime is used to increase soil pH and to neutralize soil acidity.

Limoneous—Lemon-colored; citrine.

Limoniform—Lemon-shaped.

Line—An inbred homozygous strain; a type of cultivar; an isolate; a subdivision of a physiologic race; (as a unit of measure) = 2.1167 mm or 1/12 in.

Linear—Long and narrow, resembling a line.

Linear DNA—One of several forms of DNA. The two linear strands of double-stranded DNA may be free, bound to a specific protein, or closed by a hairpin loop.

Lineate-striate—With fine, radially oriented lines.

Lineolate—Marked with fine lines.

Linguiform, lingulate—Tongue-shaped.

Linkage—Tendency of certain genes (markers) to remain together in inheritance because of their location on the same chromosome; the proximity of two or more markers on a chromosome; the closer together the markers are, the lower the probability that they will be separated during meiosis. This gives an idea of the probability that they will be inherited together.

Lip cap (of nematodes)—The anterior most cuticular annulation, disclike in shape, about the mouth and usually thicker than adjacent head annules.

Lip region (of nematodes)—That portion from the lips to the constriction (often obscure or lacking) in the body wall in the head region.

Lipase—An enzyme that hydrolyzes fats into glycerin (glycerol) and fatty acids.

Lipids—Fatty compounds, substances whose molecule consists of glycerin and fatty acids and sometimes certain additional types of compounds; generic term for oils, fats, waxes, and related products found in living tissues; esters of higher aliphatic alcohols insoluble in water but soluble in certain other solvents.

Lipo- (prefix)—Fat.

Lipochroic—Having pigment in oil drops within the cell. *See* Endochroic.

Lipolytic enzyme—An enzyme that hydrolyzes fats and lipids.

Lipoprotein—A conjugated protein containing one or more lipids.

Liposome—Lipid vesicle used to introduce biochemical molecules, such as virus particles or nucleic acid, into cells.

Lips (of nematodes)—Cuticular structures (usually six), two subdorsal, two lateral, and two subventral, surrounding the mouth opening (may be fused in pairs).

Lipsanenchyma—Primordial tissue of a basidiocarp, other than the universal veil, covering the hymenium.

Lipstick mold—*Sporendonema purpurascens*, a fungus that invades mushroom beds.

Liquefaction—Transformation of a gel to a liquid.

Lirella—A long, narrow apothecium.

Liter—2.1134 pt = 1.0567 liquid qt (U.S.) = 0.9081 dry qt (U.S.) = 0.264178 gal (U.S.) = 1,000 ml or cm^3 = 33.8147 fl oz = 61.025 $in.^3$ = 0.0353 ft^3 = 0.028378 bushel.

Lithophytic—Growing on rocks; saxicolous.

Litmus—A blue amphoteric dye from certain depside-containing lichens, used as an indicator for pH and oxidation or reduction; turns red in presence of acids and back to blue again with alkalis.

Littoral—Growing on lake or sea shores.

Lituate—Forked with the points turned slightly outward.

Liverwort—A bryophyte, sometimes used in common names for large foliose lichens.

Loam—A mellow, textural class of soil composed of about equal parts silt and sand and less than 20% clay.

LO-analysis (L, lux [light]; O, obscuritas [darkness])—A method for pollen and fungus spore analysis by which different types of surface ornamentation may be distinguished microscopically by the differences in appearance of the spore surface at upper and lower focus.

Lobate—Lobed; resembling a lobe; having large rounded divisions or lobes.

Lobed (leaf)—Shallowly or deeply divided by clefts or sinuses, as in a maple (*Acer*) leaf.

Lobulate—Having small lobes.

Local infection—Infection involving only limited or localized parts of a plant (e.g., leaf spot, fleck, scab).

Local lesion—A small, localized, chlorotic or necrotic spot produced on a leaf upon mechanical inoculation with a virus (e.g., a hypersensitive reaction).

Locular—Containing chambers or hollows.

Locule, loculus (adj. **loculiferous**)—(Of fungi) a cavity, especially an ascigerous one, in a fungus stroma enclosed by fungal, host, or fungal/host tissue; (of higher plants) one of the compartments of a compound ovary.

Loculoascomycetes—A class of the Ascomycotina characterized by bitunicate asci produced in unwalled locules (pseudothecia) in ascostromatic ascomata with an ascolocular ontogeny.

Loculospore—A spore, one per chamber, in spherical multiloculate structures of the agaric *Zerovaemyces*.

Locus (conidiogenous)—Specific area on a conidiogenous cell in which conidial formation occurs; (of genetics) the fixed position of a gene on or in a chromosome.

Lodging—Lying down; a condition in which stems (culms, stalks, etc.) approach a horizontal rather than a vertical position, as in plants beaten down by wind, rain, disease, or insect attack.

Lomasome—A vesicle derived from an intracytoplasmic membrane. *See* Plasmalemmasome.

Longevity end-point or longevity *in vitro*—The storage time after which a virus in a crude sap preparation loses its infectivity; usually determined at 0 or 20°C.

Longicollous—Having long beaks or necks.

Longitrorse—Turned longitudinally.

Longitudinal (conidioma)—Longer than broad.

Longitudinal ridges or alae (of nematodes)—Raised cuticular areas that extend through the length of the body outside of the lateral field and are present on all sides.

Longitudinal striae (of nematodes)—Superficial lines in the cuticle extending the length of the body and separating the longitudinal ridges.

Loose smut—A smut in which the teliospores are freely exposed. *See* Covered smut.

Loose smut of oats and several grasses—Disease caused by *Ustilago avenae*.

Loose smut of sorghum—Disease caused by *Sporisorium cruentum*.

Loose smut of wheat, barley, rye, and several grasses—Disease caused by *Ustilago tritici*.

Lophotrichous, lophotrichiate (of a bacterial cell)—Having several flagella at one or both ends.

Lorate—Strap-shaped; ligulate; like a narrow band.

Lorchel—The poisonous fruitbody (ascoma) of *Gyromitra esculenta*; false morel; lorel.

Low volume (LV) spray—A concentrate spray applied to uniformly cover the crop being treated, but not as a full-cover spray to the point of run-off. *See* Ultra low volume spray.

Lower fungi—Chytridiomata, Hyphochytriomycota, Oomycota, Myxomycota, Plasmodiophoromycota, and Zygomycotina. *See also* Phycomycetes.

Lubricous—Slippery; smooth.

Lucerne—A name for alfalfa (*Medicago sativa*) used in Europe, Australia, and other regions.

Lucid—Shining; clear.

Lucunose—Having lacunae.

Lumen (pl. **lumina**)—Central cavity of a cell or other structure (e.g., the canal or duct of the esophagus in nematodes); hollow center of a culm (grass stem); (of a plant cell) the protoplasmic membrane and enclosed protoplast.

Lunate, luniform, lunulate—Crescent-shaped, like a new moon.

Lungwort—The lichen *Lobaria pulmonaria*, that externally resembles a human lung; lung lichen; lungs of oak.

Lupinosis—A mycotoxicosis of sheep caused by *Phomopsis (Phoma) leptostromiformis*, teleomorph *Diaporthe woodii*; also a cause of stem rot of European yellow lupine (*Lupinus luteus*).

Lurid—Pale yellow, wan, sallow, ghostly; the color of glowing fire; brown tinged with red.

Luteofuscous—Blackish yellow.

Luteoskyrin—A carcinogenic toxin of *Penicillium islandicum*, a cause of hepatitis (yellow rice disease) in humans. *See also* Islandicin and Skyrin.

Luteous—Yellow, dull egg-yellow.

Lutous—Muddy.

Lux (lx)—A standard measure of light (metric system). The illumination impinging upon a surface of 1 m^2, each point of which is at a distance 1 m away from a standard light source of one candle. Photometric unit expressing the illuminance of 1 lumen/m^2; 1 lx = 0.093 ft-c. *See* Foot-candle.

Lycomarasmic acid—A derivative of lycomarasmin. *See* Fusaric acid.

Lycomarasmin—A dipeptide wilt toxin of *Fusarium oxysporum* f. sp. *lycopersici*.

Lymabiont, lymaphile—An organism found commonly or only in sewage.

Lymaphobe—An organism never found in sewage.

Lymaxene—An organism rarely found in sewage.

Lyocystidium—*See* Cystidium.

Lyophilization (v. **lyophilize**) (syn. **freeze-drying**)—A technique by which water is removed under high vacuum while the preparation or tissue is frozen. Used for long-term preservation of bacteria, fungi, viruses, antisera, and other types of pathogens. *See also* Freeze-drying.

Lysergic acid and derivatives—Fungi that produce hallucinogenic compounds, e.g., lysergic acid diethylamide (LSD), that occur in *Claviceps* sclerotia and are the cause of paspalum staggers. *See* Ergot.

Lysigenetic, lysigenic, lysigenous—Formed by the breaking down or dissolution of cells or tissues. *See* Schizogenous.

Lysigenome (pl. **lysigenomata**)—The name given a group of giant cells or syncytia denoting their origin from the lysis or dissolution of walls of normal cells to form a tumorlike structure; formed commonly in root tissue by gall-forming and some other kinds of nematodes.

Lysin—An enzyme or other substance capable of disrupting cells. *See also* Antibody.

Lysis—The cellular breakdown (dissolution) of tissues or destruction of cell walls by enzymes (e.g., a lysin) or viruses and subsequent rupture of cell membranes and loss of cell contents (*see* Autolysis); (of conidial liberation) secession by the dissolution of the wall of the adjacent cell. *See* Fission and Fracture.

Lysogen (adj. **lysogenic**)—A bacterium or other prokaryotic cell carrying a bacteriophage (often as prophage) to which it is not itself susceptible.

Lysogeny—The process by which viruses become integrated into a bacterial chromosome.

Lysosome—Membrane-bound cytoplasmic organelle that contains a variety of hydrolytic enzymes.

Lysozyme—A lytic enzyme found in egg-white, saliva, tears, and other substances that can dissolve the cell wall of many Gram-positive bacteria. It is used in the material (peptidoglycan) preparation of plasmids from bacteria.

Lytic phage(s)—*See* Virulent phage(s).

M

m—Meter = 100 cm = 1,000 mm = 39.37 in. = 3.2808 ft = 1.09361 yards = 0.1988 rod = 0.001 km.

Macaedium—*See* Mazaedium.

Macerate—To separate the cells of a tissue, so that they drift apart or collapse; to soften by soaking; to waste away; a soft rot.

Macergenic—Generating macerates; soft rotting.

Macro- (prefix)—Long, large, or great; commonly used in the sense of mega-. *See* Mega.

Macrocephalic—Describing the long end cells of immature ascospores; septation proceeds from the primary septum toward the poles.

Macroconidium (pl. **macroconidia**)—The larger, generally more diagnostic, conidium of a fungus that also has microconidia; a long or large conidium.

Macrocyclic—A life cycle of a rust fungus comprised of aecia, uredinia, telia, and usually spermogonia (pycnia); they may be autoecious or heteroecious. *See* Microcyclic.

Macrocyst (of Myxomycota)—An encysted aggregate of myxoamoebae; resting form of a young plasmodium, the alternative to the sporocarp in some cellular slime molds.

Macrocystidium—*See* Cystidium.

Macrofungi—Fungi with large (macroscopic) sporocarps.

Macrolichen—A large lichen of the squamulose, foliose, or fruticose habit.

Macromolecule—A molecule with a molecular weight from several thousands to many millions.

Macromorphology—The gross or macroscopic structural attributes of fruiting structures (e.g., basidiocarps) or cultures.

Macromycetes—Large fungi.

Macronematous (of conidiophores)—Hyphae morphologically very different from vegetative hyphae and usually erect. *See* Micronematous and Semimacronematous.

Macronutrients—Essential chemical elements required in relatively large quantities for the growth of plants, e.g., nitrogen (N), phosphorus (P), and potassium (K), usually in amounts greater than 1 ppm.

Macroplasmodium—A large plasmodium, e.g., *Physarum polycephalum.*

Macroscopic—Visible to the unaided eye.

Macrospore—A large spore in which there are spores of two sizes, as in *Fusarium.*

Maculate—Blotched; spotted.

Maculicole (adj. **maculicolous**)—An organism growing on spots, e.g., leaf spots.

Maculiform—In the form of spots.

Madefact—Moistened; made wet.

Madura foot, maduramycosis—*See* Mycetoma.

Maggot—The growing stage or larva of a fly.

Magnification—The number of times the apparent size of an object has been increased by the lens system of a microscope.

Malaceoid venation (of lichens)—*See* Veins.

Malacoid—Mucilaginous.

Male agaric—*Phellinus (Fomes) igniarius.*

Male-sterile (n. **male sterility**)—Having nonfunctional or no male sex organs; a condition in some plants in which pollen is not formed or does not function normally, even though the stamens may appear normal.

Malformin—Plant-malforming cyclic pentapeptide from *Aspergillus niger.*

Malignant—Describing a cell or tissue that divides and enlarges autonomously; a nonself-limiting abnormal growth.

Maltase—Enzyme that hydrolyzes malt sugar (maltose) to glucose (grape sugar).

Malthusian theory—A theory developed by Thomas R. Malthus (1766–1834), a British economist and clergyman, who theorized that the world's human population will increase faster than food productivity and thus world starvation will eventually occur unless war or disease reduces the population or the increase in population is checked by sexual restraint.

Maltoryzine—A metabolite of *Aspergillus oryzae* var. *microsporus* that is toxic to cattle.

Maltose—A carbohydrate; a disaccharide produced by the enzymatic hydrolysis of starch by diastase.

Malvaceae (adj. **malvaceous**)—Mallow family.

Mammillate, mammiform—Having nipple-shaped projections or protuberances; digitate; (of conidia) breast-shaped, apex conical with a papillate tip.

Mandibles—Horny, jawlike mouthparts of insects; used to bite and chew food.

Manna, "manna lichen"—*Shaerothallia esculenta*; may be one of the types of manna in the Bible. It can be used to make bread; thalli are unattached and blown in the wind.

Manocyst (of *Phytophthora*)—A projection, or receptive papilla, from the oogonium that undergoes fusion with the antheridium.

Mantle—Dense hyphal mass of ectomycorrhizal fungus, enclosing short feeder roots of plants connected to the Hartig net on the inside and to the extramatrical hyphae on the outside. The mantle acts as a nutrient sink.

Manual transmission—Spread or introduction of inoculum to infection courts by hand manipulation. *See* Mechanical transmission.

Manubrium (of nematodes)—The enlarged proximal portion of the spicule.

Marbled—Stained with irregular color streaks.

Marginal muscle fibers (of nematodes)—Muscle fibers of the esophagus at the apices of the triradiate lumen, having a suspensory and skeletal function.

Marginal veil (of agarics)—An incurving proliferation of the pileus margin that protects the developing hymenium. *See* Partial veil.

Marginate—Having a distinctly marked border; (of the basal bulb of an agaric stipe) having an exterior circular ridge (gytterlike rim) where the veil was attached.

Margo propius—Proper margin. *See* Excipulum proprium.

Margo thallinus—Thalline margin. *See* Excipulum thallinum.

Maritime—Confined to the seacoast.

Marker—An identifiable physical location on a chromosome whose inheritance can be monitored.

Marker rescue—The production of infective progeny virus from coinfections between a virus that lacks a specific gene function and another related or unrelated virus.

Marl—A type of soil, rich in lime, formed in the bottom of a lake or swamp.

Masked—Pertains to symptoms that are absent under certain environmental conditions but appear under other conditions; describing the condition whereby a host is infected but does not exhibit disease symptoms because the environment is not favorable for disease expression and/or development.

Masked virus—A virus carried by a plant that does not show symptoms of its presence.

Mast cell—A connective-tissue cell with many large, metachromatic granules in the cytoplasm.

Mastigonemes—Hairlike processes covering the surface of a tinsel flagellum.

Mastoid—Nipplelike.

Mating types—Compatible strains, usually designated + and − or A and B, necessary for sexual reproduction in heterothallic fungi.

Matrix—Substratum in or on which an organism is living; (of nematodes) the middle layer of the cuticle; (of fungi) mucilaginous material in which conidia and some ascospores are produced. It influences dissemination, survival, germination, etc.

Matrix protein—A term for several different types of proteins: 1) the major protein making up the structure of inclusion bodies; 2) in some viruses, the protein between the viral membrane and the nucleocapsid.

Matsutake—An edible fungus, *Tricholoma matsutake*, important in Japan; in North America "matsutake" is *Tricholoma ponderosum* and *T. murrillianum*. *See* Pine or Japanese mushroom.

Maturation—Process of becoming mature. *See* Differentiation.

Maturescent—Ripening; approaching maturity.

Maturity—The state of ripeness. Usually refers to that stage of development that results in maximum quality.

Maxam and Gilbert method—A technique for sequencing DNA using chemical base-specific modification and cleavage.

Mazaedium, mazedium—A spore mass formed by the release of spores from asci, generally with sterile elements and held together in a dry, loose, powdery mass on the fruiting surface (e.g., apothecium of *Roesleria*).

Mealybug—Small, oval insects (family Pseudococcidae, superfamily Coccoidea) with sucking mouth parts and a cottony, scalelike covering. Unlike scale insects, mealybugs possess functional legs and reproduce by producing eggs or living young. Mealybugs transmit a number of plant-infecting viruses, especially of cocoa (*Theobroma*).

Measures—*See* Weights and measures (miscellaneous).

Mechanical injury—Injury of a plant part by abrasion, mutilation, or wounding.

Mechanical transmission (or **inoculation**)—Artificial spread or introduction of inoculum (e.g., a virus) to an infection court (especially a wound) by hand manipulation accompanied by physical disruption of the host tissue; may also occur in the field when a virus is transmitted from one plant to another by leaves rubbing or root contact. *See* Manual transmission.

Medallion clamp—A clamp connection with a space between the main hypha and the "hook."

Median—In the middle.

Median effective dose (ED_{50})—The amount of a substance required to produce a response in 50% of the hosts (subjects) to whom it is given.

Median effective time (ET_{50})—The amount of time for a substance to produce a response in 50% of the hosts (subjects) to whom it is given.

Median infective dose (ID_{50})—The dose of a chemical, pathogen, etc., that, on average, will infect 50% of the hosts (individuals) to whom it is administrated.

Median lethal dose (LD_{50})- The dose of a substance that is fatal to 50% of the test population.

Median lethal time (LT_{50})—The period of time required for 50% of a group of organisms to die following a specific dose of an injurious agent, e.g., virus, radiation, drug, etc.

Median or middle bulb (of nematodes)—*See* Metacorpus.

Median survival time (ST_{50})—The period of time at which half the subjects have died following the administration of an injurious agent, e.g., virus, radiation, drug, etc.

Median tissue culture infective dose ($TCID_{50}$)—The dose (e.g., a virus) that, on average, will infect 50% of susceptible tissue culture cells.

Medium, culture medium (pl. **media**)—Chemical environment (substrate) used to provide nutrients for microbial growth and reproduction in the laboratory; may be solid or liquid, organic or inorganic, defined or undefined.

Medulla (adj. **medullary**)—Central part of an organ or tissue, as in a sclerotium; the pith of plants; loose layer of hyphae inside a lichen thallus below the cortex and algal layer; the part of fungal sporocarps composed mainly or entirely of longitudinal hyphae.

Medullary excipulum—Tissue below the generative layer of an apothecium; hypothecium. *See* Excipulum.

Meg-, mega-, megalo- (prefix)—Large, of great size. *See* Macro-.

Megagametophyte—A female gametophyte resulting from the growth of a megaspore and producing female gametes or eggs.

Megapascal—10^6 pascals; approximately 7.5×10^3 mm of mercury.

Megasporangium—A sporangium that produces megaspores.

Megaspore—Spore produced within a megasporangium and forming a megagametophyte. *See* Macrospore.

Megasporophyll—A leaf, or similar structure, that bears a megasporangium.

Mei-, meio-, mio- (prefix)—Less, smaller, fewer.

Meiocyte—A cell in which meiosis occurs. *See* Gonotocont.

Meiogyrous—Rolled inward slightly.

Meiophase—The part of a life cycle in which the diploid nucleus undergoes reduction division, giving rise to haploid nuclei.

Meiosis—Reduction division; a process that involves two nuclear divisions, during which both reductional division (separation of homologs) and equational division (separation of chromatids) occur; process by means of which the chromosome numbers are reduced by half; process by which a zygote (2n) divides twice and produces four haploid (1n) gametes. Meiosis compensates for the chromosome-doubling effect of fertilization.

Meiosporangium—Brown, pitted, thick-walled, diploid sporangium of certain Blastocladiales that produce uninucleate, haploid zoospores or meiospores. *See* Mitosporangium.

Meiospore—A spore from a meiosporangium produced following meiosis; (of ascomycetes and basidiomycetes), ascospore or basidiospore that is the product of meiosis.

Meixner test—A test for amatoxins. Juice from a fresh basidiocarp is squeezed onto a piece of newspaper or computer paper, etc., allowed to dry, and a drop of concentrated hydrochloric acid (HCl) is added. A blue color indicates the presence of amatoxins.

Melanconiales—An old name for a form-order of the Coelomycetes, containing the form-family Melanconiaceae, traditionally used for deuteromycetes that produce their conidia in acervuli. See Mitosporic fungi.

Melanin, melanose (adj. **melanoid**)—Brown-black pigment.

Melanosphaerites—Fossil fungi from the Devonian period.

Melanosporous—Black-spored.

Meline—Quince-colored.

Melioline—One of the Meliolaceae, especially species of *Meliola*.

Melleous, melline—Honey-colored.

Melzer's reagent (or iodine)—A solution, containing 2.5 g of iodine, 7.5 g of potassium iodide, and stabilized by 100 g of chloral hydrate per 100 ml of distilled water; used to detect amyloid or dextrinoid reactions of fungal spores or mycelium.

Membranaceous, membranous—Membranelike; like a thin skin or parchment; thin, compact, but pliant.

Membrane—A lipid bilayer that separates the internal contents of a cell or organelle from its surroundings; also, sometimes used for nitrocellulose or nylon fibers used in blotting.

Membrane filter—A filter made from such polymeric materials as cellulose, polyethylene, and tetrafluoroethylene.

Memnospore—A spore staying at its place of origin. *See* Xenospore.

Mendel's laws—A set of three laws formulated by Gregor J. Mendel (1822–1884), an Austrian biologist. Each law is generally true but with numerous exceptions. The laws are: 1) characters exhibit alternative inheritance, being either dominant or recessive; 2) each gamete receives one member of each pair of factors present in a mature individual; and 3) reproductive cells combine at random.

Meniscate—Bent into a half circle.

meq—Milliequivalent weight.

-mer (suffix)—Part; e.g., dimer, trimer, pentamer.

Mer-, meri-, mero- (prefix)—Part.

Merenchyma—Hyphal tissue derived by cell division in several planes. *See* Plectenchyma.

Merismatoid, merismoid (of a pileus)—Made up of smaller pilei.

Merispore—Components of a compound spore or spore-ball. *See* Sporidesm.

Merist-. meristo- (prefix)—Divisible, divided.

Meristem (adj. **meristematic**)—A layer or zone of undifferentiated plant tissue that functions principally in cell division and is thus responsible for the first phase of growth; a mass of growing cells, capable of frequent cell division; (of conidiophores) increasing in length with the laying down of cross walls and elongation of cells.

Meristem arthrospore—One of a chain of conidia that mature in basipetal succession and originate by meristematic growth at the apical region of the conidiogenous cell.

Meristem blastospore—Conidium arising apically or laterally from a conidiogenous cell (basauxic) that exhibits basal elongation.

Meristem culture—Aseptic culture of a plant or plant part from a portion of the meristem.

Meristem tip—The meristem dome of cells and one or two pairs of primordial leaves (0.5–1 mm in diameter) that comprise the explant removed from a bud and grown in tissue culture to produce a disease-free plant.

Meristem, intercalary—A meristem located between differentiated tissues, e.g., growth at the base of the internodes in grasses; a mistletoe sinker elongates by growth from a meristem near the base of the sinker.

Meristogenous (of pycnidia, etc.)—Formed by growth and division of one hypha. *See* Symphogenous.

Merogamy—Copulation between two special sex cells or gametes.

Meront (of Myxomycota)—One of the daughter myxamoebae cut off in succession by the parent myxamoeba.

Merosporangium (of Zygomycetes)—A cylindrical outgrowth from the swollen tip of a sporangiophore in which a chainlike series of sporangiospores is usually produced.

Meruloid (of the hymenophore)—Thrown into anastomosing folds, shallow pits and ridges with the hymenium extending over the edges.

Mes-, meso- (prefix)—Middle.

MES—A biological buffer (2-(N-morpholino)-ethane sulphonic acid) with a pH range of 5.2–7.2.

Mesh (screen)—Standard screens that separate solid particles into size ranges. The mesh is stated in number of openings per linear inch. The finest screen practical is the 325-mesh, which has openings 44 μm in diameter and over 105,000 openings per square inch. Quality dusting sulfur has 95% of the particles passing a 325-mesh screen. A common range of granular formulations is in the 15- to 30-μm range. Particles that pass a 60-mesh screen are considered dusts.

Mesially—Medianly.

Mesocarp—Middle layer of a fruit wall (pericarp).

Mesochroic—Having pigment in the hyphal wall.

Mesodermatous (of hyphae)—Having the outer wall and lumen of about the same thickness. *See* Leptodermatous.

Mesogenous—Borne in the middle.

Mesophiles (adj. **mesophilic**)—Organisms that grow best between about 10 to 40°C (optimum usually 20 to 35°C). *See* Thermophily.

Mesophyll—Chlorophyllous parenchyma tissue in leaves between the two epidermal layers.

Mesophytes—Plants intermediate between hydrophytes and xerophytes; grow in soils containing moderate amounts of available moisture.

Mesosome—A membranous involution of the cytoplasmic membrane.

Mesosporangium (of Zygomycetes)—A cylindrical outgrowth from the swollen tip of a sporangiophore in which a chainlike series of sporangiospores is generally produced.

Mesospore—A one-celled teliospore among two-celled ones; *or* the middle layer of a three-layered spore wall.

Mesozoic—A geologic era beginning 225 million years ago and ending 65 million years ago.

Messenger RNA (mRNA)—A chain of ribonucleotides that codes for a specific protein. Messenger RNA moves from the nucleus to the ribosomes, where protein is synthesized in the cytoplasm.

Met-, meta- (prefix)—Changed in form or position; between, among; with; after.

Metabasidium (pl. **metabasidia**)—Often defined as the part of the basidium in which meiosis takes place; also called the promycelium in certain basidiomycetes.

Metabiosis—The association of two organisms acting or living one after the other. *See* Synergism.

Metabolism—The sum total of the chemical processes (physiological activities) occurring in the body of a living organism; the process by which organisms utilize nutrients to build structural components and break down cellular material to obtain energy and simple substances for special functions.

Metabolite—Any chemical component of a reaction series or process; a product of metabolism.

Metacellulose—A cellulose in certain fungi and lichens.

Metachroic—Changing color through the appearance of a new pigment in more mature tissue.

Metachromatic granules—Cytoplasmic inclusions of concentrated cellular material.

Metachromic granule—A deeply staining body in the protoplasm of some cells (e.g., some bacteria and yeasts).

Metacorpus (of nematodes)—The swollen posterior portion of the corpus of the esophagus (sometimes called the median bulb).

Metal shadowing—A technique used to prepare viruses for electron microscopy, in which the virus particles are exposed to the vapor of a heavy metal such as gold or platinum. This technique has now largely been replaced by the *negative contrast staining method.*

Metamorphic rock—Rock that has been greatly altered from its previous condition through the combined action of heat and pressure.

Metamorphosis—Any conspicuous changes in form or structure during the growth of animals.

Metaphase—Stage of mitosis or meiosis at which the chromosomes are shortened and thickened and lie at the equator of the spindle.

Metaphysis—Same as paraphysis.

Metaplasm—Same as epiplasm.

Metathallus—Assimilative (containing the photobiont) part of a lichen thallus, especially where there is also a prothallus.

Methylated bovine serum albumin—Serum albumin used as a matrix in the separation of protein molecules in chromatography.

Methylation—The addition of methyl groups to nucleic acids or to other compounds.

Methylene blue—A photoreactive dye used as a vital stain for cells and for the detection of nucleic acids in gel electrophoresis.

Metoecious—Same as heteroecious.

Metonym—A name for which there is an older, valid name.

Metuliform—Pyramidlike.

Metuloid—A modified and encrusted cystidium that is thick-walled at maturity, as in *Peniophora.*

mg/kg—Milligrams per kilogram; used to express the amount of pesticide per amount of animal body weight needed to produce a toxic or desired effect (1 million milligrams = 1 kg = 2.2 lb).

mg—Milligram = one-thousandth of a gram.

mho—*See* Siemen.

Micaceous (of a pileus surface)—Covered with glistening, micalike particles.

Micelles—The solid particles dispersed in a colloidal system.

Micro- (prefix)—Small; minute; one-thousandth. *See* Micron.

Microacrophile, microaerophilic—An organism that makes its best growth under lowered oxygen pressure.

Microbe—Microscopic organism composed of one to many cells and capable of reproduction by division. *See* Microorganism.

Microbial control—Use of microorganisms and viruses as biological control agents for pests and diseases.

Microbial succession (metabiosis)—Natural progression of microorganisms in wounded and decaying tissue; their occurrence in composting is important.

Microbiology—The science or study of microorganisms, e.g., bacteria, actinomycetes, fungi, algae, viruses, and protozoa.

Microcephalic—Describing ascospores in which the septation proceeds from the primary septum toward the poles, so immature spores have longer end cells. *See* Septum.

Microclimate—Atmospheric environmental conditions in the immediate vicinity of a plant, including interchanges of energy, gases, and water between atmosphere and soil.

Microconidium (pl. microconidia)—The smaller conidium of a fungus that also has macroconidia; a small conidium often acting as a spermatium or male sex cell. *See* Macroconidium.

Microcrystal tests—Technique for identifying lichen substances, especially phenolic metabolites.

Microculture—A culture of an organism under the microscope, as in a hanging drop (in a van Tieghem cell), and similar microscopic techniques.

Microcyclic—A life cycle of a rust fungus comprised of spermagonia and telia, or of telia only (*see* Macrocyclic); (of conidiation) germination of spores by the direct function of the conidia without the intervention of mycelial growth..

Microcyst (of Myxomycota)—An encysted myxamoeba or swarm spore.

Microendospores—Minute cytoplasmic particles behaving like spores in *Ophiostoma ulmi* (syn. *Ceratocystis ulmi*).

Microfibrils—Aggregations of chainlike cellulose molecules into ultramicroscopic fibrils.

Microflora—Composite of microscopic plants and certain other microorganisms in a specified site or habitat.

Microfungi (micromycetes)—Fungi having small fruitbodies, requiring microscopic examination for observation.

Microgram (μg)—10^{-6} grams.

Microlichens—Small, mainly crustaceous lichens that require microscopic examination for their identification.

Micrometer (μm)—A unit of length equal to one-thousandth of a millimeter (0.001 mm or 1 micron [μm]); 0.000001 meter (10^{-6} m) or 0.00003937 in. Also a disc or slide of glass ruled with lines forming a metric scale for measuring objects under a microscope in microns.

Micromorphology—The microscopic attributes of fruiting bodies or other structures.

Micron (μm)—One millionth of a meter (m) = 0.001 mm = 0.00003937 in.; same as micrometer.

Micronematous, micronemeous—Having hyphae of small diameter; (of conidiophores) similar morphologically to vegetative hyphae but bearing conidia.

Micronutrients—Essential chemical elements required in only minute amounts (less than 1 ppm) for the growth of plants, e.g., boron, chlorine, copper, iron, manganese, molybdenum, and zinc; trace elements.

Microorganism—Microscopic organism (includes all unicellular prokaryotes and eukaryotes and some multicellular eukaryotes such as bacteria, fungi, nematodes, viruses, microscopic algae, and protozoa. Same as Microbe.

Microphylline (of lichens)—Composed of minute lobes or scales.

Microprecipitin test or precipitin test—A serological test for virus in which a positive reaction is a white precipitate.

Micropyle (adj. **micropylar**)—Minute opening between the edges of the two integuments of an ovule through which a pollen tube enters to reach the female gametophyte to effect fertilization; (of nematodes), the minute opening in the membrane of an egg through which the spermatozoon enters.

Microsclerid—*See* Cystidium.

Microsclerotium (pl. **microsclerotia**)—Often microscopic, dense aggregate of darkly pigmented, thick-walled hyphal cells specialized for survival, as in *Verticillium dahliae*; capable of germination to produce a mycelium; pseudosclerotium.

Microscopic—Too small to be seen except with the aid of a microscope.

Microsome(s)—Fraction of a cell homogenate obtained by ultracentrifugation; comprised of ribosomes and fragments of rough endoplasmic reticulum 16–150 nm in diameter.

Microsporangium—A secondary sporangium that produces microspores.

Microspore—(Of fungi) a small spore in which there are spores of two sizes (e.g., *Fusarium*); a spore from a microsporangium that grows into a male gametophyte; (of higher plants) one of the four haploid spores that originate from the meiotic division of the microspore mother cell in the anther of the flower; microspores give rise to pollen grains.

Microspore mother cell—Diploid cell in the anther that gives rise, through meiosis, to four haploid microspores.

Microsporophyll—A leaf or similar structure, that bears microsporangia.

Microtome—An instrument for making thin sections of tissue or cells.

Microtubules—Any of the minute tubular structures of an eukaryotic cell that are widely distributed in protoplasm and made up of longitudinal fibrils.

Microzoospore (of certain algae)—A microscopic, motile spore.

Middle lamella—The thin cementing layer between adjacent primary cell walls; it generally consists of pectinaceous materials (especially calcium pectate), except in woody tissues, in which pectin is replaced by lignin.

Midrib—The central, thickened vein of a leaf.

Migration pseudoplasmodium (of Acrasiales)—Migration phase after the aggregation of the myxamoebae.

Migratory—Migrating from plant to plant.

Miktohaplont—Haplont composed of cells having genotypically different nuclei. *See* Isohaplont.

Mildew—Plant disease, caused by a fungus, characterized by a thin coating of whitish, grayish, pale purplish, or black mycelial growth and spores on the surfaces of infected plant parts. *See* Downy mildew (Peronosporaceae), Powdery mildew (Erysiphaceae), and Dark mildew (Meliolales or Capnodiaceae).

Mile (statute) = 5,280 ft = 1,760 yd = 1,609.35 m = 320 rods = 0.86836 nautical mi.

Milk cap—Basidiocarp of *Lactarius*.

Millimeter (mm)—*See* mm.

Millimho (mmho)—A measure of electrical conductivity, 1 mmho = 0.001 mho. The mho is the reciprocal of an ohm.

Millimicron (mμm)—One-thousandth of a micron or 10 Å; a nanometer (nm).

Mineral soil—A soil with makeup and physical properties that are largely those of mineral matter.

Minimal medium—The simplest, chemically defined medium on which the wild type (prototroph) of an organism will grow and that must be supplemented with one or more specific substances for the growth of auxotrophic mutants derived from the wild type.

Minor element—*See* Micronutrient.

Miscibility—The physical properties of chemicals that permit a uniform mixture when two or more liquids are combined.

Miscible—Capable of being mixed and remaining mixed under normal conditions.

Miso—An oriental food product used for soups and as a flavoring agent. It is composed of rice, cereals and soybeans fermented by *Aspergillus oryzae* and *Saccharomyces rouxii*.

Mist propagation—Applying water in mist form to leafy cuttings in the rooting stage to reduce transpiration.

Mist spraying—A method of spraying in which concentrated spray is atomized into an air stream.

Mite—Minute (1/64- to 1/32-in. long) animals (order Acarina, families Tetranchidae and Eriophyidae) without evident body divisions that are six-legged as larvae and eight-legged as adults. Red spider mites commonly cover leaf surfaces with fine webbing. Eriophyid mites transmit a number of plant-infecting viruses. Species of *Tyrophagus* and *Tarsonemus* commonly infest culture collections.

Miticide—*See* Acaricide.

Mitic—Threadlike.

Mitochondrion (pl. mitochondria)—A minute, membrane-bound, long or round cellular organelle found outside the nucleus of an eukaryotic cell; it produces intracellular energy for the cell through respiration and is rich in fats, proteins, and enzymes; contains the enzyme systems required for the citric acid cycle, electron transport, and oxidative phosphorylation; it possesses DNA, messenger RNA, and small ribosomes and is capable of protein synthesis.

Mitosis—Usual process of nuclear cell division in which each chromosome duplicates, pulls apart, and produces two daughter nuclei of the same genetic constitution as the parent nucleus. Mitosis is usually, but not always, followed by cellular division (cytokinesis). See Cytokinesis.

Mitosporangium—Thin-walled diploid sporangium of certain Blastocladiales that produce uninucleate diploid zoospores (mitospores). See Meiosporangium.

Mitospore—Spore from a mitosporangium; (of ascospores and basidiomycetes) any nonbasidiosporous or ascosporous propagule; a spore produced following mitosis. See Meiospore and Mitosporic fungi.

Mitosporic fungi—Includes Deuteromycotina, Deuteromycetes, Fungi Imperfecti, asexual fungi, and other conidia-producing fungi. An artificial assemblage of fungi that have not been correlated with any meiotic states. Mitosporic fungi with teleomorphs in the ascomycetes and basidiomycetes are anamorphs or anamorphic states of these groups. Three classes of mitosporic fungi are traditionally recognized:

Hyphomycetes, in which conidia are borne on separate hyphae or hyphal aggregations (sporodochia and synnema).

Agonomycetes, which have mycelial forms that are sterile (but may produce chlamydospores, sclerotia, bulbils, and related vegetative structures.

Coleomycetes, in which conidia are produced in acervuli, pycnidia, pyenothecia, and cupulate or stromatic conidiomata.

Mitrate, mitriform—Miter-shaped.

Mixed bud—A bud that contains both rudimentary flowers and vegetative shoots.

Mixed forest—A forest containing both coniferous and broad-leaved trees.

ml—Milliliter = 1 cm^3 (approximate) = 0.001 liter = 0.061 $in.^3$ = 0.03815 fl oz.

MLO—See Mycoplasmalike organism (phytoplasma).

mm—Millimeter = 0.1 cm = 0.01 dm = 0.001 m = 1,000 μm = 0.03937 in. (about 1/24 in.).

MNPV = Bundle virion. Abbreviation for the subtype of nucleus polyhedrosis viruses in which the majority of enveloped virions contain more than one nucleocapsid.

Modal length—The length that occurs most frequently in a population of virus particles; a criterion used when grouping viruses having rod-shaped particles.

Mode infection (in virology)—Inoculation of cells or an organism with a solution that lacks virus particles; a control in virus infection experiments to ascertain any possible side effects of materials in the inoculum other than infectious particles.

Modification—A temporary change or variation in the characteristics of an organism.

Moiety—A part of a molecule having a characteristic chemical property.

Moisture, dry basis—A basis for representing moisture content of a product. It is calculated from the net weight of water lost by drying, divided by dried weight of the material. The answer is multiplied by 100 to give percent.

Mol (light)—Used in plant science when the waveband stated is 400–700 nm (PAR); 1 mol = 6.022×10^{23} photons, or 1 μmol = 6.022×10^{17} photons.

Molariform—Shaped like a molar tooth with a flattened crown.

Mold, mould—Any microfungus with conspicuous, profuse, or woolly superficial growth (mycelium and/or spore masses) on various substrates; especially an economically important saprobe. Molds commonly grow on damp or decaying matter and on the surface of plant tissues.

Mole (M)—Amount of a substance that has a weight in grams numerically equal to the molecular weight of the substance. Also called gram-molecular weight.

Molecular biology—A field of biology concerned with the interaction of biochemistry and genetics in the life of an organism. Molecular techniques in use include DNA-DNA hybridization, DNA fingerprinting, PCR (polymerase chain reaction), RFLP (restriction fragment length polymorphisms), RAPD (random amplified polymorphic DNA), and DNA sequencing. Most of these techniques are based on DNA that first must be extracted.

Molecular radius (M_r)—The radius of the space occupied by a molecule; considered a more correct term for macromolecules than molecular weight, as it allows for hydration.

Molecular weight (MW)—The weight of a molecule expressed as the sum of the atomic weights of its constituent atoms. *See* Dalton.

Molecule—A unit of matter; the smallest portion of an element or a compound composed of one or more atoms that retains the chemical identity with the substance in mass. A molecule usually consists of the union of two or more atoms; some organic molecules contain hundreds to a million or more atoms.

Mollicute—One of a group of prokaryotic microorganisms lacking a cell wall and bounded by flexuous membranes. *See* Mycoplasma.

Mollisioid—Saucer-shaped, as in the genus *Mollisia*.

Molt, moult—The shedding or casting off of a cuticle or body encasement during a phase of growth.

Mon-, mono- (prefix)—One.

Monandrous (of oospores)—Formed when only one functioning antheridium is present. *See* Polyandrous.

Monaxial—Having one stem or axis.

Monera—The kingdom that includes bacteria and cyanobacteria.

Moniliaceous (of spores, mycelium, etc.)—Hyaline or brightly colored; mucedinaceous.

Moniliales—A form-order of fungi of the Coelomycetes traditionally divided into four form-families. Conidia are produced on unorganized hyaline conidiophores (Moniliaceae), dark conidia on dark hyphae (Dematiaceae), on synnemata

(Stilbellaceae), and on sporodochia (Tuberculariaceae). Now considered to include the blastomycetes and hyphomycetes.

Moniliasis—*See* Candidiasis.

Moniliform, monilioid—Having swellings at regular intervals like a string of beads; beaded; chainlike; (of cystidia) with regularly spread constrictions; (of hyphidium) with successive swellings and constrictions, like a string of beads.

Mono- (prefix)—One.

Monoascous—Containing but one ascus.

Monoblastic (of a conidiogenous cell)—Holoblastic and blowing out (producing a blastic conidium) at only one point.

Monocarpic (of *Exobasidium* infections)—Annual and circumscribed. *See* Polycarpic and Surculicolous.

Monocentric (of a chytrid thallus)—Having a single center of growth and development. *See* Polycentric and Reproductocentric

Monocephalic (adj. **monocephalous**)—One-headed.

Monochoid (of nematodes)—The characteristics of the genus *Monochus* (predatory).

Monoclinous—Having the gametangium on the oogonial stalk. *See* Androgynous.

Monoclonal antibody—A homogeneous antibody population showing specificity to a single antigenic epitope. Produced by fusing specific antibody-producing cells (in a single clone of lymphocytes) with immortalized tissue culture cells. They are synthesized and secreted by clonal populations of hybrid cells (hybridoma) prepared by the fusion of individual B lymphocyte cells from an immunized animal (usually a mouse or rat) with individual cells from a lymphocyte tumor (e.g., myeloma).

Monocotyledoneae, monocots (pl. **monocotyledonae**, adj. **monocotyledonous**)—The subclass of flowering plants, including the grasses and cereals, having one cotyledon (seed leaf) in the embryo; characterized by parallel-veined leaves and fibrous roots. *See* Dicotyledon.

Monoculture—Continuous use of land for planting a single crop, cultivar, or noncrop species.

Monocyclic—Having one cycle per growing season; no secondary infections.

Monocystic—Having one encystment stage, as in the Saprolegniales.

Monodelphic—A female nematode possessing one genital tube or ovary.

Monoecious aphid—An aphid that spends its entire life cycle on a single plant species. *See* Dioecious aphid.

Monoecism (adj. **monoecious**)—Having male and female sex structures (flowers, gametangia) on the same plant; a fungus (particularly a rust) that has all stages of its life cycle on a single species of plant (*autoecious*); also, the stamens and pistils in separate flowers, but borne on the same individual (as in maize or corn (*Zea*). *See* Dioecism, Heterothallism and Homothallism.

Monogenic—Containing or controlled by one gene. In nematodes, producing offspring of only one sex.

Monogenocentric (of Chytridiales)—Developing one reproductive structure at the center of gravity of the thallus; genocentric. *See* Reproductocentric.

Monohybrid cross—Cross between parents differing in a single character.

Monokaryon—An individual with one haploid nucleus per cell; haplont.

Monokaryotic—Having genetically identical haploid nuclei. *See* Dikaryotic.

Monomer—A single polypeptide chain. Basic components of Ig molecules are four monomers: two H chains and two L chains.

Monomitic (of hyphal structure in basidiocarps)—Having only thin-walled, branched generative hyphae.

Monomorphic—Having only one characteristic structure, shape, and size; not pleomorphic.

Monomycelial (of an isolate)—From a single spore or hyphal tip.

Mononematous (of conidiophores)—Composed of a single thread or filament; solitary or in tufts of loose fascicles. *See* Synnematous.

Monoorchic—A male nematode having one genital tube or testis.

Monophagy (adj. **monophagous**) (of Chytridiales)—Condition of having the thallus in one host cell; opposite of *polyphagy*, in which the thallus branches and invades more than one host cell; describing an insect (e.g., aphid) that feeds on a specific type of host plant.

Monophialidic (of a conidiogenous cell)—Phialide with only one opening through which phialospores (conidia) are produced.

Monophylesis, monophyletic group—Related taxa derived from a common ancestor.

Monophyllous (of foliose lichens)—Having a single leaflike thallus.

Monoplanetic, monoplanetism (of zoospores in oomycetes)—Having one motile phase without a resting period.

Monopodial (conidiophore)—Having the main axis continue to grow at the apex always in the direction of the previous growth; lateral branches produced behind the apex growing in a similar manner; branches given off one at a time, commonly in an alternate or spiral series.

Monoreproductocentric—*See* Monogenocentric and Reproductocentric.

Monosaccharide—A simple sugar such as five- and six-carbon sugars.

Monospermous, monosporic (adj. **Monosporous**)—One-spored.

Monospore—One spore maturing on a two-spored basidium normally bearing two dispores.

Monostichate, monostichous—In one line, row or series, usually vertical.

Monostratic—Asci or basidia in one layer.

Monotretic (of conidiogenous cells)—Producing tretoconidia by extrusion of the inner wall through a single channel. *See* Tretic.

Monotrichiate, monotrichous—Describing a microorganism (bacterium) having a single polar flagellum.

Monotype, monotypic—Having only one representative specimen, as a genus having only one species or holotype.

Monoverticillate (of a penicillus)—Composed of only phialides; having one whorl.

Monoxenic culture—A culture containing an organism growing in the presence of another organism.

MOPS—A biological buffer (3-(I-mospholino)propane sulphonic acid) with a pH range of 6.2–8.2.

Morel—An edible fruitbody (ascoma) of *Morchella*.

Moribund—Being in a dying state.

Moriform—Shaped like a mulberry (*Morus*) fruit.

Morph- (prefix)—Form.

Morphogenesis—Changes in form during the growth and differentiation of cells and tissues.

Morphological subunit—The structural subunit of a virus particle as seen under an electron microscope; these are often clusters of protein subunits (capsomeres), especially in isometric particles.

Morphology—Study or science of the form, structure, and development of organisms.

Morphometrics—Body measurements.

Morphotype—Group of morphologically differentiated individuals of an unknown species or one of no taxonomic significance.

Mosaic—Disease symptom characterized by a patchy mottling of the foliage or by variegated patterns of dark and light green to yellow that form a mosaic; caused by disarrangement or unequal development of the chlorophyll content; symptomatic of many viral infections.

Mosaic fungus—A network that resembles disorganized dermatophyte mycelium as seen in skin scales cleared in potassium hydroxide (KOH); or the extracellular deposit of a dermatophyte.

Moschate—Musky.

Motile—Exhibiting or capable of independent movement from place to place.

Mottle, mottling—Disease symptom comprising light and dark areas in an irregular pattern; often symptomatic of viral diseases. *See* Mosaic.

Mottled rot—A type of white pocket rot in which the pockets are very distinct, giving a splotched or spotty appearance to the decayed wood.

mRNA—Messenger ribonucleic acid.

Mucedinaceous—*See* Moniliaceous.

Mucedinous—White or pale in color and moldlike; mucedinoid.

Mucid—Musty, moldy, or slimy.

Mucilaginous—Viscous, slimy, sticky when wet.

Muck (soil)—Highly decomposed black soil in which the original plant parts are not recognizable; similar to peat soil, often having a lower percentage of organic materials.

Mucoid, mucous—Resembling mucus or slime. Colonies produced by highly encapsulated bacteria.

Mucormycosis—Disease of humans or animals caused by one of the Mucorales (e.g., *Absidia corymbifera*); sometimes also caused by members of the Entomophthorales. *See* Phycomycosis and Zygomycosis.

Mucose—Wet or yeasty-appearing.

Mucro (of nematodes)—A stiff or sharp point abruptly terminating an organ.

Mucron (of nematodes)—A small pointed projection occurring on the tail tip.

Mucronate—Ending abruptly in a sharp point (or small pointed projection at the apex); pointed.

Mucronate hyphopodium—*See* Hyphopodium.

Mucus—Slime; gelatinous matrix of nematodes.

Muerh—Edible cultivated *Auricularia* spp., especially *A. polytricha* in China and *A. auricula* in Japan.

Mulch—A protective layer of some nonliving substance such as straw, dry leaves, wood bark chips, compost, peat moss, plastic film, gravel, small stones, ground corncobs, cocoa or peanut shells, buckwheat hulls, sawdust, etc., spread over the soil surface. Often used to catch rainfall, prevent splashing or soil crusting, retard water loss, control weeds, keep the soil temperature down in summer, protect roots against cold, keep produce clean, or improve soil structure and fertility.

Mult-, multi- (prefix)—Many; much; a great number.

Multiallelic—Having a series of alleles (more than two) at the locus or loci for incompatibility in a population of heterothallic species. *See* Biallelic.

Multiaxial—Having several or more axes with the stipe divided into many branches.

Multicellular—Many celled.

Multicistronic messenger RNA—A mRNA that contains the coding sequences for two or more proteins. *See* Operon.

Multicomponent virus—A virus whose genome, which is needed for full infection, is divided into two or more parts, each part separately encapsidated. Two or more components are thus needed to initiate an infection. This is different from a multipartite genome, in which the components may be enclosed in a single article. *See* Genome.

Multifid—Having division into a number of parts or lobes.

Multiform—Of various shapes.

Multigenic—Containing or controlled by a number of genes.

Multigenic messenger RNA—*See* Multicistronic messenger RNA

Multiguttulate—Having many oil-like drops.

Multiline—A mixed population of cultivars with two or more genotypes.

Multilocular, multiloculate—Having several to many separate cells, locules, or cavities.

Multinucleate—Having more than one nucleus per cell; polynucleate.

Multipartite genome—A viral genome split between two or more nucleic acid molecules that may be encapsidated in the same particle or in separate particles, where they are termed *multicomponent.*

Multipartite virus—Multicomponent virus (which see).

Multipileate—Having many pilei.

Multiple clamp connections (of basidiomycetes)—Several, usually three to six, clamp connections at the same septum.

Multiple fruit—A cluster of matured fused ovaries produced by separate flowers; a kind of compound fruit. Example: pineapple (*Ananas comosus*).

Multiple infection—Invasion by more than one parasite.

Multiple-cycle disease—A disease caused by a pathogen with the capacity to be disseminated to other suscepts and have secondary infection cycles during the current growing season; having the potential for causing an epiphytotic; a polycyclic disease.

Multiplicity of infection (m.o.i.)—Ratio of the number of infectious virus particles added to a known number of cells in a culture.

Multiplicity reactivation—A form of reassortment of complementation between two related viruses that have been activated; the inactivation sites must be in different parts of the genomes of the two viruses.

Multiporous—Having many pores.

Multiramose—Many branched.

Multiseptate—Having a number of septa or partitions.

Multisporous—Having many spores.

Multivesicular bodies—Small vesicles limited by a membrane that in *Monilinia (Sclerotinia) fructigena* originate from an endoplasmic reticulum and are possibly related to extracellular enzyme secretions.

Multizonate—Having many zones.

Mummification—The drying down and shriveling of fruits (e.g., brown rot of stone fruits [caused by *Monilinia fructicola, M. fructigena, M. laxa,* and *M. seaveri*] and black rot of apple fruit and other plant parts [caused by *Botrytosphaeria obtusa*]). The "mummy" may hang on the tree or fall to the ground, where it survives the winter and is a source of infection for next year's crop.

Mummy—A dried, shriveled fruit; plant part or organ partly or completely replaced with fungal structures; desiccated plant part or organ.

Mural tooth (of nematodes)—A cutting organ attached to or derived from the pharyngeal wall but formed further back in the esophagus.

Muricate—Rough with short, hard outgrowths; (of cystidia) bristly with oxalate crystals; same as capitate-incrusted.

Muriculate—Delicately muricate.

Muriform (of spores)—Having both cross (transverse) and longitudinal septa. *See* Dictyospore.

Murine, murinous—Mouse-colored, mouse-gray.

Muscardine fungi—Pathogens of silkworms and other insects: green muscardine fungus, *Metarhizium anisopliae*, and yellow muscardine fungus, *Paecilomyces farinosus*.

Muscaridin and **muscarin(e)**—Toxic quaternary ammonium compounds from *Amanita muscaria*; muscarin also comes from *Inonotus patouillardii*.

Muscazone—An insecticidal toxin from *Amanita muscaria*. *See* Tricholomic acid.

Muscicole, muscicolous—Living on or among mosses or liverworts. *See* Bryicolous.

Muscimol—Same as pantherine.

Muscose—Mossy.

Mushroom (toadstool)—A fleshy, sometimes tough, umbrellalike basidiocarp (agaric or bolete) of certain Basidiomycotina, especially one with gills. The word properly applies to all agaric fruitbodies whether edible, poisonous, tough and unpalatable, or leathery. The common field mushroom is *Agaricus campestris*, and the cultivated mushroom is *A. brunnescens* (syn. *A. bisporus*). *See* Caesar's mushroom, St. George's mushroom, Horse mushroom, Oyster mushroom, Parasol mushroom, Pine mushroom, and Straw mushroom.

Mushroom sugar—Trehalose.

Musiform—Shaped like a banana (basidiospores in *Exobasidium*).

Mutagen—A substance that causes an increased mutation rate.

Mutagenesis—The production of mutations, often caused by chemical and physical agents.

Mutagenic—Pertaining to, or causing, mutagenesis.

Mutation (n. **mutant**)—An abrupt heritable or genetic change in a gene or an individual as a result of an alteration in a gene or chromosome, or of an increase in chromosome number; also, an individual species, or the like, resulting from such a departure and the process by which such changes occur. *See* Strain, Complementation, Recombinant, and Pseudorecombinant.

Muticate, muticous—Pointless, blunt, without appendages.

Mutualism—A form of symbiosis in which two or more organisms of different species are living together for the benefit of both or all; a persistent and intimate relationship between two species of organisms in which both benefit from the association (e.g., alga and lichen).

Mutualistic—A term applied to a mutually beneficial relationship between organisms.

Mutually exclusive—Condition in which two or more parasites or pathogens cannot exist together in the same host cell or tissue.

MW—Molecular weight.

Myc-, mycet-, myceto-, myco- (prefix)—Pertaining to fungi.

Mycangium (pl. **mycangia**)—Sac- or cup-shaped fungal repository of ectodermal origin in or on an ambrosia beetle.

Mycelia sterilia—Members of the Deuteromycotina (Fungi Imperfecti), in which spores, except for chlamydospores, are not present; Agonomycetales.

Mycelial—Referring to mycelium.

Mycelial cord or strand—Discrete filamentous aggregation of hyphae that, unlike a rhizomorph, has no apical meristem; syrrotia..

Mycelial fan—A mass of mycelium, usually between bark and wood, with feathery branching strands.

Mycelial felt—A mass of mycelium that fills shrinkage cracks in decayed wood, usually becoming compactly interwoven to form tough, leathery, feltlike sheets.

Mycelioid—Resembling mycelium.

Mycelium (pl. **mycelia**)—The strands, group, or mass of interwoven, tubular hyphae making up the vegetative body (thallus) of a true fungus; "spawn." The mycelia of fungi show great variation in appearance and structure; (in bacteriology) long filaments of segmented cells.

Myceloconidium—Same as stylospore.

-mycetes (suffix)—Recommended ending for names of classes of fungi.

-mycetidae (suffix)—Recommended ending for names of subclasses of fungi.

Mycetocyte—*See* Mycetosome.

Mycetoma (madura foot, maduramycosis)—A mostly tropical disease of the foot or other part of humans that results in tumefactions characterized by mycotic granules or "grains" in infected tissues. Many different fungi (*eumycetoma*) and actinomycetes (*actinomycetoma*) are involved.

Mycetophagy (adj. **mycetophagous**)—Use of fungi as food; mycophagy.

Mycetophilous—Fond of fungi or growing on fungi.

Mycetosome—Saclike structure in the gut of Anobiid beetles lined with cells (*mycetocytes*) containing yeast cells.

Mycid—A secondary effect (eczema, urticaria, etc.) that is an allergic reaction to the spores or toxin of a dermatophyte; dermatophytid (*see* Mycosis); a mycid may be a trichophytid caused by *Trichophyton*; microsporid (*Microsporum*); or epidermophytid (*Epidermophyton*).

-mycin (suffix)—Recommended ending for names of antibiotics derived from actinomycetes.

Myco- (prefix)—Pertaining to fungi.

Mycobiont—The fungal symbiont in a lichen or mycorrhiza. *See* Phycobiont, Photobiont.

Mycobiota—Total fungal population inventory in an area under consideration; commonly used for the fungal mass present, as in a soil sample.

Mycocecidium—A gall caused by a fungus.

Mycocoenosis—Community of fungi in a particular habitat.

Mycoderm—Compact tissuelike ectotrophic mycorrhiza.

Mycodextran—Unbranched polysaccharide from *Aspergillus niger* and other fungi; nigeran.

Mycoecology—Ecology of fungi. *See* Ecology.

Mycogenous—Coming from or living on fungi.

Mycogeography—Study of the geographical distribution of fungi.

Mycohaemia, mycohemia—Condition in which fungi are present in the blood stream.

Mycoin—*See* Patulin.

Mycoliths—Masses of sand grains up to 5 or 6 cm long, joined together by mycelium especially of *Melanospora tulasnei* in soil under vines.

Mycologist—Person who studies and works with fungi.

Mycology—The science or study of fungi.

Mycolysis—The lysis of fungi, especially by the action of a mycophage.

Mycomyringitis—Fungal infection of the eardrum.

Mycomysticism—Mystical state induced by eating hallucinogenic fungi.

Mycoparasite (n. **mycoparasitism**)—The parasitism of one fungus by another; may be hyperparasitism.

Mycopathology—Study of disease caused by fungi.

Mycophage—Phagelike antibacterial substance produced by certain actinomycetes; a mycophagist.

Mycophagist—One who eats fungi.

Mycophagous (n. **mycophagy**)—Using fungi as food; mycetophagy; *or* the lysis of a fungus by a phage.

Mycophilic—Fond of fungi (or mushrooms); mycetophilous; *or* growing on fungi.

Mycophobia—Fear of mushrooms.

Mycophthorous (of a fungus)—Parasitic on another fungus; a mycoparasite.

Mycophycobiosis—An obligate symbiosis between a systemic marine fungus and a marine alga, in which the habit of the alga dominates.

Mycoplasma—A genus of prokaryotic, parasitic bacteria of the class Mollicutes, smaller than "walled" bacteria (measuring 0.1–1.0 μm in diameter), and much larger than viruses. They have an "elastic," confining, unit membrane but lack a rigid cell wall, are highly variable in shape (pleomorphic), contain both DNA and RNA, and reproduce by budding or fission; phytoplasma. Mycoplasmas induce viruslike symptoms in infected plants and are transmitted in nature by leafhoppers.

Mycoplasma viruses—Phagelike viruses isolated from mycoplasmas (prokaryotes without cell walls).

Mycoplasmalike organism (MLO)—Bacterium with apparent features of a mycoplasma without cell walls, but not proven to be a mycoplasma; found in the phloem and phloem parenchyma of diseased plants and assumed to be the cause of the disease.

Mycoplasmaphage(s)—*See* Mycoplasma viruses.

Mycoprotein—Fungal protein, e.g., commercially processed mycelium of nonpathogenic *Fusarium graminearum* A35 for human consumption.

Mycorrhiza (pl. **mycorrhizae** or **mycorrhizas**, adj. **mycorrhizal**)—A usually intimate, symbiotic association of the mycelium of a typically nonpathogenic or weakly pathogenic fungus with the roots of a higher plant; may aid in the uptake of certain nutrients by the plant host. Mycorrhizal fungi are numerous on mono- and dicotyledons, conifers, as well as some pteridiophytes and bryophytes. *See* Ectotroph, Ectotrophic, Ectendotrophic mycorrhiza, Ectotrophic mycorrhiza, Endotrophic mycorrhiza, Pseudomycorrhiza, and Vesicular-arbuscular mycorrhiza.

Mycorrhizome—A form of orchid rhizome that possesses endophytic mycorrhizae.

Mycose—*See* Trehalose.

Mycosin—A nitrogenous material like animal chitin in the cell wall of fungi.

Mycosis (pl. **mycoses**)—Fungal disease of humans, animals, or, rarely, plants (e.g., tracheomycosis). Mycoses are commonly named after the part attacked (e.g., *broncho-* [respiratory tract]; *dermato-* [skin]; *onycho-* [nails]; *oto-* [ear]; *pneumo-* [lungs]), etc., or the pathogen *blastomycosis* (*Blastomyces*), *coccidioidomycosis*, coccidioidal granuloma (*Coccidioides immitis*); also, the first limited infection of a dermatophyte. *See* Mycid.

Mycosociology—Study of fungal communities.

Mycostasis (adj. **mycostatic**)—Inhibition of fungal growth; fungistasis or fungistatic; sporostatis.

Mycostatin—Trade name for the antibiotic nystatin.

Mycosymbiont—Same as mycobiont.

Mycosymbiosis—Symbiosis of two or more fungi.

-mycota (suffix)—Ending of names of divisions of fungi.

Mycota—Same as fungi.

Mycotheca—Distributed set of dried fungal specimens.

Mycotic (especially of disease)—Caused by fungi.

-mycotina (suffix)—Ending of names of subdivisions of fungi.

Mycotope—Major fungal association of a particular type of woodland.

Mycotoxicosis—Fungus poisonings; now chiefly limited to poisoning of humans and animals by various food and feed products invaded by toxin-producing fungi and sometimes rendered carcinogenic. *See* Aflatoxins, Citreoviridin, Citrinin, Fumonisins, Islanditoxin, Luteoskyrin, Lysergic acid, Lupinosis, Maltoryzine, Ochratoxin A and B, Patulin, Roridins, Rubratoxin B, Satratoxins, Slaframine, Sporidesmin, Sterigmatocystin, Tremorgen, Trichothecenes, and Zearaleone. *See also* Phytoalexins.

Mycotoxin—Toxin produced by a fungus, especially one affecting humans and/or animals. Mycotoxins are found in seeds, feeds, or food and are capable of causing illnesses of varying severity and even death to animals that consume them.

Mycotroph—Fungus that obtains its nutrients from another fungus. *See* Mycoparasite.

Mycotrophein—"Growth factor" from fungi needed by a mycoparasite.

Mycotrophic—Describing green plants having mycorrhizae.

Mycovirus—A virus that replicates in fungal cells.

Myeloma—A tumor of an organ in the immune system.

Mylitta—A large sclerotium, e.g., that of blackfellows' bread. *See* Blackfellows' bread.

Myofilaments (of nematodes)—The thick and thin muscle elements of the contractile region of the muscle cell.

Myriocytous (of nematodes)—Having a very large number of cuboidal cells (over 8,224) in the intestinal tract.

Myriosporous—Many-spored. *See* Oligosporous.

Myrmecophilous (of fungi)—Being a covering or food for ants.

Mytiliform—Shaped like a mussel shell.

Myurous—Long and tapering like a mouse's tail.

Myx-, myxo- (prefix)—Mycus, slime.

Myxamoeba (pl. **myxamoebae**)—A zoospore (swarm cell) after becoming amoebalike; an amoeboid cell, particularly one of the myxomycetes.

Myxarioid (of basidia)—Having a stalklike portion separated by a wall from the globose metabasidial portion (as in *Myxarium*).

Myxomycetes—The true slime molds; a class of organisms characterized by amoeboid, vegetative protoplasts, multinucleate, coenocytic, saprobic plasmodia, and by brightly colored, spore-bearing capillitia.

Myxomyceticolous—Growing on myxomycetes.

Myxomycota—A division of fungi, to which the Myxomycetes and the Protosteliomycetes belong; the "primitive fungi."

Myxosporium—Same as perispore, perisporium; spore wall.

N

n.—Noun.

Nacreous—Like mother-of-pearl.

NAD—Nicotinamide-adenine-dinucleotide.

Naked bud—A bud not protected by bud scales.

Naked virus—A virus without a lipoprotein envelope.

Nameko—The edible *Pholiota nameco*, cultivated in Japan.

Nanism—Dwarfism.

Nanometer (nm)—A unit of length equal to one billionth (10^{-9}) of a meter (m) or one millimicron = 10 Å.

Napaceous, napiform—Turnip-shaped.

Narcotic—A drug that produces stupor or coma, often accompanied by hallucinations and other nervous abnormalities.

Nascent cleavage—Proteolytic cleavage of a polypeptide while it is being synthesized by ribosomes.

Nascent RNA—Oligoribonucleotides in the process of being transcribed from template DNA or RNA.

Nassace, nasse—Fingerlike projection of the inner part of a bitunicate ascus into the inner tunica; internal apical beak.

Nastic movements—Growth movements of flattened organs, such as leaves and petals, as a result of differences in growth or turgor of cells on the upper and lower surfaces of these organs.

Native bread—*See* Blackfellows' bread.

Natural (self) pruning—Natural abscission of branches and twigs of woody plants.

Natural immunity—Immunity based on qualities natural to the organism.

Natural openings in plants—Stomata, lenticels, hydathodes, and nectarthodes.

Natural preservatives—Preservative substances such as tannins and resins, which occur naturally in wood and bark and tend to reduce attacks by fungi and insects.

Natural selection—The process in nature by which the best-fitted types of organisms survive and produce offspring, the less competitive being eliminated; an important feature of Darwin's theory of the mechanisms of evolution.

Naturalized plant—A plant introduced from one environment into another in which the plant has become established and more or less adapted to a given region by growing there for many generations.

Navel—*See* Umbilicus.

Navicular, naviculate—Boat-shaped; cymbiform; with a sharply tapered end and a broad end.

NCPPB—Abbreviation for Natural Collection of Plant Pathogenic Bacteria maintained in the Ministry of Agriculture, Fisheries and Food in the United Kingdom.

Nearest neighbor sequence analysis—A technique for studying the relationship between nucleic acids. A determination is made of the relative frequencies with which pairs of the four bases occur in adjacent positions and this can distinguish between nucleic acids with similar or identical base compositions.

Nebulose, nebulous—Cloudy; dark.

Neck—(Of plants) peduncle; culm area just below the head and above the uppermost leaf of a grass plant; (of pteridophytes and bryophytes) the tapering portion of an archegonium; a sperm enters an archegonium through a canal extending lengthwise through the neck; (of nematodes) the anterior portion of the body occupied by the esophagus (neck region); (of bacterial viruses) the region in tailed phage particles that links the head to the tail.

Necral layer—Layer of horny, dead fungal hyphae (above or below the algal layer) with indistinct lumina in or near the cortex of lichens.

Necro- (prefix)—Dead.

Necrogenic—Generating necrosis (cell death).

Necrophyte (adj. **necrophagous**)—An organism living on dead material; saprobic. *See* Saprophyte.

Necrosis (pl. **necroses**, adj. **necrotic**)—Localized or general death and disintegration of plant cells or plant parts, usually resulting in tissue turning brown or black due to oxidation of phenolics; commonly a symptom of fungus, nematode, virus, or bacterial infection; a symptom of disease or injury.

Necrotroph—A parasite that typically kills and obtains its energy from dead host cells. *See* Biotroph.

Nectar—The sticky, often sweet, secretion, (especially of spermogonia [pycnia] of rust fungi) in which spores or spermatia may be freed, and that has an attraction for insects; a sweet fluid secreted by the nectaries of plants and the chief raw material of honey.

Nectarthode—An opening at the base of a flower from which nectar exudes.

Nectary (pl. **nectaries**)—A floral, nectar-secreting gland; (of nematodes) plant cells near the heads of sedentary individuals that have been stimulated to enlarge by salivary secretions and from which the nematodes derive their sustenance (nurse cells or giant cells).

Needle blight (of conifers)—Fungal attack on evergreen foliage of any age and therefore active anytime there is coincidence of high relative humidity or free moisture and spores. Blighted foliage is also cast after spore release. *See* Needle cast.

Needle cast (of conifers)—Fungal attack resulting in a large, premature drop of needles; generally caused by fungal species of Rhytismatales (syn. Phacidiales). Only young needles of the current year are generally attacked.

Negative contrast staining—Staining procedure used to prepare virus particles for examination in an electron microscope. *See* Metal shadowing.

Negative-sense strand (negative strand)—Nucleic acid complementary to the plus strand.

Negative-strand virus—Virus whose genome is negative-sense RNA. Five families of negative-strand viruses have been recognized.

Nematicide—A chemical or physical agent that kills or inhibits nematodes.

Nematode, nema—Eelworm; an unsegmented, wormlike animal (phylum Nematoda), parasitic in or on plants and animals, or free-living in soil, decaying matter, or water; a generally microscopic, tubular roundworm with a cuticle, a hydrostatic skeleton, abundant in many soils. Practically all plant-parasitic nematodes pierce plant cells with a stylet and suck juices. Nematodes play an important role in providing wounds by which bacteria and fungi may enter, as well as transmitting microorganisms and viruses, into plants.

Nematogenous (of conidiogenous cells)—Cells that arise at all different levels from single hyphae; lanose.

Nematology—The science or the study of nematodes.

Nematophagous—Feeding on nematodes, as in nematophagous fungi or nematodes.

Nematotoxin—A metabolic product toxic to nematodes.

Nemeous—Filamentous; threadlike.

Nemin—A principle from a nematode-free filtrate of *Neoplectana glaseri*, that causes *Arthrobotrys conoides* to differentiate traps.

Nemoral—Inhabiting woods or groves.

Nemorose, nemorous—Woody; shady.

Neomycin—An antibiotic produced by *Streptomyces fradiae*.

Neoplasm—An aberrant new growth of abnormal cells or tissues. *See* Tumor.

Neosexual—Organisms having a sexual but not a protosexual cycle. *See* Protosexual.

Neotope—An antigenic site that arises only after two or more molecules interact with one another.

Neotype—A specimen (strain, culture) selected as the nomenclatural type of a taxon when all the original material is lost or destroyed. *See* Type, Holotype, and Lectotype.

Nephroid—Kidney-shaped; reniform.

Nepovirus (NEPO)—Member of the tobacco ringspot virus group; usually transmitted by nematodes; *NE*matode-transmitted *PO*lyhedral-shaped viruses.

Nerve ring (of nematodes)—The circumesophageal commissure, or the center of the nervous system of nemas, encircling the esophagus; composed largely of nerve fibers and associated ganglia.

Nervicole, nervicolous—Growing on veins of leaves or stems.

Nervisequent—Following the veins.

Net necrosis—Irregular, anastomosing, necrotic lines giving a netted appearance, e.g., potato leaf roll virus in a potato tuber.

Net venation—Vein arrangements in leaves in which the veins branch frequently, forming a network.

Netted—Same as reticulate.

NETU—A coined term referring to *NE*matode-transmitted, *TU*bular-shaped (rod-shaped) viruses.

Neutral soil—A soil in which the surface layer is neither acid nor alkaline in reaction. *See* pH.

Neutralization—Inactivation of infectious virus by reaction with its specific antibody, thereby blocking sites on the virus that normally adsorb to susceptible cells.

Neutralizing antibody—An antibody that inhibits virus infectivity.

Nidose, nidorose—Having a foul smell.

Nidulant—Lying free in a cavity.

Nidulate—Nested; nestling; as if borne in a nest.

Nietsuki—A product of a failure in the drying of the basidiocarps of *Lentinula edodes*.

Nig-, nigri-, nigro- (prefix)—Black.

Nigeran—*See* Mycodextran.

Night-break lighting—Low light intensity from incandescent or fluorescent lamps, used at night to change the photoperiod.

Nigrescent, nigricant—Turning blackish.

Nigrolimitate—Outlined in black; black-lined.

Nigropile—Covered with black hairs.

Nigropunctate—Covered with black dots.

Nigrostrigose—Covered with black bristles.

Nimbospore—Spore with a gelatinous, apparently many-layered wall.

Nitid, nitidous—Smooth and clear; lustrous.

Nitrate reduction—The reduction of nitrates to nitrites or ammonia.

Nitrification—The conversion of ammonia and ammonium compounds into nitrites and nitrates through bacterial action in soils.

Nitrocellulose—A nitrated derivative of cellulose used as a powder or made into membrane filters of defined porosity. Used to bind nucleic acids in Northern and Southern blotting procedures and proteins in Western blotting.

Nitrogen activity index (AI)—An index applied to ureaformaldehyde compounds and mixtures containing such compounds. The AI is the percentage of cold-water-insoluble nitrogen (CWIN) that is soluble in hot water (i.e., is not hot-water-insoluble nitrogen [HWIN]). AI = (% CWIN − % HWIN) divided by % CWIN × 100.

Nitrogen assimilation—The incorporation of nitrogen into organic cell substances by living organisms.

Nitrogen fixation—The conversion of atmospheric, gaseous nitrogen (N_2) into oxidized nitrogen compounds in soils or plant roots by certain soil bacteria and blue-green algae characteristic of *Rhizobium*, the genus of root nodule bacteria, and of *Azotobacter* spp.

Nitrogenous—Relating to or containing nitrogen.

Nitrogenous bases—Nitrogen-containing compounds found in DNA and RNA that, in sequence, specify precise genetic information. The nitrogenous bases in DNA are adenine, thymine, cytosine, and guanine. In RNA, they are adenine, uracil, cytosine, and guanine. *See* DNA and RNA.

Nitrophilous—Preferring habitats rich in nitrogen; chionophilous.

Nitrophobous—Preferring habitats poor in nitrogen.

Niveous—Snow-white.

nm—Nanometer = 10^{-9} meter; 1 mµm.

Nocturnal—Active or occurring in the night.

Node—Enlarged joint on a stem that is usually solid; site from which a leafy bud and branch arises.

Node cell—A stigmatocyst in a hypha. *See* Hyphopodium.

Nodose—Having many or large nodelike swellings or joints.

Nodose septum—*See* Clamp connection.

Nodular bodies (of dermatophytes)—Rounded bodies composed of massed hyphae.

Nodulation (adj. **noduled**)—The process of forming nodules, especially root nodules containing symbiotic bacteria.

Nodule—A lump, knot, or tubercle; often on the roots of certain plants; an enlargement within which nitrogen-fixing bacteria live.

Nodulose, nodulous—Having very small rounded knobs, knots, or nodules; (of spores) having broad-based, blunt, wartlike excrescences.

Nodum (pl. **noda**)—Used in phytosociology for particular well-defined plant communities.

Noisiness—Errors that may occur during the replication of a virus's genome.

Nomen (Latin)—Name.

Nomen ambiguum—A name having different senses.

Nomen confusum—A name of a taxonomic group based on two or more different entities, e.g., an impure culture.

Nomen conservandum—Name made valid by a decision of the International Botanical Congress.

Nomen conservandum propositium—Name put up for conservation.

Nomen dubium—A name of uncertain sense or application.

Nomen hybridum—Name formed by combining words from different languages.

Nomen monstrositatis—A name based on an abnormality.

Nomen novum—A new name; a replacement.

Nomen nudum—A name for a taxon lacking an adequate diagnosis or description and hence not validly published.

Nomen oblitum—A forgotten name; one not used as a senior synonym in the past 50 years.

Nomen provisorium—Provisional name.

Nomen rejiciendum—A rejected name.

Nomen species (n. sp.)—Named species; taxonomic category; (of bacteria) a type species.

Nomenclature—The scheme (believed to be a system) by which names are attached to objects, including microorganisms. The naming of organisms is governed by the International Code of Botanical Nomenclature. The Code is comprised of principles, rules (termed articles), and recommendations.

Nonacid-fast—Bacterial cells decolorized by 25% sulfuric acid after staining with carbol fuchsin.

Nonaggressive strain—Strain of fungus limited in pathogenicity.

Nonamyloid (of spore coverings)—Remaining hyaline or turning yellowish in Melzer's reagent. *See* Amyloid and Dextrinoid.

Nonfissitunicate (of asci)—Asci in which discharge does not involve separation of the wall layers.

Noninfectious disease—A disease (or disorder) that is caused by an abiotic agent and not by a pathogen; cannot be transmitted from a diseased plant to a healthy plant. *See* Physiogenic (physiological) disease.

Nonionic—Lacking a positive or a negative charge; electrically neutral.

Nonpathogenic—Incapable of causing disease.

Nonpermissive cells—Cells in which a specific virus will not infect and replicate.

Nonpersistent transmission (syn. **stylet-borne transmission**)—Type of insect transmission in which the virus is acquired by the vector after very short acquisition feeding times and is transmitted during very short inoculation feeding periods. The vector (e.g., aphid) remains viruliferous for only a short period (e.g., a few minutes to perhaps 4 h) unless it again feeds on an infected plant. *See* Persistent and Semipersistent transmission.

Nonpreference—*See* Vector preference.

Nonproducer cells—Cells carrying all or part of a viral genome but not producing virus particles; they are usually transformed by the virus.

Nonsense codon—*See* Stop codon.

Nonseptate—Without cross walls.

Nonstructural protein—Protein encoded by a viral genome but not involved in the structure of the virus particle. The protein is usually functional during replication.

Nontarget organism—A plant or animal other than the one against which a pesticide is applied.

Normal length (of virus particles)—The most common length, believed to be the intrinsic natural length of a virus that consists of rod-shaped particles.

Normal saline solution—Sodium chloride (8.5 g) and water (1,000 ml).

Northern blotting—A procedure analogous to Southern blotting (which see) but involving the transfer of RNA to nitrocellulose or activated paper sheets.

Notate (of surfaces)—Marked by straight or curved lines.

Notifiable disease—A plant disease about which agricultural authorities in a given country must be notified.

No-till (stubble culture)—A cultural system most often used with annual crops, in which the new crop is seeded or planted directly in a field on which the preceding crop plants were cut down, had the tops harvested, or were destroyed by a nonselective herbicide. The old crop is not removed or incorporated into the soil as is common in preparing a plant bed.

Novobiocin—An antibiotic that inhibits DNA synthesis.

Nozzle—A device for metering and dispersing a spray solution.

Nubilate, nubilous, nubilated—Cloudy and/or semiopaque when viewed by transmitted light.

Nucellus—Tissue (megasporangium) found inside the integuments of the young seed (ovule), enclosing the megagametophyte and in which the embryo sac develops; in some species it is absorbed as the seed matures.

Nuciform—Nut-shaped.

Nuclear cap (of Blastocladiaceae)—Body at one side of the nucleus of a zoospore or gamete.

Nuclear magnetic resonance (NMR) spectroscopy—The measurement of energy absorption by magnetic atomic nuclei placed in a strong magnetic field.

Nuclear membrane—A double membrane that surrounds the nuclear contents of an eukaryotic cell.

Nuclear sap—Liquid present within a living nucleus.

Nuclease—An enzyme that can hydrolyze the internucleotide linkages in a nucleic acid.

Nucleic acid—A compound of high molecular weight that consists of pentose (ribose or deoxyribose), phosphate, and nitrogen bases (purines and pyrimidines) joined in a long chain of repeating units (nucleotide complex). Nucleic acids are present in all living cells. The infectious parts of plant viruses consist of nucleic acids. All known nucleic acids fall into two classes, DNA and RNA.

Nucleocapsid—The nucleic acid (RNA or DNA) of a virus particle or virion enclosed by a protein capsid.

Nucleoid—(Of bacteria) structure containing nuclear material (DNA) that, during multiplication, divides and passes to the daughter cells; (of viruses) electron-dense region.

Nucleolus (pl. **nucleoli**)—A small, dense, usually spherical protoplasmic body within the nucleus of most kinds of cells.

Nucleoprotein—Viruses or another complex consisting of nucleic acid and protein.

Nucleoside—The combination of a sugar (ribose or deoxyribose) joined to a base molecule. *See* Nucleic acid.

Nucleosome—Structures found in large DNA genomes and chromosomes that comprise DNA and histones (chromatin). Nucleosomes alternate with protein-free stretches of nucleic acid. Beadlike arrangement of nucleocapsid.

Nucleotide(s)—The building blocks of DNA and RNA consisting of a phosphate group joined to a five-carbon sugar (ribose or deoxyribose) that, in turn, is joined to a base; a phosphate ester of a nucleoside. Thousands of nucleotides are linked to form a DNA or RNA molecule.

Nucleotide phosphohydrolase—An enzyme that removes a phosphate group from the triphosphate end of nucleotides.

Nucleotide sequence—*See* Sequence.

Nucleus (pl. **nuclei**)—The dense, usually spherical or ovoid, protoplasmic body present in most living cells of plants and animals and essential in all synthetic, developmental, and reproductive activities of a cell; contains the chromosomes that transmit hereditary characteristics.

Nudicaulous—Naked-stemmed.

Nummiform—Corn-shaped.

Nurse cells (in *Scleroderma*)—Hyphae supplying nutrients to spores after becoming detached from the basidia; (of nematodes) normal-sized host parenchyma cells with thickened walls, dense cytoplasm, an enlarged nucleus, and no central vacuole, formed near the head of certain nematodes (e.g., *Tylenchulus* spp.) that require these cells for further development. *See* Giant cell and Nectaries.

Nut—A dry, indehiscent, one-seeded fruit with a hard, woody pericarp (shell), generally one produced from a compound ovary, such as walnut and pecan (*Carya*).

Nutant—Nodding; drooping.

Nutlet—Seedlike fruit segment of the mint and related families. The fruit splits longitudinally into four sections; each section, shaped like a quarter section of an apple, is a nutlet.

Nutrients, plant—Essential elements available to plants through soil, air, and water that are utilized in metabolism and growth.

Nutrilite—Any organic compound needed in small amounts for the nutrition of an organism.

Nutriocyte (of *Ascosphaera*)—Inflated part of the ascogonium that later develops into a spore cyst.

Nymph—Juvenile insect (e.g., aphid, grasshopper) that superficially resembles the adult.

Nystatin (Mycostatin)—An antifungal antibiotic from *Streptomyces noursei* widely used against *Candida albicans* infections of humans.

O

"o"—Distance of dorsal esophageal gland opening in a nematode from the stylet knobs × 100.

"O", "ovary"—Percent of the nematode body occupied by each gonad of the female reproductive system, i.e., length of the gland divided by the length of the nematode × 100; value usually superscripted left or right of "V" or both, depending on gonadal arrangement.

O antigens—Antigenic complex of the body (or soma) of a bacterium, that may be masked by more labile antigens overlying the soma. O antigens are heat stabile and resist 100°C for several hours.

Oak-moss, oak moss—The lichen *Evernia prunastri*, a source of perfume.

Ob- (prefix)—Inversely, oppositely, or reversely.

Obclavate—Shaped like a club upside down; inversely clavate; thickened toward the base.

Obconic, obconical—Like an inverted cone, having a very narrow base and a broad apex.

Obcordate, obcordiform—Reversely heart-shaped.

Obcrenate—Same as denticulate (which see).

Obcuneate—Wedge-shaped with the thin edge at the base.

Obcurrent—Running together and adhering at the point of contact.

Obdeltoid—Triangular; with the apex inward or downward; reversely deltoid.

Objective—The system of lenses in a compound microscope nearest the object being viewed.

Oblanceolate—Relatively narrow but broadening from the base to the tip; reversely lanceolate.

Oblate—Flattened at the poles, like an orange.

Obligate—Necessary, essential, obliged.

Obligate anaerobe—Organism that grows only in an environment that has no oxygen.

Obligate parasite—A parasite that can grow and multiply in nature only on or in living tissue and that cannot be cultured on an artificial medium. *See* Parasite and Facultative parasite.

Obligate saprophyte or saprobe—An organism that can develop only on dead organic matter.

Obliterate—Indistinct; inconspicuous; almost effaced.

Oblong (of spores)—Short, cylindric, and less than twice as long as broad with somewhat truncate ends.

Oblong-ellipsoid, oblong-elliptical (of spores)—Rounded-oblong; having long, nearly parallel sides and ends almost hemispherical.

Obovate, ovovoid (of spores, etc.)—Reversely ovate (ovoid), narrowest at the base; egg-shaped in two dimensions.

Obpyriform—Reversely pear-shaped with the broad end near the base.

Obrotund—Somewhat round.

Obrute—Covered, buried, overwhelmed.

Obsolete—(Of organs or parts) rudimentary or absent; (of terms) no longer in use.

Obsubulate—Very narrow; pointed at the base and a little wider toward the tip.

Obtrite—Broken; crushed; rubbed.

Obturbinate—Inversely top-shaped.

Obtuse—Rounded or blunt; greater than a right angle.

Obvallate—Surrounded; walled in on all sides.

Obvolute—Wrapped or rolled up; with the margins of one structure overlapping those of another.

Obvolvent—Enveloping.

Occluded—Closed; describing a virus producing an occlusion body; often describes the lumina of hyphae or pseudoparenchymatous cells.

Occlusion—Block or plug that stops flow of liquids (as in vessels).

Occlusion body (of virology)—Large virus-coded protein crystal containing virions, preferred to the more general term "inclusion body."

Occultate—Hidden.

Ocellate, oculate—Having rounded spots, dots or circular patches; like little eyes.

Ocellus (pl. **ocelli**)—An eyelike spot (pigment spot) of nematodes; an eyespot functioning as a lens and concentrating light rays on a sensitive spot.

Ochraceous, ocherous—A yellowish buff, cinnamon-buff to brownish yellow color; ochre-yellowish.

Ochratoxin A,B—Toxins produced by *Aspergillus ochraceus* and *Penicillium viridicatum*, the cause of nephrotoxicosis in sheep, cattle, and pigs; toxins that are carcinogenic and have been found in coffee.

Ochre codon—*See* Stop codon.

Ochroalba—A combination of ochre color and white.

Ochrosporous—Having yellow or yellow-brown spores.

Oct-, octo- (prefix)—Eight.

Octad—Group of eight cells arising from a division of a single original cell; a typical ascus.

Octonate—In eights.

Octophore—*See* Haerangium.

Octopolar (of incompatibility systems)—Having three loci, as in *Psathyrella coprobia*. *See* Tetrapolar.

Octoseptate—Having eight cross walls.

Octospore—One spore of an eight-spored ascus.

Octosporous—Having spores in eights.

Ocular micrometer disc—A ruled glass plate used in the eyepiece of a microscope for measuring minute objects.

Odium (pl. **odia**)—An asexual spore formed by the fragmentation of vegetative hyphae into short cylindric segments.

Odontium (pl. **odontia**)—A labial tooth (teeth) of nematodes situated on the esophageal wall but formed in the esophagus and moved forward during the molt.

Odontoid—Dentate, toothlike; hymenial surface with short, often compound teeth.

Odontostyle, odontostylet (of nematodes)—*See* Stylet.

Odorate—Scented.

Odor—*See* Smell.

Oedema—An abnormal condition, usually on leaves, characterized by small areas of watersoaking usually later accompanied by slight swelling (intumescence or blister formation) of the spots. *See* Edema.

Oedocephaloid—Having an enlarged or swollen head or tip, as the conidiophores of *Cunninghamella, Oedocephalum*, certain hymenomycetes, etc.

-oid (suffix)—Like; having the shape of.

-oideae (suffix)—Added to the stem of a legitimate name of an included genus in botany and the type genus in bacteriology to form the name of a subfamily.

Oidiomycin—An antigen prepared from *Candida albicans*, especially for skin testing.

Oidiophore—A structure, usually a hypha or portion of one, that produces oidia.

Oidium, oidiospore (pl. **oidia**)—Used in varying, often imprecise or incorrect ways; can mean 1) spermatia formed on hyphal branches, especially in heterothallic hymenomycetes; 2) flat-ended asexual spores formed by the breaking up of a hypha into cells; arthrospore; 3) a powdery mildew in which the sexual state is unknown (teleomorphs are Erysiphales).

Oidization—Dikaryotization by the fusion of an oidium with a haploid hypha.

Oil—A fatty substance that is liquid at room temperature.

Okazaki fragments—Short pieces of DNA, with attached RNA primers, produced during DNA synthesis. The primers are later replaced by DNA, and the fragments are joined together.

Old man's beard—*See* Beard moss.

Oleaginous—Oily.

Oleiferous hypha—Hypha that does not carry latex but commonly has resinous substances.

Oleocystidium—A cystidium with an oily, resinous exudate.

Oleoresin—Resin, pitch; a viscous, aromatic mixture of terpenes, resin acids, and fatty acids produced by various conifers.

Oleoso-locular—Fungal spores having cells like drops of oil.

Oligocytous (of nematodes)—Having a small number of rectangular cells (up to 128) in the intestinal epithelium.

Oligogenic—Applied to a character determined by only a few genes. *See* Polygenic.

Oligonucleotide—A short-chain nucleic acid molecule.

Oligonucleotide probe—A short DNA probe whose hybridization is sensitive to a single base-mismatch.

Oligopeptide—A short chain of amino acids.

Oligosporous—Having few spores. *See* Myriosporous.

Oligotrophic—Poor in nutrients. *See* Eutrophic.

Olivaceous—The color of a green olive (yellowish green).

Olivaeform—Olive-shaped.

-olus, -ole (suffix)—Used to form diminutives.

Omnivorous (of parasites)—Attacking many different hosts; feeding on substances of both animal and vegetable origin.

Omphalodisc—An orbicular, conical-shaped disc; an apothecium with a central knob of sterile hyphae.

Onchiostyle (**odontostyle**) **of nematodes**—*See* Stylet.

Onchium (pl. **onchia**)—Pharyngeal tooth (teeth) of nematodes that is formed "in place" in the stoma, as compared to an odontium developed in the esophagus and then moved forward.

Oncogenic—Generating tumors (galls).

Oncom morah, oncom hitah—Japanese fermented soybean (*Glycine max*) products in which the principal fungi are *Neurospora sitophila* (anamorph *Chrysonilia sitophila*) and *Rhizoctonia oligosporous*, respectively.

One-step growth curve—A single cycle of replication of a virus in a synchronously infected cell culture.

Ontogenesis, ontogenetic—Pertaining to the origin and development of an organism.

Ontogeny—Life history or development of an individual, as opposed to that of the race (*phylogeny*).

Onychomycosis—*See* Mycosis.

Oo- (prefix)—Egg.

Oocyte (of nematodes)—A cell giving rise to an egg.

Oogamy—Heterogamy in which the gametes are a large, nonmotile egg and a small, motile sperm.

Oogenesis—The development of the oogonium after being fertilized; the formation of the egg, its preparation for fertilization and development.

Oogonium, oogone, (pl. **oogonia**)—1) One-celled female sex organ (gametangium) of oomycetes, containing one or more eggs (gametes or oospheres); 2) (of nematodes) a cell derived from the cap cell giving rise to an oocyte.

Oomycota (**oomycetes, peronosporomycotina, peronosporomycetes**), **Chromista**—A class of the Mastigomycotina. Typically aquatic, saprobic, or parasitic fungi that produce oogonia, antheridia, and oospores; (of nematodes) the first stage in the differentiation of an egg cell from a primordial germ cell.

Ooplasm (of Peronosporales)—Protoplasm at the center of the oogonium that becomes the oosphere. *See* Gonoplasm and Periplasm.

Ooplast (of Saprolegniaceae)—A large cellular inclusion in the oospore.

Oosphere (of oomycetes)—A large, naked, nonmotile female gamete; an unfertilized ovum, a female reproductive cell that, after fertilization, develops into the oospore. A *compound oosphere* has many functional nuclei.

Oospore—Thick-walled, resting spore in the oomycetes that develops from a fertilized oosphere or by parthenogenesis.

Ooze—Viscous fluid exuding from diseased plants; droplets or strands composed of bacteria or fungus spores mixed with host fluids; found on the surface of lesions or on the cut ends of diseased stems, fruit, or leaves.

Opal codon—*See* Stop codon.

Opalescent—Reflecting an iridescent light like an opal.

Opaline—Clear.

Opaque—Dull; not shining.

Open reading frame (ORF)—A set of codons for amino acids (usually encoding a protein) uninterrupted by stop codons.

Open-circular DNA—One of the three forms that double-stranded DNA can take. This circular DNA has one or both strands not covalently closed.

Operator—Region in the chromosome contiguous with an operon and controlling its functioning.

Operculate—Opening by an apical circular flap or lid of tissue for the discharge of spores, as in some asci, sporangia, and conidiomata; produced by circumscissile rupture of overlying tissues. *See* Inoperculate.

Operculiform—Lid-shaped.

Operculum (pl. **opercula**)—1) A hinged cover or lid on a sporangium or an ascus; 2) the lid covering a moss capsule; 3) (of nematodes) less sclerotized areas of the poles of hard-shelled eggs of some animal parasites allowing the escape of the larva.

Operon—A cluster (two or more) of functionally related genes in the chromosome, whose function is subject to a common control mechanism. The genes are regulated and transcribed as a unit.

Ophiobolin (cochliobolin)—An antifungal and antibacterial antibiotic from *Cochliobolus miyabeanus*, anamorph *Bipolaris oryzae*, and *C. heterostrophus*, anamorph *B. maydis*, anti-*Trichomonas vaginalis*; phytotoxic to rice.

Opine—Condensation product of an amino acid and a keto acid.

Opisthodelphic (of nematodes)—Describing a female having the uterus directed posteriad.

Opisthokont, opisthokontus—Having one or more flagella or cilia at the posterior end.

Opplete—Filled.

Opportunistic pathogen—A pathogen that is naturally saprobic and often common but on occasion able to cause disease in a host plant, human, or higher animal, rendered susceptible by one or more predisposing factors.

Opposite—Bearing two leaves or two buds at a node on opposite sides of a stem.

Opsonin—A substance in the blood serum that renders microorganisms susceptible to ingestion by phagocytes.

Optical density—*See* Absorbance.

Optical diffraction—The bending of light waves around the edges of an obstacle; used in electron micrographs to determine regular structures, as the bending of light is inversely proportional to the repeat distance in a regular structure.

Optimal proportions—In precipitin tests, if antibody and antigen are present in such a ratio that the maximum combination between antibody combining sites and antigenic determinants can occur, they are said to be present in optimal proportions or "at equivalence." At this ratio, maximum precipitation occurs.

Optimum—Most favorable, e.g., optimum temperature is that most favorable for growth and reproduction of a culture, spore germination, invasion of a host, etc.

Oral capsule or cavity—The stoma or buccal capsule.

Oral or buccal aperture, opening, or orifice (of nematodes)—The anterior entrance into the stoma.

Oral toxicity—The relative toxicity of a chemical when introduced through the mouth. *See* LD_{50}.

Orbicular, orbiculate—Spherical or ringlike; round

Orchinol and **hircinol**—Dihydrophenanthrenes produced by orchids as a response to infection by mycorrhizal fungi.

Orculiform—Cask-shaped. *See* Polarilocular.

Order—Taxonomic category ranking between a family and a class. An order is composed of related families.

ORF—*See* Open reading frame.

Organ—One of the major parts of a plant body—leaf, stem, root. An organ is composed of various cells and tissues and adapted to perform certain specific functions.

Organ culture—A form of tissue culture in which the organization of the tissue is maintained.

Organ or organ Z (of nematodes)—A muscular structure in certain species of *Xiphinema*, located between the oviduct and uterus.

Organelle—Membrane-delimited structure or body within a cell (e.g., Golgi apparatus, mitochondria) having a specialized function.

Organic compound—A chemical compound that contains carbon, as distinguished from a noncarbon containing, or inorganic, compound.

Organic matter—Any plant or animal material that is decomposed, partially decomposed, or undecomposed.

Organic phosphorus insecticide—A synthetic compound derived from phosphoric acid. These insecticides are primarily contact killers with relatively short-lived effects. They decompose in water, by pH extremes, high temperatures, and microorganisms. Examples are malathion, diazinon, chlorpyrifos, Aspon, phorate, isoferphos, trichlorfon, dimethoate, and fenthion.

Organic soil—A soil containing more than 20% organic matter.

Organism—A living plant, animal, or microbe.

Orifice—An opening in a nozzle tip, duster, or granular applicator through which spray, dust, or granules flow.

Ornamental plant—A plant grown for accent, attraction, beautification, color, pleasure, screening, specimen, and other aesthetic reasons.

Ornamented (of organs, spores, etc.)—Having surface fibrils, scales, striations, ridges, warts, or other appendages or variously sculptured; not smooth.

Ornate—Adorned.

Ornithocoprophilous—Preferring habitats rich in bird droppings.

Orthographic variant—A variant spelling; under the Code the original spelling must be retained but typographic or orthographic errors should be corrected; orthographic variants should not be listed as synonyms, but when two or more generic names are so similar as to be confused, they are treated as orthographic variants or, when based on different types, as homonyms.

Oscule—An old term for a pore in a rust spore.

-ose (suffix)—Full of; given to; like.

-osis (suffix)—Condition of; state caused by; abnormal or diseased condition.

Osmiophilic—Reacting to the presence of osmium tetroxide.

Osmophily (adj. **osmophilic, osmophilous**)—Growing under conditions of high osmotic pressure, as some yeasts on concentrated sugar solutions.

Osmosis (adj. **osmotic**)—Diffusion of fluids (usually water) through a differentially permeable membrane, from the side of the higher concentration of water to the side of the lower concentration of water.

Osmotic concentration—Concentration of osmotically active particles in a solution.

Osmotic pressure—The negative pressure that influences the rate of diffusion of water through a semipermeable membrane such as a cytoplasmic membrane; the maximum pressure that can be developed in a solution that is separated from pure water by a rigid membrane permeable only to water.

Osmotrophic—Exhibiting absorptive nutrition.

Ostiolar cells—Cells that delimit the aperture (ostiole) of perididate uredinia.

Ostiolate—Having an ostiole.

Ostiole, ostiolum—Small, more or less circular, schizogenous, paraphysis-lined pore or cavity in the papilla or neck of a spermagonium (pycnium), perithecium, or pycnidium through which spores are discharged after secession from conidiogenous cells.

Ostiole buffer—Tissue formed in the ostiolar region of perithecia and pycnidia that enlarges and breaks open the host tissue.

Ostropalean (of asci)—*See* Ascus.

Otomycosis—An infection of parts of the ear. *See* Mycosis.

Ouchterlony gel diffusion test—Serological test in which antibody and antigen are placed in wells in agar and allowed to diffuse toward each other. A positive reaction shows as a band of precipitate between the antibody and antigen wells. Serological relationships can be determined by the bands interacting with each other.

Oudin tube—Simple diffusion in agar. Antigen in solution is placed over agar containing antibody in a test tube. Antigen diffuses into agar and forms a precipitin band at equivalence.

Ounce, avoirdupois (oz av) = 0.911458 troy oz = 28.349527 g = 0.0625 lb = 437.5 grains.

Ounce, troy (troy oz) = 1.097 oz av = 31.10348 g = 480 grains.

Outbreeding—The formation of heterokaryons from plasmogamy between compatible homokaryons from genetically different parental sources.

Outer bark—Exterior, nonliving portion of bark in woody plants.

Ov-, ovi- (prefix)—Egg.

Oval—Broadly elliptical to nearly circular; egg-shaped, with the larger end at the base.

Ovariicolous—Living in ovaries.

Ovariiform—Egg-shaped.

Ovary—The female reproductive structure of higher plants that produces or contains one or more ovules, each ovule contains an egg; the basal, generally enlarged part of a pistil, within which seeds develop. The mature ovary is a fruit.

Ovate, oviform, ovoid—Shaped like a hen's egg, with one end narrower than the other.

Ovate-acuminate—Egg-shaped (in section) but tapering to a point.

Overseason—Live over from one planting season to the next.

Oversummer—To survive the summer.

Overwinter—To survive the winter.

Oviduct (of nematodes)—A short, usually tubular, thick-walled part of the female reproductive system between the ovary and spermatotheca or uterus.

Ovijector, ovejector (of nematodes)—The elongated, generally heavily muscularized vagina in some animal parasites that aids in the expulsion of eggs.

Oviparous (of nematodes)—Producing eggs that hatch after expulsion from the body.

Ovipositor—Egg-laying organ of a female insect, located at the end of the abdomen.

Oviposit—To deposit or lay eggs; in insects with an ovipositor.

Ovoid—Resembling an egg in shape.

Ovoviviparous (of nematodes)—Production of thinly shelled eggs that hatch in the mother's uterus.

Ovulate—Somewhat egg-shaped.

Ovule—A rudimentary seed; an enclosed structure, consisting of a female gametophyte (egg cell), nucellus, and one or two integuments; after fertilization it becomes a seed.

Ovum (pl. **ova**)—The reproductive cell of the female nematode; an egg.

Oxidase—Enzyme of qualitative significance in some organisms; accelerates the oxidation of some substance.

Oxidation—A chemical reaction in which oxygen combines with another substance or in which hydrogen atoms or electrons are removed from a substance. *See* Reduction.

Oxidation-reduction reaction—A chemical reaction in which one substance is oxidized (loses electrons, or loses hydrogen ions and their associated electrons, or combines with oxygen) and a second substance is reduced (gains electrons, or gains hydrogen ions and their associated electrons, or loses oxygen).

Oxidative phosphorylation—Utilization of energy released by the oxidative reactions of respiration to form high-energy ATP bonds.

Oxidative respiration—The chemical decomposition of foods (glucose, fats, and proteins) requiring oxygen as a terminal electron acceptor and yielding carbon dioxide, water, and energy. The energy is commonly in ATP (which see).

Oxybiont (adj. **oxybiotic**)—Microorganisms capable of using atmospheric oxygen during growth.

Oxydated (of crustose lichens)—Oxidized; having thalli tinged by rust-red iron oxides.

Oyster cap fungus or mushroom—The edible *Pleurotus ostreatus*.

oz av—Ounce avoirdupois. *See* Ounce avoirdupois.

oz troy—Troy ounce. *See* Ounce, troy.

Ozone (O_3)—A highly reactive form of oxygen (a photochemical oxidant) that in relatively high concentrations may injure plants, humans, and animals.

P

Pachydermate, pachydermatous (of hyphae)—Having the outer wall thicker than the lumen; thick-skinned.

Pachypleurous—Thick-walled.

Paddy straw mushroom—*See* Straw mushroom.

Palaeceous—Chaffy or chafflike.

Palatability—Agreeableness or attractiveness of a feedstuff to animals or how readily they consume it.

Palea—The upper and smaller of the two bracts that enclose a grass flower.

Paleomycology—Study of fossil fungi.

Paleozoic—A geologic era beginning about 570 million years ago and ending about 225 million years ago.

Paliform—Stake-shaped; palisadelike.

Palisade cells (of lichens)—Terminal cells of the hyphae of a fastigiate cortex.

Palisade fungi—Basidiomycotina.

Palisade parenchyma tissue—Parenchyma tissue found immediately beneath the upper epidermis of leaves, composed of elongate, tubular cells arranged upright like posts in a palisade fortification and packed with chloroplasts. Found in dicots but not in monocots.

Palisade plectenchyma—Plectenchyma in the cortex of a lichen thallus composed of hyphae arranged perpendicular to the surface.

Palisoderm (of basidiomata)—An outer layer composed of several strata of cells or hyphal tips. *See* Derm.

Pallescent—Turning pale.

Pallet—Rectangular or square platform, usually wooden, designed for ease of mechanical handling and transportation of material (containers) placed on the platform.

Pallid—Light-colored; pale.

Palmate leaf—A compound leaf having three or more lobes or leaflets radiating from a common center but not reacting to the point of insertion; shaped like a hand with the fingers spread. *See* Chiroid and Digitate.

Palmately veined—Type of net venation in which the main veins of a leaf blade branch out from the apex of the petiole like the fingers of a hand.

Palmicole, palmicolous—Growing on palm trees.

Paludal, paludose—Growing in wet places, e.g., marshes or swamps.

Palumbine—Dull grayish blue; dove-colored.

PAN (peroxyacyl nitrates)—Toxic air pollutants produced by photochemical reactions in daylight air, originating from the exhausts of internal combustion engines, and injurious to plants.

Panama disease of banana—Vascular disease of banana and abaca (*Musa*) caused by *Fusarium oxysporum* f. sp. *cubense*.

Pandemic—A widespread and destructive epidemic (epiphytotic) that occurs over an extended geographical area or areas.

Pandurate, panduriform—Fiddle-shaped.

Panicle—A loose, branched, open-flower cluster common in the grass family (Poaceae); a compound raceme.

Paniculate—Branched; having the form of a panicle.

Pannose, panniform—Like felt or woolen cloth in appearance or texture; ragged.

Pantherine—A metabolite of *Amanita pantherina* that is toxic to humans and flies (*Musa* sp.); muscimol; amanita factor B.

Papain—A proteolytic enzyme produced in the juice of fruit and leaves of pawpaw or papaya (*Carica papaya*); used to hydrolyze immunoglobulin to Fab and Fc fragments.

Paper chromatography—*See* Chromatography.

Papery—Thin basidiocarp that may be more or less coriaceous when fresh but like paper when dry.

Papilionaceous—Variegated; mottled; marked with different colors as the lamellae of certain species of *Panaeolus*.

Papilla, papillum (pl. **papillae**, adj. **papilloid**)—Minute, rounded, blunt, or nipple-shaped projection; (of fungi) the tip of a sporangium through which zoospores escape.

Papillae—(Of nematodes) tactile, sensory organs found on various body regions (labial, cephalic, cervical, etc.); (of plants) localized wall thickenings on the inner surface of plant cell walls at sites penetrated by fungi; synonyms or part synonyms include callosity, lignituber, callus, and "sheath."

Papillate, papillose—Bearing a papilla, i.e., a hump or swelling; nipplelike protuberance; (of basidiocarp surfaces) with small, rounded or hemispherical projections.

Papillulate—Having very small papillae.

Papillum—See Papilla.

Papulose—Covered with pustules or pimples.

Papulospore—Asexual spore as in *Papulaspora sepedonioides*; a closed, hyphal coil composed of short, somewhat enlarged cells, each of which is capable of germinating and the whole acting as a spore.

Papyracea—Pertaining to papyrus or paper.

Papyraceous—Resembling paper or parchment; papery.

Par-, para- (prefix)—Beside, near.

PAR—Abbreviation for photosynthetically active radiation; i.e., in the 400–700 nm spectral waveband.

Paraboloid—The surface generated by the rotation of a parabola about its axis.

Paracapillitium (of Lycoperdales)—A capillitium composed of thin-walled, hyaline, septate hyphae. A true capillitium has thick-walled, brown, aseptate hyphae.

Paracortex—A cortex of pseudoparenchymatous cells with hyaline walls.

Paracrinkle—A symptom of mild leaf crinkle in virus infections.

Paraderm—A pseudoparenchymatous derm, in which original hyphal elements have become more or less isodiametric.

Paragynous (of Pythiaceae)—Having the antheridium contact the oogonium at iits side.

Parallel evolution—Evolution in a similar direction in different groups of organisms; similar structures are produced but do not show phylogenetic relationships.

Parallel venation—Type of leaf venation, in which the principal veins are parallel to each other and to the longitudinal axis of the leaf, typical of monocots.

Parameter (in microbiology)—Points on or near the limits of a boundary; these variable points or boundaries represent known values.

Paramorph—Any *variant* of a species, especially one that cannot be more accurately defined due to a lack of information.

Paraphysate—Having paraphyses.

Paraphysis (pl. **paraphyses**)—Sterile, upward growing, basally attached hyphal filament or cell in a hymenium, especially in the Ascomycotina and certain Basidiomycotina, where they are generally clavate or filiform in shape, branched or unbranched, and free at the apex; often produced among fertile conidiophores or conidiogenous cells; the free ends frequently make an epithecium over the asci. *See* Hamathecium and Apical paraphyses; in basidiomycetes *see* Hyphidium.

Paraphysoid network—Branched and anastomosing paraphysoids surrounding asci in some *ascolocular ascomycetes.*

Paraphysoids—Threads of hyphal tissue between asci, like delicate paraphyses but without free ends; also, sterile accessory hymenial structures allied to basidia, cystidia, or hyphae. *See* Hamathecium, Basidiole, and Hyphidium.

Paraplectenchyma—Plectenchyma composed of cells with isodiametric lumina and unthickened walls.

Paraplectenchymatous—Somewhat interwoven or webbed tissue formed by interlacing of mycelium.

Paraprosenchyma—Tissue with elongated and spherical cells.

Parasexual cycle—A mechanism in filamentous hyphae in which a recombination of hereditary properties is based on the mitotic cycle. Its essential features are: production of diploid nuclei in a heterokaryotic haploid mycelium; multiplication of the diploid nuclei along with haploid nuclei in a heterokaryotic mycelium; the sorting out of a diploid homokaryon; segregation by crossing-over at mitosis; and haploidization of the diploid nuclei. *See* Protosexual.

Parasexualism (adj. **parasexual**) —Recombination of genetic characters based on mitosis without sexual processes.

Parasite—An organism (fungus, bacterium, nematode, insect, etc.) virus, or viroid living with, in, or on another living organism (host) and obtaining food from it; may benefit the host in return, but more frequently causes disease in the host; a biotroph or necrotroph. *See* Facultative parasite, Facultative saprophyte, Obligate parasite, Parasymbiont, and Pathogen.

Parasitic—Living on or in another organism and deriving nourishment from it.

Parasitism—An association in which one organism (the parasite) grows at the expense of another (the host).

Parasol mushroom—The edible Macrolepiota (*Lepiota*) *procera.*

Parasymbiont—Fungus or lichen living symbiotically on a lichenized fungus.

Parasymbiosis—An association of two organisms that is harmless to both but not mutually beneficial.

Parathecium—*See* Ectal excipulum.

Paratope—The antigen-combining site of an immunoglobulin (antibody) molecule.

Paratype—Every specimen (strain, isolate) other than the holotype or isotype(s) on which the first of a new species or other group is based. *See* Type, Holotype, and Syntype.

Parenchyma—Physiologically active, soft tissue of higher plants composed of thin-walled, often isodiametric cells that commonly store food or perform other functions, usually retain meristematic potential, and have intercellular spaces between them.

Parenthesome—A curved double membrane that may be perforate, imperforate, or vesiculate; on each side of a dolipore septum; septal spore cap.

Parietal—The layer next to the wall in certain fungal fruiting bodies; attached to the wall, e.g., of asci in a perithecium.

Parietin, physcion—Bright yellow-orange to red.

Paris inch = 27.9 mm.

Paris line = 2.2558 mm.

Parmuliform—Shield-shaped with margins slightly upturned.

Part spore—One-celled spore resulting from the breaking up of a two- or more-celled ascospore.

Parthen-, partheno-—Virgin.

Parthenocarpy—Natural or artificially induced development of fruit without sexual fertilization (pollination). Such fruits are seedless.

Parthenogamy, parthenomixis—Copulation between two "female" cells, especially of an archicarp; state of an oospore formed with a diploid nucleus.

Parthenogenesis (adj. **parthenogenetic**)—A type of asexual reproduction. Development of an egg (female gamete) into a new individual without fertilization by a sperm (male gamete); apomitic development of haploid cells.

Parthenospore—An oospore (aboospore) or zygospore (azygospore) produced by parthenogenesis.

Partial fusion line—*See* Spur precipitation line.

Partial veil or inner veil (of agarics)—A layer of fungal tissue extending from the stem (stipe) to the margin of the cap (pileus) and enclosing the gills (lamellae); when it ruptures, it usually leaves a ring (annulus) on the stem.

Partible—Easily or ultimately separating.

Particle size, soil—The effective diameter of a particle measured by sedimentation, sieving, or micrometric methods.

-partite (suffix)—Of viruses; consisting of parts; e.g., bipartite, tripartite, etc.

Partition coefficient—The constant ratio of the concentration of a solute in the upper phase to its concentration in the lower phase when the solute is in equilibrium distribution between two liquid phases.

Partridge wood—Wood invaded by a pocket rot fungus, e.g., caused by *Xylobolus frustulatus*, syn. *Stereum frustulatum*; or wood of *Caesalpinia* (= *Poinciana*).

Parts per million (ppm)—milligrams per liter; 1 ppm = 1 mg/liter.

Paspalum staggers—*See* Ergot; caused by *Claviceps paspali*.

Passage—Experimental infection of a host with a parasite that is later reisolated; method used to increase the virulence of the parasite and to separate a specific virus from a mixture of viruses.

Passive hemagglutination—A serological test used to detect a virus-specific antigen by coating red blood cells with viral antigen. If viral antibody is present the red blood cells agglutinate.

Passive immunity—Protection gained by injecting blood or serum containing antibodies.

Pasteur effect—Increased respiration and decreased fermentation in the presence of oxygen and the opposite in the absence of oxygen.

Pasteur filter—Unglazed porcelain tube for sterilization by filtration.

Pasteur pipette—Short length of glass tubing with one end drawn out into a sealed capillary, the other end plugged with cotton and sterilized. The tip of the capillary is broken before use.

Pasteurization—Freeing a medium of selected pathogenic microorganisms in soil, or other propagating media, using heat. The treatment does not materially change the natural characteristics of the substance treated.

Pasteurized soil—*See* Soil pasteurization.

Pasture—An area of domesticated forages, usually improved, on which animals are grazed.

Patch—Distinctly delimited area of plants in which most or all are affected by disease; (of scales or remnants of the universal veil) a flat, closely applied piece.

Patellate, patelliform—Like a round plate or disk with a well-defined edge; dishlike.

Patent—Spreading; stretching out.

Pateriform—Saucer-shaped.

Path-, patho- (prefix)—Suffering, hence disease.

Pathogen, pathogene—An organism or agent (e.g., fungus, bacterium, nematode, virus, or viroid) capable of causing disease in a particular host (*suscept*) or range of hosts. Most pathogens are parasites, but there are a few exceptions.

Pathogenesis—The sequence of processes in disease development from the time of infection to the final reaction in the host; production and development of disease.

Pathogenesis related (PR) proteins—Proteins that accumulate within cells in plant tissues as a reaction to the hypersensitive response to fungal or viral infection or to certain chemical treatments.

Pathogenic—Causing, or capable of producing, disease.

Pathogenicity—The ability of a pathogen to cause (incite) disease; the state or condition of being pathogenic.

Pathology, plant—Science or study of diseases, their nature and effects on plants, causes, and their control.

Pathotoxin—A toxin or poison produced by a pathogen and/or its host within the infected plant that functions in the production of disease but is not itself the initial inciting agent; vivotoxin.

Pathotype—Variant or subdivision of a given pathogenic species; pathogenic reaction on a given host; type (or reference) culture of a bacterial pathovar.

Pathovar (pv.) (of bacteria)—A type of subspecies; strain or group of strains of a bacterial species differentiated by pathogenicity in one or most hosts (species or cultivars).

Patulin (clavicin, clavitin, claviformin, expansin, mycoin, penicidin)—An antifungal and antibacterial antibiotic from *Aspergillus clavatus, Penicillium claviforme, P. expansum, P. patulum*, etc.; toxic to plants and animals (carcinogenic to mice) and the cause of neurotoxicosis in cattle; apple and pear juice may contain it (from *P. expansum*).

Patulous—Slightly spreading.

PBS—*See* Phosphate-buffered saline.

PDA—Potato-dextrose agar growth medium; a medium containing potato broth, dextrose (glucose) sugar, and agar.

Peach scab—Disease caused by *Cladosporium carpophilum*, teleomorph *Venturia carpophila*.

Pear scab—Disease caused by *Venturia pyrina*, anamorph *Fusicladium pyrorum*.

Pearl—The process of grinding off the hull, bran, aleurone, and germ of barley or rice to yield a pellet of endosperm.

Peat—An unconsolidated soil mass, high in semicarbonized organic materials consisting of partially decomposed plant tissue formed in water of marshes, bogs, or swamps, usually under conditions of high acidity.

Peat mold—*Chromelosporium fulvum*. Found on debris associated with heat-sterilized soil, sand, and peat moss in heated greenhouses.

Peck, U.S. = 0.25 bushel = 2 gal = 8 qt = 16 pt = 32 cups = 8.80958 liter = 537.605 cu in.^{3}.

Pecky cypress—Decay of bald-cypress (*Taxodium distichum*) by *Laurilia (Stereum) taxodii*.

Pectic enzymes—Components of the macerating enzymes of a number of fungal parasites. *See* Pectinesterases and Polygalacturonases.

Pectic substances—Complex organic compounds found chiefly in the middle lamella of cells.

Pectin—A methylated polymer of galacturonic acid found in the middle lamella and the primary cell wall of plants; jelly-forming substance found in fruit.

Pectinase (adj. **pectolytic**)—An enzyme that breaks down pectin.

Pectinate—Comb-shaped; having narrow parallel projections like teeth in a comb.

Pectinesterases (syn. **pectinmethylesterases**)—Specific enzymes that saponify methyl ester groups of pectinic acid.

Ped—A unit of soil structure such as a soil aggregate, crumb, prism, block, or granule commonly found in noncleaned lots of seed. Such "balls" of soil can contain pathogens.

Pedate—Footlike; having a foot.

Pedicel, pedicle—Small slender stalk; stalk bearing an individual flower, inflorescence, or spore.

Pedicellate, pediculate—Having a stalk or pedicel; borne on a pedicel.

Pediform—Foot-shaped.

Pedigree—A record of ancestry.

Pedogamy—Pseudomixis between mature and immature assimilative cells, e.g., copulation between a yeast mother cell and its bud.

Pedogenesis—Reproduction in young or immature organisms; also, soil formation.

Peduncle—Stalk or main stem of an inflorescence; part of an inflorescence, or a fructification; stalk.

Peg—*See* Hyphal peg.

PEG—*See* Polyethylene glycol.

Pellet—The material concentrated at the bottom of a centrifuge tube after centrifugation; three-dimensional colony in a liquid culture.

Pellicle, pellicula—A film or skinlike aggregation of microorganisms (commonly bacteria and/or yeasts) on the surface of a liquid medium; any surface with a thin, compact skinlike growth; (of basidiocarps of agarics) a delicate outside membrane or film that is easily peeled off or flaked away; outermost living layer below any nonliving secreted material containing the plasma membrane and underlying epiplasm or other membranes.

Pellicular veil—Very thin partial veil of a sporophore that lacks a stipe. *See* Cortina.

Pelliculate, pelliculose—Like a thin crust, as in the hymenial layer in Thelephoraceae; provided with a pellicle.

Pellis—Cellular cortical layers of a basidioma not belonging to the veils; cuticle. Three cortical layers have been recognized: *supra-*, *medio-*, and *sub-*.

Pellucid—Translucent or transparent.

Pellucid-striate (of a pileus)—Having a somewhat transparent top so the gills are seen through it as rays or striae.

Peloderm (of nematodes)—Caudal alae (bursa) that meet posterior to the tail tip.

Peloton—*See* Vesicular-arbuscular mycorrhiza.

Peltate—Shield-shaped like a round plate with a stalk from the center of the underside, like a leaf of garden nasturtium (*Tropaeolum majus*).

Penatin (corylophillin, notatin, penicillin B)—Antibacterial antibiotic from *Penicillium chrysogenum* (syn. *P. notatum*).

Pendant—Hanging down or suspended by a slender dorsal attachment.

Pendulous—Hanging down; pendant.

Penetration—The initial invasion (entrance) of a host by a pathogen.

Penetration peg—A thin hyphal strand developing on the underside of an appressorium and penetrating between closed guard cells into the substomatal cavity, or through the cuticle and epidermal cell wall into an epidermal cell; infection peg.

Penicidin—*See* Patulin.

Penicillate, penicilliform—Arranged like a little brush; especially the penicillus (conidiophore conidiogenous cells and conidia) of *Penicillium* and related genera. *See* Penicillus.

Penicillic acid—An antibacterial tetronic acid from *Penicillium puberulum*, *P. aurantiogriseum* (syn. *P. cyclopium*), etc.

Penicillin BB—*See* Penatin.

Penicillin dihydro F—Gigantic acid.

Penicillin F—Flavicin.

Penicillin N—Cephalosporin N.

Penicillinase—Bacterial enzyme that inhibits the action of penicillin.

Penicillins—Group of antibiotic substances produced by certain species of *Penicillium* and other microorganisms, generally effective against Gram-positive bacteria and of low toxicity to humans.

Penicillus—The small brushlike conidioma of *Penicillium* and related genera composed of a stipe bearing a cluster of conidiogenous cells (phialides) and other elements such as rami and metulae.

Pentaploid—Having five sets of chromosomes (5n).

Penton—Capsomer with five subunits, usually surrounded by five hexons, said to be "pentavalent."

Pentose—A sugar with five carbon atoms, e.g., ribose.

Penultimate—Next to the last.

Peplomer—Glycoprotein (spike, knob) located on a virus envelope.

Pepsidase—Enzyme catalyzing the liberation of individual amino acids from a peptide.

Pepsin—Gastric proteolytic enzyme used to hydrolyze immunoglobin to $F(ab')_2$ and Fc fragments.

Peptide—A compound consisting of two or several amino acids joined by peptide bonds.

Peptone—Partially hydrolyzed protein.

Per- (prefix)—Through or sometimes completely.

Percolation, soil water—The downward movement of water through soil.

Percurrent—(Of conidiogenous cell, conidiophore) growing straight, either through the open end left when the first conidium becomes detached or through a terminal pore; (of a gasteromycete basidiocarp) extending the entire length of the columella.

Perennial canker—*See* Canker.

Perennial—A plant that continues growth more or less indefinitely from year to year and usually produces seed each year; (of fungi) a fruitbody (e.g., basidiocarp) that persists in an active, living state for more than one growing season.

Perfect flower—A flower having both stamens and pistils (carpels); a hermaphroditic flower.

Perfect state (stage or **phase)**—State in the life cycle of a fungus in which sexual spores (e.g., ascospores, basidiospores, oospores, zygospores) are formed after nuclear fusion or by parthenogenesis; sexual state or teleomorph; capable of sexual reproduction. *See* Teleomorph.

Perforate—Pierced through.

Perforation lysis—Process that initiates degradation of resistant fungal propagules in soil.

Perfossate—Hollowed out.

Perfuse—Completely covered.

Pergameneous, pergamenous, pergamentaceous—Like paper or parchment in texture.

Peri- (prefix)—Near, around, about.

Perianth—The external envelope of a flower, consisting of sepals and petals (calyx and corolla).

Pericarp—Outer tissue layer (modified walls) of a ripened ovary (fruit); a covering; also the entire fruitbody.

Periclinal—Curved in the direction of, or parallel to, the surface or the circumference; cell division in the plane parallel to the surface. *See* Anticlinal.

Periclinal thickening (of phialides)—Unstainable, thickened, apical region surrounding the phialidic channel, referred to as the annulus in *Alternaria*.

Pericycle—The layer of cells in a stem or root located between the endodermis and the vascular cylinder. Branch roots arise from the pericycle.

Periderm—(Of fungi) the membrane surrounding a sorus; (of higher plants) plant tissue external to the cortex and/or phloem of stems and roots, composed of cork and the cork cambium, more or less impervious to water, solutes, and organisms and serving to protect underlying tissues.

Peridermioid—Having the appearance of the anamorphic or form genus *Peridermium*, the aecial state of most rust fungi on conifers.

Peridermium—A blisterlike, tongue-shaped aecium, as in the form-genus *Peridermium*.

Peridial cells (especially of aecia)—Cells of the peridium.

Peridiate—Having a peridium.

Peridiole, peridiolum (especially of Nidulariaceae)—A small, seedlike or egglike division of the gleba with a distinct wall, frequently acting as a unit for distribution.

Peridium (pl. **peridia**)—The outer cellular structure (wall) or limiting membrane of a sporangium, perithecium, rust sorus, or other fruitbody.

Périgord (French) truffle—*Tuber melanosporum. See* Truffle.

Perigyny—A condition in flowers in which the pistils and stamens are usually fused with the calyx and in which the pistil is seated in a concave receptacle.

Perineal pattern (of nematodes)—Cuticular patterns in the perineal area (area of vulva and anus) of *Meloidogyne* females; composed of fine, wavy, fingerprint-like striations that are more or less species characteristic.

Perineum (of nematodes)—The region surrounding the anus.

Perinuclear space—The region between the two membranes of the nucleus or close to the outer membrane.

Periodic acid-Schiff's (PAS) reagent—A sensitive stain for glycoproteins, often used to detect these substances after SDS-gel electrophoresis.

Peripheral (n. **periphery**, adv. **peripherally**)—Of, relating to, or involving the surface of a body; around the margin of an area or structure; (locules) situated in the outer regions of the stroma and often subtended or supported by a considerable amount of sterile stroma below; (of a rust sorus) around the margin, e.g., peridium or paraphyses.

Periphysis (pl. **periphyses**)—A hairlike, sterile projection from, in or near, an ostiole of a perithecium, pycnidium, or spermogonium (pycnium). *See* Hamathecium.

Periphyton—Assemblage of organisms growing on free surfaces of submerged objects in water and covering them with a slimy coat.

Periplasm (of Peronosporales)—The outer, nonfunctional, spongy layer of protoplasm surrounding the oogonium and antheridium. *See* Gonoplasm and Ooplasm.

Perispore, perisporium—Sheath or layer outside the true spore wall; spore membrane or wall.

Perisporial sac—A perispore that forms a loose envelope around a spore, as in *Coprinus*.

Peristome—A ring of teeth surrounding the opening of a moss capsule and by its hygroscopic movements scattering the spores; a circular edging variously adorned and marked around an opening, especially of basidiocarps of certain gasteromycetes.

Perithecium (pl. **perithecia**)—Flask-shaped to more or less globose, thin- or thick-walled, ascocarp of the Pyrenomycetes containing asci, ascospores, and paraphyses. Spores are expelled or otherwise released through a pore or slit (ostiole) at the tip of the neck or beak.

Peritrichous, peritrichiate—Having hairs or flagella distributed over the whole surface.

Permanent wilting percentage—The amount of water in the soil at the time plants become permanently wilted. *See* Wilting Point.

Permeability (adj. **permeable**)—The quality or condition allowing a fluid or substance in a fluid to pass or diffuse through a membrane.

Permissive cells—Cells in which infection results in the production of infectious progeny virus.

Permutate—Changed completely.

Peronate—Sheathed; having a boot or covering, especially of the lower part of a stipe covered by a volva or universal veil.

Peroxidase—Enzyme that catalyzes the decomposition of peroxides. Used as a reporter molecule in diagnostic tests such as ELISA.

Perrumpent—Breaking through. *See* Erumpent.

Persicine—Peach-colored.

Persistent (n. **persistence**)—Said of circulatory viruses that remain infectious within their insect or other vectors for long periods without inducing lysis and are transmitted via salivary fluids; (of fungus spores) nondeciduous; (of conidiogenous cell) carried away at the base of the conidium at secession; (of teliospore pedicels) remaining firmly attached to the spore after liberation; (of a pesticide) persisting on plants, seed, or soil for weeks or months after application during which it retains at least some of its pesticidal activity; (of interascal tissues) still evident at maturity.

Persistent transmission—Type of insect or nematode transmission in which a pathogen (e.g., a virus or mycoplasma) is acquired by a vector only after a long (minutes, hours, or days rather than seconds) acquisition feeding period. There is usually a latent period, following the acquisition feed, before the vector can transmit the pathogen. The vector often remains inoculative throughout its life span. The pathogen sometimes multiplies within the vector. Nematode vectors can transmit immediately after acquisition and lose the capacity to transmit when they molt. Nonmolting dagger and stubby-root nematodes retain the capacity to transmit longer than do needle nematodes. *See* Nonpersistent transmission, Semipersistent transmission, Circulative, and Propagative.

Perthophyte, perthotroph—A "saprophyte" that kills cells in advance of actual mycelial invasion by means of secreted toxic materials; a necrophyte on dead tissues of living hosts.

Pervious—Having an open passage; (of lichen scyphi) open or perforate basally.

Pest—Any organism (e.g., some bacteria, fungi, insects, mites, nematodes, rodents, weeds) injuring or detrimental to a beneficial plant, plant product, animal, or human (or human property or that annoys humans), or to the environment.

Pesticide—Any chemical or physical agent that destroys, prevents, mitigates, repels, or attracts pests (e.g., acaricide or miticide, bactericide, fungicide, herbicide, insecticide, nematicide, rodenticide, etc.).

Petal—One of the structural units of the corolla of a flower, frequently conspicuously colored.

Petaloid—Resembling a petal of a flower in shape, texture, and/or color, narrowed somewhat at the base.

Petiole—The stemlike part of a leaf; the stalk that attaches a leaf blade to a stem.

Petraceous, petrose—Growing among stones.

Petri plate (or **dish**)—*See* Culture plate.

Petrophilous—*See* Saxicolous.

pH—A measure of acidity and alkalinity within the range of 0 to 14 in which pH 7 is neutral; numbers less than 7 indicate increasing acidity; those more than 7, increasing alkalinity. One pH unit change is equal to a 10-fold change in the hydrogen ion concentration because the scale is logarithmic, i.e., pH = –log (H+).

Phaeo- (prefix)—Dark-colored, brown or swarthy, especially of spores. *See* Deuteromycotina (Mitosporic fungi).

Phaeodictyosporous—Having dark muriform spores.

Phaeomycosis, phaeohyphomycosis—General term for mycoses of humans and animals caused by dematiaceous fungi.

Phaeophragmious, phaeophragmosporous—Having dark spores with cross walls.

Phaeosporous—Having dark, one-celled spores.

-phage (suffix)—One that eats.

Phage—A general term for viruses isolated from prokaryotes, including bacteria, blue-green algae (cyanobacteria), and mollicutes (mycoplasma and spiroplasma). Phages are tailed, cubic, rod-shaped (filamentous) or pleomorphic in form and are further subdivided according to biochemical and morphological properties. A given phage type infects a limited range of susceptible hosts, usually within the established taxonomic divisions among prokaryotes. *See* Bacteriophage, Cyanophage, and Mycoplasmaphage.

Phage type—A subdivision (usually of a species) determined by the susceptibility of a bacterium to one or more of a series of bacteriophages that are themselves called *typing phages*. The type is often seen as a pattern of lytic zones in the places in which phage has been seen on a plate sown with the strain under test.

Phagocyte—An amoeboid cell capable of ingesting microorganisms or other foreign particles; also known as macrophage.

Phagotrophic—Feeding by engulfing particles versus absorption, as in Myxomycetes versus Eumycota.

-phagous (suffix)—Eating, feeding on.

Phalacrogenous—Of conidiogenous cells that arise at the same level from single hyphae and form a turflike or velvety layer.

Phalloid—A member of the Phallales.

Phallotoxins—Cyclic heptapeptides toxic to humans from *Amanita phalloides*. *See* Amatoxins.

Phaneroplasmodium—Plasmodium composed of a well-differentiated advancing fan and thick strands of granular protoplasm exhibiting ecto- and endoplasmic regions. See Plasmodium.

Pharyngeal bulb (of nematodes)—A muscular swelling of the esophageal wall around the buccal capsule.

Pharynx (of nematodes)—The muscular esophagus, by some authors.

Phase—A recognizable stage in the growth of bacteria.

Phase variation—Separation of a bacterial species into strains having somewhat different characters.

Phase-contrast microscope—A compound microscope with an annular diaphragm in the front focal plane of the substage condenser and a phase plate at the rear focal plane of the objective to make visible differences in phase or optical path in transparent or reflecting media.

Phaseoliform—Bean-shaped.

Phaseollin—A phytoalexin from bean (*Phaseolus vulgaris*).

Phasmid (of nematodes)—Porelike structure, usually minute and usually occurring in pairs opposite each other, in the lateral field on the tail of members of the class Secernentea; the function is believed to be sensory (sometimes called precaudal glands).

Phellem—Cork; a protective tissue composed of nonliving cells with suberized walls produced by the phellogen.

Phelloderm—A secondary tissue produced by cork cambium and found on the inner surface of the cork cambium (resembles the cortical parenchyma in morphology); the innermost layer of the periderm.

Phellogen—Cork cambium; lateral meristem forming the periderm, a protective tissue in stems and roots. Phellem (cork) is produced toward the surface of some plants and phelloderm toward the inside.

Phenetic—A classification based on overall similarity, as determined by an unprejudiced assessment of all known observable characters; the word phenetic has no evolutionary implications.

Phenol (adj. **phenolic**)—A toxic acidic compound, C_6H_5OH, used as a disinfectant or protein denaturant.

Phenol coefficient—The ratio between the greatest dilution of a test compound capable of killing a test organism in 10 min but not in 5 to the greatest dilution of phenol giving the same result.

Phenolase—An enzyme capable of degrading phenolic compounds.

Phenology—The study of the timing of periodic phenomena such as flowering and the start or ending of growth, especially as related to seasonal changes in temperature or photoperiod.

Phenolytic—A compound containing one or more phenolic rings.

Phenotype—External, visible physical characteristics of an organism determined by its genotype and modified by the environment.

Phenotypic mixing (or genomic masking)—A process in which an individual progeny from a mixed viral infection contains structural proteins derived from both viruses or has the genome of one virus encapsidated in the structural proteins of the other; the latter situation is called genomic masking or transcapsidation.

Pheromone—Any of a class of hormones, odor-attractive substances secreted to the outside by an individual as a sexual or aggregation stimulus and received by a second individual of the same species; first discovered in insects.

Phialide—An enteroblastic conidiogenous cell that produces a basipetal succession of conidia through one (monophialide) or more (polyphialide) open ends in which neither wall contributes toward formation of the conidium, often with collarettes surrounding the openings. The thin-walled conidia often hang together in chains or in slimy clumps.

Phialidic (conidiogenous cell)—Enteroblastic and producing conidia (phialoconidia), usually in large numbers, in basipetal succession through one opening or several openings.

Phialiform—Saucer- or cup-shaped.

Phialospore, phialoconidium, phialidic conidium—Conidium produced by a phialide.

-phile, -philic, -philous (suffix)—Inhabiting, living upon; also, a liking for a substance or state.

Phlebioid—Hymenial surface with (usually radial) folds.

Phloem—Food-conducting tissue in plants; complex vascular tissue consisting of sieve tubes (through which synthesized nutritive and other materials move from the leaves), companion cells, fibers, and phloem parenchyma; inner bark.

Phloem, secondary—Phloem formed by the vascular cambium.

Phomin—Cytostatic antibiotic from *Phoma* sp.; cytohaclasin B.

-phore (suffix)—Bearer.

-phoric, -phorous (suffix)—Bearing.

Phorophyte—Host tree of an epiphyte.

Phosphatase—An enzyme that catalyzes the hydrolysis and synthesis of phosphoric acid esters and the transfer of phosphate groups from phosphoric acid to other compounds; an enzyme that splits phosphate from its organic compound.

Phosphate-buffered saline (PBS)—A solution containing 8 g NaCl, 0.2 g KCL, 0.2 g KH_2PO_4, and 0.15 g Na_2HPO_4 widely used in serology and some culture media.

Phosphodiester bond—Link formed between the nucleotides of polypeptide chains by covalent bonding of the phosphoric acid with the 3′-hydroxyl group of one ribose or deoxyribose sugar and the 5′-hydroxyl group of the next ribose or deoxyribose ring. *See* Nucleic acid.

Phospholipase—An enzyme that catalyzes the hydrolysis of a phospholipid and is used in the study of cytoplasmic membranes.

Phosphoprotein—A protein with one or more amino acids phosphorylated by a protein kinase.

Phosphorylation—The addition of a phosphate group, for example H_2PO_4, to a compound.

Phosphotungstic acid—A negative stain used in electron microscopy consisting of dodecatungstophosphoric acid dissolved in water to give a 1–2% solution. It is adjusted to about pH 7 with NaOH.

Photic characters—Characteristics relating to light.

Photo- (prefix)—Pertaining to light.

Photobiont—Photosynthetic symbiont of a lichen that may be a green alga (phycobiont), or a cyanobacterium (bactobiont, cyanobiont).

Photochemical oxidants—Highly reactive compounds formed by the action of sunlight on less toxic precursors.

Photodegradation—Degradation due to light, usually sunlight.

Photogenic—Glowing in the dark; phosphorescent.

Photolysis—Photochemical process that cleaves water molecules into H and OH fragments.

Photon correlation spectroscopy—A technique for studying the movements of macromolecules, organelles, and cells from their scattering of laser light; can be used to estimate the diffusion coefficient of virus particles.

Photon flux density—Radiometric irradiance per unit area per unit time at specific wavelengths (400–700 nm waveband for plants). Units micromoles $m^{-2}s^{-1}$ or microeinsteins $m^{-2}s^{-1}$.

Photoperiod—The relative duration of night and day to which plants are exposed; the length of day or period of daily illumination required for the normal growth and sexual reproduction of some plants.

Photoperiodism—Growth and developmental responses of plants (e.g., flower initiation and bulbing) to the daily length of light and to differing photoperiods.

Photophilous—Preferring well-illuminated habitats. *See* Heliophilous, Anheliophilous.

Photophobous—Preferring a shady habitat.

Photophosphorylation—The production of ATP by the addition of a phosphate group to ADP using the energy of light-excited electrons produced in the light reactions of photosynthesis. Photo = light; phosphorylation = adding phosphorus. *See* ADP and ATP.

Photoreactivation—The enzymic repair of DNA damaged by UV light. The enzymes involved are activated by exposure to long wavelength light.

Photosporogenic—Requiring light for sporogenesis.

Photosynthesis—The complex fundamental process by which green plants make carbohydrate food (sugar) from water and carbon dioxide in the presence of chlorophyll(s), using light energy and releasing oxygen.

Phototaxis—Movement, such as of zoospores, influenced by light.

Phototroph—A microorganism capable of utilizing light energy for metabolism.

Phototropism—A growth movement (curvature) induced by the stimulus of light. The response, which is auxin-regulated, is a bending toward the strongest light.

Phragmobasidium—A basidium in which the metabasidium is divided by primary, often cruciate septa (as in *Tremella*) or transverse (as in *Auricularia*).

Phragmospore—A spore that differs from an amerospore and didymospore in having two to many transverse septa. *See* Mitosporic fungi.

Phyco- (prefix)—Pertaining to algae.

Phycobiont—The algal part of a lichen; photobiont. *See* Mycobiont.

Phycocyanin—A blue pigment found in blue-green and in red algae.

Phycoerythrin—A red pigment found in red and blue-green algae.

Phycology—The science or study of algae.

Phycomycetes (adj. **phycomycetous**)—An outdated class of fungi (now treated as Chromista, Chytridomycota, and Zygomycota) that included the chytrids, oomycetes, and zygomycetes; a form class not showing phylogenetic relationships. Fungi whose mycelium has few or no cross walls and that reproduce sexually by the union of two sex cells; used for "lower fungi" (those other than Ascomycotina, Basidiomycotina, and Deuteromycotina). *See* Oomycete.

Phycomycosis—General term for a disease of humans and animals caused by a phycomycete. *See* Mucormycosis and Zygomycosis.

Phycophilous—Growing with or on algae.

Phycosymbiodeme—Joined lichen thalli with a single mycobiont but having different photobionts.

Phycosymbiont—Same as phycobiont.

Phycotrophic (of fungi)—Obtaining food from algae.

Phycotype (of fungi)—Each of the morphologically distinct structures derived by symbiosis between a single mycobiont and different photobionts, especially in which each is usually free living but rarely forms composite thalli (hyphae of fungi) essentially specialized, regularly free living cephalodia.

Phyll-, phyllo (prefix)—Leaf.

Phyllidium—A lichen propagule formed by abstriction of a leaf- or scalelike part of the thallus.

Phyllody—Transformation of normal floral parts (petals) into leaflike structures; caused by certain pathogens such as mycoplasmas, insects, or mites.

Phylloplane—The leaf surface; the nonparasitic flora of the leaf surface. *See* Rhizoplane and Spermoplane.

Phyllosphere—Zone immediately surrounding a leaf.

Phylogenetic—Indicating natural evolutionary relationships.

Phylogeny—History of the evolution of a plant or animal group in relation to other groups; the study of "family trees." *See* Ontogeny.

Phylum (pl. **phyla**)—One of the major categories in classification of the animal kingdom; same as a division in classification of plants.

Physcion—*See* Parietin.

Physiogenic (physiological) disease—A disease (or disorder) produced by some unfavorable genetic, physical, or environmental factor (e.g., excess or deficiency of light, water, soil nutrients, chemical, physical, or other injury, etc.). *See* Noninfectious disease.

Physiologic form—Subspecies group that differs in behavior or other characteristics but not in morphology. *See* Physiologic race.

Physiologic race—Subdivision within a species of fungus or other organism; members of a race are alike in morphology but differ from other races in virulence, symptom expression, biochemical and physiological properties, or host range. *See* Biotype.

Physiological specialization—The occurrence of physiologic races within a species or variety of a pathogen.

Physiology, plant—The science or study of the functions and activities of living plants.

Phyto-, -phyte—Combining forms meaning plant.

Phytoalexin—An antibiotic or fungistatic metabolite, arising from host-parasite interaction, inhibitory to microorganisms attacking plants; a substance that inhibits the development of a microorganism; formed mainly when host plant cells come in contact with the parasite (response to certain stimuli), as opposed to preformed toxins. Phytoalexins are low-molecular weight, microbial compounds that are both synthesized by and accumulated in plants after exposure to microorganisms.

Phytoalternarin, A,B,C—Host-specific toxins produced by *Alternaria kikuchiana* the cause of black spot of Japanese pears (*Pyrus serotina*).

Phytochrome—A reversible, photosensitive protein pigment in the cytoplasm of green plants, associated with the absorption of light, that influences flowering, germination, growth, development, and other physiological processes independent of photosynthesis (e.g., in the photoperiodic response).

Phytolysine—An enzyme from *Dothiora (Plowrightia) ribesia* that macerates plant tissue.

Phytonicide—A chemical substance produced by higher green plants that can inhibit the growth of microoganisms.

Phytopathogenic—Capable of causing disease in plants.

Phytopathology—Plant pathology; science or study of plant diseases.

Phytoplankton—Collective term for the plants and plantlike organisms present in plankton.

Phytoplasma—A mycoplasma that causes disease of plants.

Phytosanitation—Any measures involving the removal or destruction of infected plant material likely to form a source of reinfection by a pathogen.

Phytosociology—Scientific study of plant communities.

Phytotoxicity (adj. **phytotoxic**)—Poisonous, injurious, or lethal to plants; usually describing a chemical.

Phytotoxin—A substance (toxin) injurious to plants; or a toxin produced by a plant.

Piedra—*See* White piedra of humans.

Pigment—A colored compound, such as chlorophyll; molecules that are colored by the light they absorb. Some plant pigments are water-soluble and found mainly in the cell vacuole (e.g., anthocyanin).

Pigmentation—Coloration.

Pile-. pilei-, pileo- (prefix)—Hat or cap.

Pileate—(Of basidiocarps) with a reflexed, shelflike portion having a sterile upper surface; (of an agaric) having an umbrellalike cap.

Pileipellis—*See* Pellis.

Pileocystidium—A cystidium or cystidiumlike cell on the surface of a pileus.

Pileolus—A small pileus or cap.

Pileus (pl. **pilei**)—Hymenium-supporting part of a fruitbody of a nonresupinate higher fungus, especially a hymenomycete; the cap of certain types of basidiocarps.

Pili-—Same as fimbriae.

Pilose—Covered with long, soft hairs giving a plushlike texture.

Pilus (pl. **pili**)—Filamentous, nonmotile, somatic or sex appendages found on many Gram-negative bacteria.

Pin mold—*Mucor* spp. and other zygomycetes.

Pin rot—A white wood rot with small empty pockets, giving it a lacy appearance.

Pinch (n. **pinching back**)—To remove the tip of a stem (apical meristem), an extra flower bud, or terminal bud (using fingernails, knife, or shears) to stimulate branching.

Pine moss—Species of *Alectoria* and *Bryoria*.

Pine or Japanese mushroom—Usually matsutake; rarely, *Tricholoma murrillianum*.

Pinna (pl. **pinnae**)—One of the primary divisions of a pinnate leaf or frond.

Pinnate—Having branches, lobes, leaflets, or veins arranged in a featherlike manner; attached or arranged on two sides of a stem.

Pinnote sporodochium (pseudopinnotes of *Fusarium*)—Minute sporodochia near the surface of the substrate having no stroma, the spores forming a continuous slimy layer.

Pinnotes—Slimy gelatinous spore mass of *Fusarium* having a fatty or greaselike appearance.

Pinocytosis—A form of active transport of water and dissolved or suspended molecules across a cell membrane involving the internalization of a fluid-filled vacuole.

Pint, U.S. = 16 fl oz = 32 tablespoons = 2 cups = 0.125 gal = 473.167 ml = 1.04 lb water = 28.875 in.^{3} = 0.473167 liter = 0.01671 ft^3.

Pinwheel inclusions—Shape of some virus-induced, cytoplasmic inclusion bodies seen in cross- and thin-section in an electron microscope. They are composed of sheets of virus-coded protein and found in cells of plants infected with potyviruses.

Piperate—Peppery; pungent.

PIPES—A biological buffer (peperazine-N,N'-bis-2-ethanesulphonic acid) used in the pH range of 6.2–8.2.

Pip-shaped (of spores)—Shaped like an apple seed.

Piricularin—Phytotoxin from *Pyricularia oryzae*.

Piriform—Pear-shaped. Same as pyriform.

Pisatin—A phytoalexin from pea (*Pisum sativum*)

Pisiform—Having the size and shape of a pea.

Pistil—The seed-producing organ of the flower in a seed plant; typically consists of an ovary, and one or more styles, and stigmas. A simple pistil consists of a carpel, while a compound pistil has two or more partly or wholly fused carpels.

Pistillate flower—One that contains pistils (female parts) but no stamens (male parts).

Pit—A thin or open place in a cell wall.

Pitch—In a filamentous virus particle, the axial distance between adjacent turns of a row of capsids.

Pith (adj. **pithy**)—Soft, spongy (loosely packed), thin-walled, parenchyma cells in the center of certain plant stems and roots; primary tissue; (of a stipe) soft tissue in the interior that often disappears leaving the stipe hollow.

Pith ray—A band of interfascicular parenchyma.

Pitted—Having depressions or cavities, often due to dying of small localized areas in fleshy or woody tissues beneath healthy-appearing external tissue.

Pityriasis versicolor ("tinea versicolor")—A superficial skin disease of humans caused by *Malassezia furfur*, syn. *Pityrosporum orbiculare*.

pl.—Plural.

Placenta (pl. **placentae**)—A small mass of ovary tissue in a fruit to which the seeds (ovules) are attached.

Placental—Pertaining to the part of the ovary from which ovules arise.

Placodioid (of a crustose lichen thallus)—Disc-shaped with plicate lobes at the circumference.

Placodiomorph—Two-celled spore with a thickened septum that may have a pore.

Placodium (pl. **placodia**) of Diatrypaceae—The horny, sclerotic rind layer around the perithecial mouth.

Placoid—Same as placodioid.

Placomycetoid—Pileus with a diameter to stipe ratio of <1. *See* Campestroid.

Plage—Smooth, pale or colorless spot on a surface; (of basidiospores) especially a smooth place above the hilar appendage.

Plane—Having a flat or even surface.

Planetism (of oomycetes)—Condition of having motile stages.

Plankton—Collective term for the passively floating or drifting flora and fauna (*zooplankton*) of a body of water; consists largely of microscopic organisms.

Plano- (prefix)—Motile *or* flat.

Planoconvex (of a pileus)—Convex but somewhat flattened; a flat zygote.

Planocyte, planont—A motile cell.

Planogamete—A motile gamete; zoogamete.

Planogamic copulation—Fusion of motile spores resulting in a motile zygote or planozygote.

Planose—Plane.

Planospore—A zoospore.

Planozygote—A motile zygote.

Plant disease—*See* Disease, plant.

Plant formation—*See* Formation.

Plant growth regulator—A substance that affects plants or plant parts through physiological rather than by physical action by accelerating or retarding growth, prolonging or breaking a dormant condition, promoting decay, or inducing other physiological changes.

Plant pathology—The science or study of plant disease; also phytopathology.

Plant-disease interaction—The concurrent parasitism of a host by more than one pathogen in which the symptoms or other effects produced are of greater magnitude than the sum of the effects of each pathogen acting alone, an example of synergism.

Planthopper—Small, leaping, homopterous insects (family Delphacidae) with piercing-sucking mouth parts that are known to transmit a number of plant-infecting viruses and mycoplasmalike organisms.

Plaque—A circular clear zone of lysis produced by a virulent bacteriophage in a colony of bacteria on an agar medium, also applied to similar areas in a fungal colony; (of nematodes) an infected condition of the cuticle with the general appearance of deep annulations divided longitudinally.

Plaque assay—An assay in which the concentration of infective particles in a virus solution is recorded as the number of plaques induced in a bacterial colony (lawn) or cells; also used to distinguish between different strains or distinct viruses by the features of the plaque.

Plaque forming units (pfu)—The number of plaques formed per unit of volume or weight of a virus suspension.

Plaque mutants—Mutants that produce plaques that differ in size or appearance from those produced by the wild-type virus.

Plaque picking—The selection of individual plaques considered to be formed by a single infection event; thus clones of a virus can be selected and studied further.

Plasm-, plasmo-, plasmato- (prefix)—Anything formed or molded.

Plasma—*See* Blood plasma.

Plasmalemma—Outer three-ply membrane bounding the protoplast next to the cell wall; cell, cytoplasmic, or plasma membrane composed of phospholipids and proteins that regulates the exchange of materials between the cell and its environment.

Plasmalemmasome—Intracytoplasmic vesicle (formed by invagination of the plasmalemma) filled with tubular diverticula. *See* Lomasome.

Plasmatoogosis (of Pythiaceae)—Budlike outgrowth (shaped like a prosporangium) in host tissue.

Plasmid—Generally a small, covalently closed, circular piece of nonchromosomal, double-stranded DNA, found in certain bacteria and fungi, that carries generally nonessential genetic information and is self-replicating. (Some plasmids are linear molecules.) Plasmids are used in recombinant DNA experiments as acceptors of foreign DNA.

Plasmid vector—A plasmid used for cloning "foreign" DNA, often manipulated to contain desirable features.

Plasmodesma (pl. **plasmodesmata**)—A fine cytoplasmic strand passing through the cell wall of adjoining cells, interconnecting the two living protoplasts.

Plasmodic granules—Microscopic, usually dark-colored particles on the surface of the peridium and frequently on the spores of the Cribrariaceae of the Myxomycetes; dictydine granules.

Plasmodiocarp (of myxomycetes)—A sessile and veinlike sporangium, resembling part of the larger veins of a plasmodium.

Plasmodium (pl. **plasmodia**) (of myxomycetes)—A naked, multinucleate, motile mass of protoplasm bounded by a plasma membrane and lacking a cell wall, generally reticulate, that feeds in amoeboid fashion and is the result of fusion of uninucleate amoeboid cells, characteristic of the vegetative growth phase. Several types have been distinguished: *protoplasmodium* (undifferentiated microscopic plasmodium that gives rise to a single sporangium); *aphanoplasmodium* (a plasmodium composed of a network of undifferentiated strands of nongranular protoplasm); *phaneroplasmodium* (a plasmodium composed of a well-differentiated advancing "fan" and thick strands of granular protoplasm with ecto- and endoplasmic regions); *aggregate plasmodium* or *pseudoplasmodium* (structure formed by the aggregation of myxamoebae or cells before reproduction); and *filoplasmodium* (the netlike pseudoplasmodium of the Labyrinthales).

Plasmogamy—Fusion between two cells or plasmodial cytoplasms without nuclear fusion (karyogamy) or a precursor to karyogamy. *See* Karyogamy.

Plasmolysis—Contraction and shrinkage of the cytoplasm in a living cell from the cell wall due to loss of water by exosmosis.

Plastid—Any of various cytoplasmic organelles (chloroplasts, leucoplasts, etc.) of cells that serve in many cases as centers of special metabolic activities.

Plating efficiency (efficiency of plating)—The number of plaques divided by the total number of virions in the inoculum.

Platyphyllous, platylobate—Broad-lobed.

Plect-, plecto- (prefix)—Twisted.

Plectenchyma (adj. **plectenchmatous**)—A thick tissue formed by intertwining and adhering of hyphae; synchyma.

Plectoid—Resembling the genus *Plectus*.

Plectonematogenous—Of conidiogenous cells arising from intertwined ropelike strands of hyphae; funiculose.

Pleio-, pleo- (prefix)—More.

Pleiosporous—Many-spored.

Pleioxeny—The condition in which parasites can invade several species of host plants; plurivorous parasitism.

Pleomorphic—Able to assume various shapes (and perhaps sizes); having more than one independent form or spore stage in the life cycle; polymorphic; (of dermatophytes) changes in culture due to "degeneration."

Pleomorphic phages—Enveloped, double-stranded, DNA-containing phages without apparent capsid structure.

Pleomorphism—The condition of being pleomorphic.

Pleont—Any one of the two or more states of a pleomorphic fungus.

Plerome (procambium strands)—Inner tissue in root meristem, first cells of the stele; plant tissue inside the cortex.

Plerotic (of oospores)—Filling the oogonium, as in the Pythiaceae; also, an oospore that occupies more than 65% of the volume of the oogonium.

Pleur-, pleuro- (prefix)—Relating to a side; at the side.

Pleuracrogenous—Formed at the end and on the sides.

Pleurobasidium—A basidium relatively broad at the base with bifurcated spreading "roots," as in *Pleurobasidium*.

Pleurocystidium—A cystidium on the face of a gill or tube.

Pleurogenous—Borne or formed along the side.

Pleurosporous—Having spores on the sides, e.g., a basidium of the rust fungi (Uredinales).

Plexus—A network.

Pliant—Easily bent; not firm or rigid.

Plica—A pleat.

Plicate—Folded into pleats; bearing folds or ridges; reflexed portion of a basidiocarp that is pleated radially; a more severe folding than undulate.

Plow layer—The surface soil layer ordinarily moved in tillage.

Plow pan—A hardpan layer of soil formed by continual plowing at the same depth.

Plow-plant—The practice of plowing and planting a crop in one operation without additional seedbed preparation.

Plumose, plumate—Finely feathery.

Plumule—Bud- or shoot-forming meristem of an embryo or that portion of the young shoot above the cotyledons. *See* Epicotyl.

Plur-, pluri- (prefix)—Several, many.

Pluricellular—Many-celled.

Pluriciliate—Having many cilia.

Plurifurcate—Many-forked.

Pluriguttulate—Having many guttules.

Plurilocellate—Having many hollows.

Plurilocular, pluriloculate—(Of stromata) having several cavities (locules); (of ascospores) many-celled.

Pluriperforate—Having several openings.

Plurivorous—Attacking a number of hosts or substrates; not specialized.

Plus strand—*See* Positive-sense strand (positive strand).

PMSF (phenyl methanesulphonyl fluoride), gamma-toluenesulphonyl fluoride—An inhibitor of proteases used in studies on viral proteins to avoid proteolysis.

Pneumomycosis—Mycosis of the lungs.

Pocket plum, bladder plum, plum pocket—Plum fruit swollen then "mummified" by *Taphrina pruni*.

Pocket rot—Localized decay of trunks or roots of trees by wood-destroying fungi.

Poculiform—Cup-shaped.

Podetium (pl. **podetia**)—A stalklike elevation in some lichens that arises from the thallus and supports an apothecium. *See* Pseudopodetium.

Podzol—Type of light colored, relatively infertile, acidic soil of cool, coniferous forests poor in lime and iron.

Poison—A substance that, when absorbed or taken orally, can cause illness, death, retardation of growth, or shortening of life; also called *biocide*.

Pol A mutant—A mutant of a bacterium that affects the DNA repair enzyme, DNA-dependent DNA polymerase II.

Pol gene—A gene that codes for a polymerase.

Polar (of bacteria, spores, etc.)—At the ends or poles.

Polar mutant—A mutant having effects on genes transcribed downstream.

Polar nuclei—Two centrally located nuclei in the embryo sac that fuse with a second sperm nucleus within the embryo sac to form a triploid nucleus. In certain seeds this fusion develops into the endosperm.

Polar transport—The directed movement within plants of compounds (usually hormones) mostly in one direction; this overcomes the tendency for diffusion in all directions.

Polar-diblastic, polaribilocular—Same as polarilocular.

Polarilocular (of ascospores)—Bicellular with the two cells separated by a central perforated septum; orculiform, polaribilocular, polar-diblastic.

Polarity—Morphological and physiological direction in plants.

Pole (unit of length) = 16.5 ft = 1 rod; also 1 sq rod or 30.25 yd^2.

Poleophilous (of lichens)—Lichens that thrive in urban areas; town loving.

Poles—The opposite ends of a mitotic spindle.

Pollen, pollen grains—Minute yellow bodies produced within the anthers of flowering plants or the cones of other seed plants; the male (micro) gametophyte.

Pollen mother cell—A 2n cell that divides twice (once by meiosis and once by mitosis) to form a tetrad of four pollen grains.

Pollen tube—A tubelike structure containing the tube nucleus in the microspore that helps guide the sperm from the pollen grain through the stigma and style of a flower to the embryo sac. A pollen grain and a mature tube are a male (micro-) gametophyte.

Pollination—The transfer of pollen from a stamen (or staminate cone) to a stigma or seed cone.

Poly (A)—Polyadenic acid.

Poly (A) polymerase—An enzyme that adds adenylate residues to the 3′ end of RNA

Poly- (prefix)—Many; a large number.

Polyacrylamide gel electrophoresis (PAGE)—Electrophoresis in a gel composed of polyacrylamide made by cross-linking acrylamide, usually with N,N′-methylene-bis-acrylamide. It is used for the separation of nucleic acid or protein molecules according to their molecular size and thus in estimating relative molecular weights.

Polyadenylation—The addition of adenylate residues usually to the 3′ end of RNA molecules.

Polyandrous (of oospores)—Formed when more than one functioning antheridium is present. *See* Monandrous.

Polyascous—Having many asci, especially asci in one hymenium not separated by sterile bands.

Polyblastic (of a conidiogenous cell)—Holoblastic and blowing out at more than one point; producing blastic conidia at several points.

Polycarpic (of *Exobasidium* infections)—Systemic, or circumscribed, and perennial. *See* Monocarpic.

Polycentric—Having several to many centers of growth and differentiation and more than one reproductive organ. *See* Monocentric and Reproductocentric.

Polycephalous—Many-headed.

Polychotomous—Having an apex that divides simultaneously into more than two branches.

Polycistronic—A nucleic acid coding for more than one cistron.

Polyclonal—Derived from many clones.

Polyclonal antibody—A preparation containing antibodies against more than one epitope of an antigen. *See* Monoclonal antibody.

Polycyclic—A disease of which many cycles occur in one growing season, resulting in many secondary infections.

Polycytous (of nematodes)—Having a large number of hexagonal cells (128–8,224) in the intestinal epithelium.

Polydactyloid venation (of a lichen)—An arrangement of veins in which the strands of tissue on the lower surface are confluent toward the tips of the lobes.

Polydelphic (of nematodes)—Describing females having three or more gonads.

Polyembryony—The presence of more than one embryo in a developing seed.

Polyenergid—Coenocytic.

Polyethylene glycol (PEG)—A polymer, $HOCH_2(CH_2OCH_2)_x CH_2OH$, available in molecular weights from 200 (PEG 200) to 20,000 (PEG 20,000); a coacervate that will bind water and can be used to concentrate solutions by withdrawing water from them. PEG 6000 is used in precipitating viruses and DNA.

Polyetic—Requiring many years to complete one life or disease cycle.

Polygalacturonases—Class of enzymes that catalyze the uncoupling of pectic chains into smaller units, ultimately into galacturonic acid residues.

Polygenic—Describing a character controlled by a number of genes. *See* Oligogenic.

Polygonal, polygonous—Having many angles.

Polyheads—Phage heads normal in diameter but exaggerated in length due to mutations.

Polyhedral, polyhedron—A sphaeroidal particle or crystal (e.g., inclusion body) having many sides or plane faces.

Polylysogeny—Bacterial strains lysogenic for several different phages. *See* Lysogen.

Polymerase—An enzyme that catalyzes the formation of RNA or DNA by the addition of ribonucleotide or deoxyribonucleotide triphosphates.

Polymerase chain return (PCR)—The selective amplification of DNA by repeated cycles of: 1) heat denaturation of DNA; 2) annealing of two oligonucleotide primers that flank the DNA segment to be amplified; and 3) the extension of the annealed primers with the heat-sensitive DNA polymerase; may be used to produce probes for virus diagnosis and in the amplification of low copy number sequences; a technique that allows a sequence of interest to be amplified selectively against a background of an excess of irrelevant DNA.

Polymerize—To subject to or undergo a chemical reaction in which two or more similar molecules combine to form larger molecules of repeating structural units.

Polymorphic—Having different forms; pleomorphic.

Polymyarian (of nematodes)—An arrangement of the somatic musculature, in which a large number of longitudinal rows of muscle cells is present between each two chords; best viewed in transverse body section.

Polynucleate—Having more than one nucleus per cell.

Polynucleotide kinase—An enzyme that phosphorylates the 5'-OH terminal ends of RNA and DNA nucleotide chains.

Polynucleotide ligase—A generic term for enzymes that catalyze the linking or repair of DNA or RNA strands. *See* DNA ligase and RNA ligase.

Polypeptide—A chain of amino acids linked together by peptide bonds from the synthesis or partial hydrolysis of a protein.

Polyphagous (n. **polyphagy**)—Feeding on many kinds of food, especially nematodes capable of feeding on a large number of host plants and decaying organic matter; the act of an insect (e.g., aphid) feeding on various secondary host species. *See* Monophagous, Primary host, and Secondary host.

Polyphialide, polyphialidic (of a conidiogenous cell)—Phialide with more than one opening; a cell with more than one conidiogenous locus in which conidia are produced. *See* Monophialidic.

Polyphyllous (of foliose lichens)—Having many connected leaflike lobes.

Polyplenetic, polyplanetism (of zoospores of oomycetes)—Condition of having motile and resting phases in turn; two or more flagellate phases interspersed with mobile apanosporic phases in the zoosporic part of the life history; motile phases may be monomorphic or dimorphic.

Polyploid virus—A virus, the particles of which contain a variable number of genomes depending on such factors as host cell and cultural conditions. Polyploidy is found in viruses of various groups.

Polyploid, polyploidy—Having more than two sets of 2n chromosomes per nucleus.

Polypore—One of a large group of wood-decaying fungi (Polyporaceae) with a basidiocarp bearing a hymenophore consisting of united tubes opening by pores on the lower surface.

Polyporoid (of the hymenophore)—In the form of united tubes opening by pores.

Polyprotein—A large precursor protein later cleaved to give two or more functional proteins.

Polysaccharide—A large, long-chain, organic molecule (e.g., cellulose, glycogen, starch) consisting of many monosaccharide units (simple sugars).

Polysome, polyribosome—A cluster of ribosomes attached to a messenger RNA (mRNA) during the translation of that RNA.

Polysporic, polysporous—Many-spored.

Polystichous—Occurring in many rows.

Polytomous—Dividing into many branches, usually one at a node.

Polytretic (of a conidiogenous cell)—Tretic with several channels or pores.

Polytrichous—Having many hairs.

Polytypic—Species based on more than one type; a species containing two or more subspecies each based on a different type.

Polyxeny—*See* Pleioxeny.

Pome—Simple, fleshy, indehiscent fruit derived from several carpels, the receptacle fleshy; outer pericarp fleshy and the inner pericarp papery, e.g., apple, pear, and quince fruits.

Pomiform—Shaped like an apple.

POPOP—A secondary solute—1,4-bis-2-(5-phenyloxazolyl)benzene—used in scintillation fluids to shift wavelength and enhance radioisotope detection by liquid scintillation counting.

Pore—A small opening; in Polyporaceae and Boletaceae, the mouth of a tube.

Pore fungi—Polyporaceae and Boletaceae.

Pore size distribution—The volume of the various sizes of pores in a soil expressed as a percentage of the bulk volume (soil plus pore space).

Poricin—An antitumor antibiotic from *Oxyporus (Poria) corticola.*

Poroid (of the hymenophoral surface of polypores)—With pores that are the openings of united tubes.

Porosity—That percentage of the total bulk volume of soil not filled by solid particles.

Porospore—Spore developed by expansion through a small, porelike opening of fixed dimension.

Porraceous—Leek-green.

Porrect—Extended; protracted.

Portal—Place of entrance.

Positive-sense strand (positive strand)—For RNA, the strand that functions as the messenger (mRNA), and for DNA, the strand with the same sequence as the mRNA.

Postemergence—The period after the appearance of a specified weed or crop plant.

Postemergence damping-off—*See* Damping-off.

Postemergence spray—A pesticide applied after crop plants have emerged from the soil.

Posteriad—Directed backward; opposed to anteriad.

Posterior (of a lamella)—The end at or toward the stipe; at or in the direction of the back.

Posterior bulb—The basal bulb of nematodes.

Post-transcriptional cleavage—Cleavage of polyprotein or RNA into functional units, usually monocistronic mRNAs. *See* Subgenomic RNA.

Post-transcriptional processing—Alterations in the structure of an mRNA after it has been transcribed from either DNA or RNA.

Postuterine sac (of nematodes)—A rudimentary or vestigial ovary that may serve to store spermatozoa (often called postvulval uterine sac).

Postvulval-uterine sac (of nematodes)—*See* Postuterine sac.

Potassium phosphotungstate—A negative stain for electron microscope samples usually made up as a 2% solution of phosphotungstic acid with the pH adjusted to pH 6.8 with KOH.

Potting mixture (soil mix)—A combination of several ingredients, e.g., soil, peat moss, sand, perlite or vermiculite, etc., designed for starting seeds, cuttings, or growing plants in containers.

Pound avoirdupois (lb av) = 16 oz av = 14.5833 troy oz = 453.5924 g = 0.4535924 kg.

Powdery mildew—Fungus (or disease) that forms a superficial white coating on the surface of leaves, stems, fruits, buds, and flowers; generally refers to members of the Erysiphales or a type of disease caused by these fungi. *See* Downy mildew.

Powdery scab (of potato)—Disease caused by *Spongospora subterranea*.

ppb—*See* Parts per billion.

ppm—*See* Parts per million.

PPO—A primary solute (2,5-diphenyoxyazole) or scintillator used in scintillation fluids.

Praemorse, premorse (of the stipe base)—As if broken off abruptly; truncate.

Praticole, praticolous—Living in meadows.

Preadult (of nematodes)—The last juvenile stage before adulthood; the fourth-stage juvenile or larva.

Precambrian era—*See* Proterozoic.

Precipitation (syn. precipitin) reaction—In immunology, the formation of a visible fine precipitate complex that occurs when adequate quantities of soluble antibodies (precipitins) and soluble antigens react to form an insoluble lattice.

Such complexes are detected in a test tube as a sediment and in agar gels as a white line appearing where the antigen and antibody interact. *See* Immunoprecipitation.

Precipitin—The reaction in which an antibody causes precipitation of soluble antigens. *See* Antigen.

Predacious fungi—Fungi that parasitize amoebae, nematodes, and other small terrestrial or aquatic protists or animals.

Predator—An animal that feeds by preying on other animals.

Predispose—To make prone to infection and disease.

Predisposition—The tendency of nongenetic factors (living and nonliving) acting before infection, to increase the susceptibility of a plant to disease; commonly results from pathogenic or environmental stresses.

Preemergence—The period before the emergence of a specified weed or crop plant.

Preemergence damping-off—*See* Damping-off.

Preemergence spray—A pesticide applied after planting but before crop plants have emerged from the soil.

Preoccupied name—A later homonym; a name identical to one given previously to a different taxon.

Prerectum (of nematodes)—A differentiated portion of the posterior portion of the intestine of dorylaims and related forms; separated from the main intestine by a constriction in the lumen; the granules of its cells are smaller, less numerous, and much lighter in color than those of the intestine.

Preservative—A substance used for killing or stopping the growth of microorganisms in or on any kind of substrate.

Prevalid (of names or authors)—Published before 1753, the starting point for the nomenclature of organisms under the Code; devalidated. *See* Nomenclature and Valid.

Primary—First; first-formed.

Primary culture—The establishment of cells in culture from tissue.

Primary cycle—Of plant disease; the first cycle to begin in a given year.

Primary homothallism—*See* Homothallism.

Primary host—Plant on which the sexual forms of an aphid mate and lay eggs to overwinter. *See* Secondary host.

Primary infection—The first infection by a pathogen after going through a resting or dormant period. *See* Secondary infection.

Primary inoculum—Propagules or vegetative structures of a pathogen, usually from an overwintering source, that causes initial rather than secondary outbreaks of disease.

Primary mycelium (of basidiomycetes)—Haploid mycelium formed directly by the germinating basidiospore.

Primary response—First immune response to an immunogen not previously recognized, predominantly IgM antibody.

Primary root—Root that develops directly from the mesocotyl (radicle) of an embryo rather than from a crown or node.

Primary septum—Septum found in direct association with nuclear division, by constriction or mitosis, separating the daughter cells and having a pore that may be modified as a dolipore (in basidiomycetes) or be associated with Woronin bodies (in ascomycetes). *See* Septum.

Primary squamules—First formed squamules of *Cladonia* from which podetia arise.

Primary symptom (syn. **local symptom**)—Symptom(s) produced soon after infection at the site of entry, in contrast to a secondary symptom, which follows more complete invasion (colonization) of the host. *See* Secondary symptom.

Primary tissue—Tissue developed by an apical meristem during growth in length.

Primary universal veil—Same as protoblem.

Primary wall—Cell wall layer lying next to the middle lamella, or intercellular layer.

Primase—An enzyme that synthesizes the RNA primers for DNA synthesis on a DNA template. *See* Okazaki fragments.

Primer—Small fragment of nucleic acid with a free 3′-hydroxyl group necessary for the initiation of DNA and, sometimes, RNA synthesis. *See* Primase.

Primordial—First in appearance; pertaining to the earliest stages of development.

Primordial covering (or cuticle)—Same as blematogen.

Primordial hyphae—Very intensely colored hyphae of the epicutis in *Russula*.

Primordial shaft—The monoaxial basidiocarp initial especially in the Clavariaceae.

Primordial tissue—Undifferentiated tissue of a basidiocarp initial. *See* Lipsanenchyma.

Primordial veil—Same as protoblem.

Primordium (pl. **primordia**, adj. **primordial**)—The rudimentary or initiating portion from which any organ, structure, or individual is formed; first in order of appearance; pertaining to the earliest stages of development.

Primospore—A spore very much like an ordinary cell of the organism.

Priorable—*See* Legitimate.

Pristine—Early; original; primitive.

Private idiotope—*See* IdI.

Private pesticide applicator—A person certified to use or supervise the use of a restricted use pesticide.

Pro- (prefix)—Before.

Probasidium (pl. **probasidia**)—That portion of the basidium in which karyogamy takes place.

Probe—A radioactive nucleic acid used to detect the presence of a complementary strand by hybridization; a specific sequence of DNA or RNA used to detect complementary sequences by hybridization.

Probit analysis—A statistical method that involves transformation of dosage-response data as percentages or proportions to probits that has the effect of changing the normal sigmoid curve characteristic of this type of data into a straight line by stretching the linear scale on which the percentages or proportions are measured.

Probolae (of nematodes)—Prominent, sometimes ornate, specialized structures on the lips.

Procambium—A primary meristem that gives rise to vascular bundles.

Procapsid—A viral capsid lacking nucleic acid and considered to be a stage in virion formation.

Process—An outgrowth or a projection from a surface.

Procorpus (of nematodes)—The cylindrical portion of the corpus anterior to the metacorpus (of the esophagus).

Procumbent—Nearly prostrate; spreading.

Prodelphic (of nematodes)—Describing a female having the ovary anterior to the vulva; the uterus directed anterior to the vulva.

Prodiploidization cell—The single cell in an ascocarp in which diploidization is possible.

Prodiploidization hypha—Every growing hypha capable of being diploidized. *See* Flexuous (receptive) hypha.

Productive infection (in virology)—Infection of a cell in which complete virus particles are found.

Profile (soil)—A vertical section of the soil through all the horizons and extending into its parent material.

Progametangium (in the Mucorales)—A swollen lateral branch forming a gametangium and suspensor cell.

Progamous (of zygomycetes)—A group of sex hormones.

Progeny—The young or offspring.

Progressive evolution—Evolution from simple toward more complex and more highly specialized structures.

Prohybrid (of fungi)—A mycelium having additional nuclei through hyphal fusions and nuclear migrations.

Prokaryote (adj. **prokaryotic**)—An organism lacking membrane-limited nuclei or organelles and not exhibiting mitosis (e.g., bacteria and blue-green algae); organisms with a genome of a simple, circular, double-stranded DNA free in the cytoplasm and not in organelles. *See* Eukaryote.

Prolate (of a spore, sporocarp, etc.)—Elongated (drawn out) toward the poles.

Proliferation—The rapid and repeated production of new cells, tissues, or organs; development of an abnormal number of flowers or fruits in the position normally occupied by a single organ; process of elongation of conidiogenous cells often associated with conidium production.

Proliferin—An antitubercle bacillus antibiotic from *Aspergillus proliferans*.

Promeristem—A rounded cone of meristematic tissue at the growing point.

Promitosis—A special type of intranuclear division during the growth stage of Plasmodiophoraceae; cruciform division.

Promoter—A region of DNA or RNA that is recognized by RNA polymerase in initiation of transcription.

Promycelium (pl. **promycelia**)—Basidium of the rusts and smuts; initial short and short-lived hypha produced upon teliospore germination from which promycelial spores (now called *sporidia*) are produced.

Pronase—A nonspecific proteolytic enzyme isolated from *Streptomyces griseus*.

Pronate—Inclined to grow prostrate.

Propagative virus—A virus that multiplies (replicates) in its insect vector.

Propagule—Any part of an organism capable of initiating independent growth when separated from the parent body.

Proper exciple (margin)—Same as excipulum proprium.

Properdin—A factor in serum associated with natural resistance to infection.

Prophage—The genome of a phage perpetuated in the host cell by integration into the host chromosome or by plastid formation. *See* Temperate phage.

Prophase—An early stage of mitosis in which the chromosomes become distinct and in which the nuclear membranes disappear; also prophase one and prophase two of meiosis.

Prophialide—Primary sterigma. Same as metula.

Prophylactic—Preventive treatment for protection against disease.

Prophylaxis—Prevention.

Prosenchyma (pl. **prosenchymata**)—Plectenchma in which the single hyphal elements are still recognizable as such.

Prosenchymatous (stroma)—Made up of elongated hyphal filaments still recognizable as such; often refers to undifferentiated tissue filling a fruitbody, especially a young perithecium.

Prosoplectenchyma—Plectenchyma composed of cells with elongated lumina. *See* Euthyplectenchyma.

Prosorus (of Chyridiales)—A cell giving rise to a group of sporangia, the sorus.

Prosporangium (in Oomycota)—The initial sporangiumlike body that extrudes a vesicle (sporangium) in which zoospores undergo development and from which they escape; presporangium.

Protandry (of nematodes)—A state in a hermaphroditic system in which the gonad produces sperms that are stored; they later fertilize the eggs subsequently developed by the same gonad.

Proteases—Enzymes that digest proteins by first cleaving polypeptide chains into small fragments; involved in activating and degrading proteins and widely used experimentally for peptide mapping and structural studies.

Protectant (of pesticides)—An agent, usually a chemical, that tends to prevent or inhibit infection by a pathogen.

Protection—Placement of a chemical or physical barrier between the pathogen and the host that prevents infection.

Protective fungicide—Fungicide used to protect an organism against infection by a fungal pathogen.

Protective layer—Layer of suberized cells formed below the abscission layer.

Protein—Complex, high molecular weight, organic, nitrogenous substance (polymer compound), built up of amino acids joined by peptide bonds.

Protein A (staphylococcal protein A)—Protein found in extracts of *Staphylococcus aureus* that reacts serologically with a majority of human sera. This reaction is due to binding of Protein A to the Fc fragment of human and some other mammalian IgG.

Protein kinase—Enzyme that catalyzes the phosphorylation of proteins usually in the presence of cyclic AMP or cyclic GMP.

Protein subunit—A small protein molecule that is the structural and chemical unit of the protein coat of a virus; a capsomere.

Proteinase—An enzyme that hydrolyzes proteins to polypeptides.

Proteolytic cleavage—The splitting or digesting of polyproteins into simpler structural compounds.

Proteophilous fungi—Fungi associated with ammonia-rich soils.

Proterandry—A condition in flowers in which the stamens mature and shed their pollen before the stigma of the same flower matures.

Proterogyny—A condition in flowers in which the stigma matures and is pollinated by pollen of another flower before the stamens of the same flower shed their pollen.

Proterospore—Spore formed at the start of the sporulation period of *Ganoderma*. It can germinate easily without passing through the gut of a fly larva.

Proterozoic—The earliest geologic era, beginning about 4.5–5 billion years ago and ending some 570 million years ago; also called the Precambrian era.

Prothallus (pl. **prothalli**)—The gametophyte of ferns and similar plants; (of lichens) *see* Hypothallus.

Prothecium—A primitive or rudimentary perithecium, as in the Gymnoascaceae.

Protista—*See* Protoctista.

Protist—Unicellular or a cellular organism, distinct from multicellular plants or animals, of the Kingdom Protoctista (Protista) including protozoans and many algae.

Proto- (prefix)—First; primitive; primordial.

Protoaecium, protoperithecium, protouredinium—Haploid structures that, after diploidization or dekaryotization, become fruiting structures.

Protobasidium—*See* Basidium.

Protoblem—A loose flocculent mycelial layer covering the universal veil, as in *Amanita*; primordial or subuniversal veil.

Protoconidium—*See* Hemispore.

Protoctista (Protista)—Kingdom of eukaryotic microorganisms but not including the Kingdoms Animalia, Fungi, and Plantae. Protoctista currently includes the Myxocycota, Oomycota, and related groups, with some authors instead accepting Chromista and Protozoa.

Protoderm—A primary meristem that gives rise to epidermis.

Protohymenial—Having a primitive hymenium.

Protolog, protologue—Everything associated with a species, genus or other name given by an author(s) on its first publication, i.e., diagnosis, description, geographical data, synonymy, citation of specimens, discussion, comments, illustrations, and references.

Protomers—Protein subunits that form a capsomere.

Protonema—A branching filament forming an early stage in the gametophyte generation of a moss.

Protonym (in nomenclature)—A name effectively but not validly published after the starting point for the group.

Protoperithecium (pl. **protoperithecia**)—A young walled perithecium before ascus formation.

Protophloem—The conductive tissue of actively growing parts of the plant. Its sieve tubes function for a brief period and are replaced by metaphloem elements.

Protophyte—*See* Antithetic.

Protoplasm—Living material within a cell in which all vital functions of nutrition, secretion, growth, and reproduction depend; essential semifluid, viscous, translucent colloid of all plant and animal cells.

Protoplasmodium—An undifferentiated microscopic plasmodium that gives rise to a single sporangium. *See* Plasmodium.

Protoplast—The organized living unit of a cell exclusive of its wall; the cytoplasmic membrane and the cytoplasm, nucleus, and cytoplasmic organelles are a part of it. *See* Sphaeroplast.

Protoplast fusion—A tissue culture procedure for somatic hybridization used in cell manipulation studies; the joining of two protoplasts or the joining of a protoplast with any of the components of another cell such that genetic transfer may occur. *See* Genetic engineering.

Protosexual (of yeasts and other organisms)—Describing organisms having diploid or dikaryotic cells that produce haploid or unisexual cells in the absence of fruiting structures or sexual spores; in contrast to parasexual, which is redefined to cover organisms having both protosexual and sexual cycles, and *neosexual*, for organisms having a sexual but not a protosexual cycle.

Protospore—A multinucleated mass of cytoplasm cut out by primary cleavage planes, followed by further cleavage to form the uninucleate spores of *Phycomyces* and other Mucorales, and the sporangiospores of *Coccidioides*; (of Synchytriaceae), a one-nucleate portion of protoplasm that becomes the sporangium.

Protostigma—*See* Basidium.

Protostom (of nematodes)—The middle portion of the stoma between the cheilostom and telostom; sometimes subdivided into a prostom, mesostom, and metastom.

Protothecium—An incompletely differentiated ascoma without asci nor ascospores.

Prototroph—A strain of microorganisms having the same nutritional requirements as the parent strain; an organism that is nutritionally independent and able to synthesize the growth factors it needs from similar substances.

Prototunicate (of asci)—*See* Ascus.

Protoxylem—Undifferentiated wood element in a fibrovascular bundle; the first-formed xylem, with annular, spiral, or scalariform wall thickenings.

Protozoan (pl **protozoa**)—Microscopic, motile animal consisting of one cell or a colony of like or similar cells.

Protuberant—Bulging out, prominent.

Protuberate (conidia)—Having short projections not qualifying as appendages.

Protype—Same as neotype.

Provenance—The natural origin of a tree or group of trees. In forestry, the term is generally considered synonymous with geographic origin.

Proximal, proximate—Nearest; next to or nearest the point of attachment or origin; opposite of distal.

Prozone—Suboptimal precipitation or agglutination that occurs in antibody or antigen excess depending upon which component is of variable concentration in the assay.

Pruinate, pruinose (of a surface)—Finely powdered, floury, or frostlike.

Pruinulose—Somewhat powdery.

Pruniform—Plum-shaped.

Pruning—The judicious removal of leaves, shoots, twigs, branches, or roots of a plant to control its size and shape or to increase its usefulness, vigor, or productivity.

Pseudo- (prefix)—False; spurious.

Pseudoacanthohyphidium—Hyphidium with a few apical, cylindrical prongs.

Pseudoaethalium (of myxomycetes)—A dense cluster of separate sporangia that appears as an aethalium.

Pseudoamyloid (of spores)—Reacting actively and strongly to iodine in Melzer's reagent and staining deep purplish brown. *See* Dextrinoid.

Pseudoangiocarpous, pseudoangiocarpic (n. **pseudoangiocarp**) (of a basidiocarp)—Having the hymenial surface first exposed but later covered by an incurving pileus margin and/or excrescences from the stipe.

Pseudobulb—The thickened or bulblike stems of certain orchids borne above the ground or substratum; (of nematodes) a swelling of the esophageal musculature in which the lumen does not widen to form a cavity; a nonvalvate esophageal bulb.

Pseudocapillitium (pl. **pseudocapillitia**) (of myxomycetes)—A sterile structure (plates, tubes, or threadlike bodies) in the fruitbody that has had no direct connection with the sporogenous protoplasm.

Pseudoclamp—A clamp tip not in union with the basal cell.

Pseudocleistothecium—*See* Discrete body.

Pseudocoelom (of nematodes)—The body cavity that is not lined by epithelium (mesoderm) and in which the various internal organs are suspended in a fluid. The body fluid apparently functions as a respiratory and circulatory system.

Pseudocolumella (of Physaraceae)—Lime-knots in a columellalike mass in the center of the sporangium.

Pseudocyphella (pl. **pseudocyphellae**)—An opening in the cortex of lichens in which the medulla is exposed to open air but lacks specialized cells around the cavity.

Pseudocystidium (pl. **pseudocystidia**)—A cystidiod terminal part of a conducting lactiferous hyphal element in agarics (e.g., *Stereum* and *Xylobolus*), turning into and sometimes protruding slightly from the hymenium (*see* Cystidium); (of *Entomophthora*) an organ penetrating the cuticle of an insect, allowing emergence of conidiophores.

Pseudodiblastic (of ascospores)—Having oil drops at the poles that superficially resemble polarilocular spores.

Pseudoepithecium—An amorphous or granular layer overlying paraphyses in an apothecium in which their tips are immersed, but not forming a separate tissue.

Pseudoidia—Separated hyphal cells capable of germination.

Pseudoisidium (pl. **pseudoisidia**)—Outgrowth of a lichen thallus resembling an isidium.

Pseudomixis, pseudogamy—Type of fertilization in which two copulating vegetative cells are not closely related and not special sexual cells.

Pseudomonad—A general term for aerobic Gram-negative rods that have the general characters of members of the bacterial genus *Pseudomonas*.

Pseudomorph—An indefinite stroma composed of plant parts bound together by plectenchyma that fills the interspaces.

Pseudomycelium (of *Candida*, etc.)—Loosely united catenulate groups of cells.

Pseudomycorrhiza—A situation in which the fungus parasitizes shortened, lateral roots of conifers.

Pseudoostiole, pseudoostiolum—A false ostiole; one with an orifice formed lysigenously without a lining of periphyses.

Pseudoparaphyses (of ascomycetes)—*See* Hamathecium; (of basidiomycetes)—*See* Hyphidium.

Pseudoparenchyma (adj. **pseudoparenchymatous**)—Isodiametric or oval fungus cells organized into a tissue in which the individual hyphae have lost their identity; aggregate of closely interwoven hyphae forming a definite body.

Pseudoparenchymatous (stroma)—Resembling the parenchyma of flowering plants.

Pseudoperidium—A false peridium; covering membrane of the aecium of rust fungi (Uredinales).

Pseudoperithecium (of Laboulbeniales)—Peritheciumlike structure in which the asci and ascospores become free.

Pseudophialide—A cell bearing a sporangiolum in the Kickxellaceae.

Pseudopinnotes—A slimy gelatinous mass of conidia with no evident conidiophores present. *See* Pinnote sporodochium.

Pseudoplasmodium—*See* Aggregate plasmodium.

Pseudopod, pseudopodium (pl. **pseudopodia**) (of myxomycetes)—Protoplasmic protrusion from a myxamoeba or plasmodium.

Pseudopodetium (of lichens)—Podetiumlike structure of vegetative origin on which the ascogonia arise.

Pseudopycnium (pl. **pseudopycnia**)—A pycnidium- or spermogoniumlike structure formed of hyphal tissue found in certain deuteromycetes (mitosporic fungi); now considered obsolete.

Pseudorecombinants (n. **pseudorecombination**)—New strains of a virus that result from the *in vitro* mixing of segments (reassortment) of genome nucleic acids during the replication of viruses with divided genomes in mixed infections.

Pseudorhiza (pl. **pseudorhizae**)—A rootlike extension of the stipe structure as in *Xerula (Collybia) redicata.*

Pseudosclerotial plate—A hard, dark plate, formed by certain fungi that colonize decaying wood, more or less impervious to water, solutes, or organisms and composed of large, thick-walled or encrusted fungal cells, affording protection to mycelium behind it; appearing as a black line (zone line) in transverse view.

Pseudosclerotium (pl. **pseudosclerotia**)—Sclerotiumlike structure; compacted mass of intermixed substrata (soil, wood, stones, etc.), possibly with host tissue but held together by mycelium to form definite bodies as in *Polyporus tuberaster. See* Stone-fungus.

Pseudoseptate (pl. **pseudosepta**)—With the appearance of having septa that do not reach across a spore from one side wall to the other.

Pseudoseptum, distoseptum—A protoplasmic or vacuolar membrane that appears to be a septum as in *Corynespora*; (in the Blastocladiales) a septum with pores.

Pseudosetae, false setae—Upturned free ends of context hyphae in the hymenium of *Duportella*.

Pseudospore (of Acrasales)—An encysted myxamoeba with no cell wall; a chlamydospore, as in *Rhizoctonia rubi*.

Pseudostem (of gasteromycete basidiocarps)—Spongy tissue in which hyphae are not oriented parallel to the stipe axis.

Pseudostroma (pl. **pseudostromata**)—A mass of fungal cells or closely interwoven hyphae combined with host cells to produce a stromalike structure; a false stroma.

Pseudothallus (pl. **pseudothalli**)—A false thallus.

Pseudothecium (pl. **pseudothecia**)—Aprotoperithecium; an ascocarp having bitunicate asci in one to many unwalled locules in a stroma that is sometimes very much reduced and may resemble a perithecium (as in the loculoascomycetes). *See* Euthecium.

Pseudotype—The genome of one virus enclosed in the capsid or outer coat of another, resulting from a mixed infection.

Psi—Pounds per square inch.

Psilocin and **Psilocybin**—Hallucinating indole derivatives from *Psilocybe mexicana*.

Psychrophile (adj. **psychrophilic**)—Organisms that are able to grow at low temperatures (below 100°C) with an optimal temperature usually below 200°C.

Psyllids—Jumping plant lice (Homoptera) of the family Psyllidae; some species are vectors of mycoplasmalike organisms.

PTA—Phosphotungstic acid.

Pterate—Alate; having wings.

Ptyophagous (of endotrophic mycorrhizae)—Young vigorous hyphae that rupture at the tips and extrude plasmal masses (*ptyosomes*) that are then digested by the host cells.

Puberulent—Minutely downy; somewhat hairy; very finely hairy.

Pubescence (adj. **pubescent**)—An outer covering of short, soft, silky hairs or down, as on the surfaces of leaves, stems, and fruiting bodies, etc.; also, the state of being so covered.

Public idiotope (IdX)—*See* Cross-reactive idiotope (CRI)

Puffball—The fruitbody of the Lycoperdales that emit clouds of dusty basidiospores when disturbed. *See* Giant puffball.

Pugoniform—Dagger-shaped.

Pullulation—A budding, as in yeasts.

Pulpose—Pulpy; fleshy.

Pulveraceo-delitescent—Covered with a layer of powdery granules.

Pulverulent—Powdery as though dusted over.

Pulvillus—A cushionlike group of cells.

Pulvinate—Cushion-shaped; circular and usually strongly convex.

Pulvinoid—More or less cushion-shaped.

Pulvinulus (pl. **pulvinuli**)—A small raised area.

Punctate—Marked with dots, very small spots, scales, or hollows.

Punctation (of nematodes)—Minute dots, ovals, pits, or depressions in deeper layers of the cuticle, occurring in transverse or longitudinal rows and sometimes arranged in various patterns.

Puncticulate, puncticulose—Minutely punctate.

Punctiform—Dotlike, under 1 mm in diameter, but seen with the naked eye.

Punctulate—Marked with small points.

Pungent—With an acrid or biting flavor.

Punk—Same as touchwood or amadou; a lumberman's term for the fruitbody of a hymenomycetous, wood-destroying fungus.

Punk knot—A dense sterile mass of mycelium of a wood-decaying fungus in a trunk or limb, issuing from the interior of the stem to the surface along a channel once occupied by a branch.

Punky—Describing decaying wood; soft and rather tough.

Pupa (pl. **pupae**)—Quiescent, nonfeeding stage between the larva and the adult of certain insects.

Pupate—To be inactive, as with certain insects in a cocoon or case, before ultimate maturation into an adult.

Pure culture—A culture in which only one species or biotype of an organism is growing.

Pure line—Plants in which all members are descended by self-fertilization from a single homozygous individual.

Pure live seed (PLS)—Percentage of the content of a seed lot that is pure and viable.

Purging agaric—*See* Female agaric.

Purification—The separation of molecules, cellular components, or virus particles in a pure form, free from other cellular constituents.

Purine—A heterocyclic compound containing fused pyrimidine and imidazole rings. The purine bases of nucleic acid are adenine and guanine. *See* Nucleic acid.

Puromycin—A broad-spectrum antibiotic produced by a strain of *Streptomyces*.

Purpurascent—Becoming purple.

Pustular, pustulate—Blisterlike; bearing blisters.

Pustule—A small, blisterlike or pimplelike, frequently erumpent elevation from which erupts a fruiting structure of a fungus that produces spores.

Pustuliform—Blisterlike; pimplelike.

Pustulose—Covered with small papillae or blisterlike structures.

Putrefaction—Anaerobic decomposition of organic substances, especially proteins, by microorganisms, producing disagreeable odors.

Putrescent—Soon decaying, becoming soft and mushy; not persistent.

pv.—*See* Pathovar.

Pycnidiospore—Conidium produced within a pycnidium; the term is now obsolete.

Pycnidium (pl. **pycnidia**, adj. **pycnidial**)—Asexual, usually cup- or flask-shaped to globose, ostiolate, thin-walled, brown fungus fruitbody of the Sphaeropsidales lined inside with conidiophores and conidiogenous cells that produce conidia (pycnidiospores) and composed of textura angularis, usually two to three cells thick but sometimes much thicker; pycnidial conidioma.

Pycniospore (of Uredinales)—Haploid (1n), sexually derived spore (*spermatium*) formed in a pycnium (*spermogonium*).

Pycnium (pl. **pycnia**)—Older term now replaced by spermogonium. Haploid (1n), flask-shaped fruitbody of the rusts (Uredinales) that contains spermatia (pycniospores) and filaments (receptive hyphae) that extend through the ostiole; designated by spore state O.

Pycnosclerotium—A more or less hard-walled structure that resembles a pycnidium but has no spores.

Pycnosis (of Microthyriaceae)—Process by which part of the stromatic thallus is arched up and thickens while an ascigerous hymenium forms under it.

Pycnothecium (of Microthyriaceae)—An ascoma formed by pycnosis.

Pycnothyrium—Superficial, shield-shaped, hemispherical, conidioma, sometimes comprised of only a radiate upper wall, sometimes also with a basal wall; pycnothyrial conidioma characteristic of Microthyriaceae.

Pygmaeous—Dwarfy; pygmylike.

Pyreniform—Nut-shaped.

Pyrenocarp—A perithecium.

Pyrenoid—Centers of starch formation on certain chloroplasts, especially of algae.

Pyrenomycetes—The class of Ascomycota that produce unitunicate asci in perithecia.

Pyriform—Pear-shaped. Same as piriform.

Pyrimidine—A heterocyclic organic compound containing nitrogen atoms at positions 1 and 3. *See* Nucleic acid.

Pyrophilous—Growing on burned ground, steam-sterilized soil, etc.; carbonicolous.

Pyroxylophilous—Growing on burned wood.

Pyxidate—Boxlike; provided with a lid.

Q

Q,—The ratio of length to breadth of elongate spores of agarics; spores are elliptical or ovoid when $Q = <2$; ellipsoidal-oblong, fusoid, cylindrical, etc., when $Q = >2$.

qt (quart, U.S., dry) = 2 pt = 1.1012 liter = 0.125 peck = 0.03125 bushel = 0.038889 ft^3 = 67.2 in.^{3}.

qt (quart, U.S., liquid) = 2 pt = 4 cups = 32 fl oz = 64 tablespoons = 0.25 gal = 0.946333 liter = 57.749 in.^{3} = 0.3342 ft^3.

Q-technique (of bacteria)—A method in assessing resemblances between two taxa by comparing all the characters. Numerical taxonomy deals principally with Q-techniques. *See* R-technique.

Quadi-, quadri (prefix)—Four, fourfold.

Quadrate (of spores)—Square or nearly so.

Quadricoccus—Composed of four round cells.

Quadrifid—Divided or cleft into four parts.

Quadrigonal (of spores)—Having four angles.

Quadripartite—Divided into, or consisting of four parts.

Quadrisporous—Having four spores.

Quarantine—State and/or federal legislative laws controlling the transport, import, export, and/or sale of plants or plant parts, usually to prevent spread of pathogens, insects, mites, weeds, or other pests; holding of imported plants or plant parts in isolation for a period to ensure their freedom from diseases and pests.

Quaternate—Arranged in fours.

Quercina—Inhabiting oaks.

Quick leaf-dip (syn. **quick-dip**)—*See* Epidermal-strip test.

Quiescent (n. **quiescence**)—Dormant; quiet; a rest period caused by external conditions unfavorable to germination or growth.

Quinine fungus—*Fomitopsis (Fomes) officinalis.*

Quinones—Any of various (usually yellow, orange, or red) quinonoid compounds, including several that are biologically important as coenzymes, hydrogen acceptors, or vitamins.

Quinque- (prefix)—Five.

Quinquefid—Divided or cleft into five parts.

R

R plasmids—Plasmids containing genes for drug resistance particularly common in enterobacteria but also present in other Gram-negative and Gram-positive bacteria. Some R plasmids code for sex pili and can thus influence phage host range.

Rabbit reticulocyte lysate—A cell-free system prepared from lysed rabbit reticulocytes used for the translation of eukaryotic mRNAs.

Race (or strain)—A subgroup or biotype of pathogens within a species, variety, or pathovar distinguished by behavior (differences in virulence, symptom expression, or, to some extent, host range) but not by morphology; different in virulence to cultivars (varieties) of the same host species; a genetically and often geographically distinct mating group within a species. See Physiologic race.

Raceme—Type of inflorescence in which the main axis is elongated and unbranched; the flowers are borne on pedicels that are about of equal length.

Racemose—Racemelike; having racemes.

Rachiform (of conidiogenous cells)—Having a rachis. See Raduliform.

Rachilla—Internal axis of a spikelet of grasses and sedges.

Rachis—Elongated main axis of a grass or cereal inflorescence (head); an extension of the petiole of a compound leaf that bears the leaflets; (of fungi) a geniculate or zigzag, holoblastic, spore-bearing extension of a conidiogenous cell resulting from sympodial development; (of nematodes) a central strand of non-nucleated tissue, to which oocytes are attached in the ovary and whose function is uncertain.

Racket or racquette cell (of dermatophytes)—A hyphal cell with a swelling at one end that more or less resembles a tennis racket with a very long handle. See Hypha.

Radial (of lichen thalli)—Radially symmetrical in transverse section.

Radial face—The wood surface exposed when a stem is cut along a radius from pith to bark, and the cut parallels the long axis of the majority of the cells.

Radial immunodiffusion—A serological test in which liquid antigen (or antibody) is placed in wells cut in agar gel containing the other reactant, and allowed to diffuse out into the gel. The resulting antigen-antibody complex forms a halo or ring of precipitate around the well.

Radial muscle fibers (of nematodes)—Muscle fibers of the esophagus, acting to dilate the lumen of the esophagus.

Radial section—Section of a stem or root cut longitudinally on a radius.

Radial symmetry—Type of floral symmetry in which the flower may be separated into two approximately equal halves by a longitudinal cut in any plane passing through the center of the flower; that is, a flower built upon a wheel plan, rather than on a right-and-left plan, e.g., rose.

Radiate—To extend outward from the center of a circle, as spokes in a wheel.

Radiate-lineate—Marked with radiating lines.

Radiately ridged—Ridges extending out from a center.

Radicating (of stipes)—With an elongated and rootlike underground portion; rooting.

Radiciform—Root-shaped.

Radicine—Rootlike.

Radicle—The part of the embryonic axis that becomes the primary root; the root primordium of an embryo; the first part of the embryo to start growth during seed germination.

Radioactivity radiation—The three major types emitted by a radioactive substance: 1) decay in which an energetic helium ion is ejected; 2) decay in which an energetic negative ion is emitted; and 3) decay that involves the ejection of protons. There are a number or radioisotopes used in biological experiments.

Radioimmunoassay—A sensitive and versatile system of protocols similar to those of enzyme-linked immunosorbent assay (ELISA), with solid-phase binding of either antibodies or antigens, but substituting radiolabeled antibodies or antigens for enzyme-IgG conjugation and recording data by means of a radiation (gamma or beta) counter. *See* Radioimmunoprecipitation.

Radioimmunoprecipitation—A serological test in which one of the reactants, usually the antibody, is labeled with a radioisotope, usually ^{125}I. The amount of isotope precipitated with the antigen-antibody complex can be measured very accurately for quantifying immunoprecipitation. The amount of antibody in the immunoprecipitate can also be measured using radiolabeled protein A.

Radula spore, radulaspore, radulospore—A slimy spore borne over the surface of ascospores as in *Nectria coryli* while still in the ascus. *Dry radula spore* is the same as *sympodulospore.*

Raduliform (of conidiogenous cells)—The elongating conidiogenous axis resulting from holoblastic sympodial conidial development, clavate or somewhat inflated rather than zigzag. *See* Rachiform.

Raduloid (of the hymenophore)—With broad, flattened, toothlike projections.

Ragi—A starter for arrack, etc., composed of small balls of rice flour containing *Mucor, Rhizopus,* yeasts, and bacteria.

Raised (in bacteriology)—Colony growth thick, with abrupt or terraced edges.

Ram-, rami-, ramo- (prefix)—Branch.

Ramal, rameal, rameous—Of, pertaining or belonging to, or growing on a branch.

Ramicole, ramicolous—Growing on branches.

Ramiferous, ramigerous—Bearing branches.

Ramiform—Shaped like a branch.

Ramify—To branch; to separate into divisions; to split into branches or constituent parts; to send forth branches or extensions.

Ramoconidium—An apical branch of a conidiophore that secedes. It functions as a conidium (e.g., *Cladosporium* and *Subramaniomyces*).

Ramose—Branched; branchy.

Ramulus (pl. **ramuli**)—A branchlet as in most Clavariaceae.

Ramus (pl. **rami**) (of a penicillus)—A cell bearing a verticil of metulae and phialides.

Ramycin—*See* Fusidic acid.

Range—(Of a plant pathogen) geographical region or regions in which it is known to occur; (of land and native vegetation) vegetation that is predominantly grasses, grasslike plants, or shrubs suitable for grazing by animals.

Range management—Producing maximum sustained use of range forage without detriment to other resources or uses of land.

Rangiferoid—Branched and pronged like a reindeer's horn.

Rank—In taxonomy, the word is used to indicate relative position in an orderly sequence; similar to *category*.

Rapaceous—Turnip-shaped.

Raphe—(Of fungi) specialized longitudinal dehiscence mechanism, only known in *Chaetomella*; (of higher plants) the ridge on seeds formed by the fusion of the stalk of the ovule with the seedcoat, in those seeds in which the funiculus is sharply bent at the base of the ovule.

Raphides—Needle-shaped crystals; as in some lichen thalli.

Rate zonal gradient—A technique in which a sample containing macromolecules is centrifuged through a gradient of an inert material (e.g., sucrose or glycerol), and the constituents of the sample are separated as bands on their sedimentation rates.

Rate—The amount of active ingredient of a pesticide applied to a given surface; also called *dose* and *dosage*.

Ratoon—Second crop produced from living roots after the first crop has been harvested (as in sugarcane).

Ray—The corolla of the marginal flowers of a head (inflorescence) in the family Asteraceae; also a vascular medullary ray, the tissue or narrow group of cells (usually parenchyma, but sometimes also tracheary cells) that extends radially in the secondary xylem and phloem of a woody plant.

Ray fungi—Actinomycetes (bacteria).

Rays (of nematodes)—Muscular genital papillae or tactile organs located within the bursa (primarily in animal parasites).

Razor-strap fungus—*Piptoporus* (*Polyporus*) *betulinus*, the birch fungus.

RDE (Receptor-destroying enzyme)—An enzyme that destroys the specific receptor by which a virus can attach to a susceptible cell.

RDE—Receptor-destroying enzyme; one that destroys the specific receptor by which a virus can attach to a susceptible cell.

rDNA—Recombinant DNA. A new DNA molecule produced by enzymatic insertion of one piece into another *in vitro*.

Re- (prefix)—Back, backward; again.

Reaction (soil)—*See* pH.

Reactivation—The activation of a virus from a latent stage or the "rescue" of a defective virus. *See* Rescue.

Reading frame—A sequence of codons in RNA or DNA beginning with the initiation codon AUG. *See* Open reading frame.

Readthrough—The reading of an mRNA through a stop codon. A suppressor tRNA causes the insertion of an amino acid into a growing polypeptide chain in response to the stop codon.

Reannealing—Linking by hydrogen bonds of complementary strands of nucleic acid after melting. *See* Hybridization.

Reassortment—Production of a hybrid virus that contains parts derived from the genomes of two viruses in a mixed infection. *See* Pseudorecombination.

Receptacle (pl. **receptacla**)—The more or less expanded terminal portion of the stem on which flower parts are borne; enlarged upper end of a pedicel or peduncle to which flower parts are attached; (of fungi) an axis having one or more organs; any hymenium-supporting structure; the inflated tips of certain brown algae within which gametangia are borne.

Receptive body—A small, branched or unbranched process from a stroma (as in *Stromatinia* (*Sclerotinia*) *gladioli*) able to be "spermatized" by microconidia.

Receptive hypha—Specialized hypha that protrudes from the top of a spermogonium (pycnium) and functions in sexual reproduction; flexuous hypha, trichogyne, and possibly other similar structures.

Receptor—(in physiology) An end organ or group of end organs of sensory or afferent neurons, specialized to be sensitive to stimulating agents; sensitive to a distinct (specific) signal molecule.

Recessive—One of a pair of contrasting characters that is masked, when both are present, by the other (dominant) character; also refers to the genes determining such characters.

Recessive gene—A gene that is not expressed in the phenotype of the heterozygote. *See* Dominant gene.

Reclinate—Turned or bent downward.

Recognition (of symbionts)—The process by which two compatible potential symbionts initiate a symbiotic relationship; discrimination between molecules or pathogens.

Recognition factors—Specific molecules or structures on the host (or pathogen) that can be recognized by the pathogen (or host).

Recombinant (n. **recombination**)—A new strain of a virus, bacterium, fungus, or other microorganism that occurs as a result of the breakage and renewal of covalent links in a nucleic acid chain, so that genes are swapped or replaced with different alleles; a cell or clone of cells resulting from recombination; formation in daughter cells of gene combinations not present in either parent; the mixing of genotypes that results in sexual reproduction; the exchange of genetic material from two or more virus particles into recombinant progeny virus during a mixed infection. *See* Reassortment.

Recombinant DNA—A new hybrid DNA molecule produced by enzymatic insertion of one piece into another from a different source *in vitro*. *See* Genetic engineering.

Recondite—Hidden; not readily seen.

Rectal glands (of nematodes)—Three to six large glands opening into the rectum of some Secernentea; in the genus *Meloidogyne*, they produce the gelatinous matrix into which eggs are deposited.

Rectum (of nematodes)—Posterior or hind gut of the female, opening into the anus or the cloaca in the case of males; a narrow, dorsoventrally flattened tube that is lined with cuticle and separated from the intestine by a sphincter muscle.

Recurrence—Return of symptoms after a period of absence.

Recurved—Curved backward, inward, or downward; to curve in an opposite direction; (of a pileus) convexo-expanded.

Red bread mold—*Chrysonilia sitophila*, teleomorph *Neurospora sitophila*. A common nuisance in bakeries.

Red rice—*See* Ang-kak.

Red rust—The summer or urediniospore spore state of rusts, especially of cereals, (e.g., *Puccinia graminis* and its *forma speciales*); (of tea) the alga *Cephaleuros*.

Red truffle—*Melanogaster variegatus*. *See* Truffle.

Reducing sugar—Sugar with a free carbonyl group, such as fructose, formed from hydrolysis of a complex sugar.

Reduction—A chemical process involving the removal of oxygen, the addition of hydrogen, or a gain of electrons. A corresponding oxidation always accompanies reduction. *See* Oxidation.

Reduction division—Stage of meiosis in which two daughter nuclei each receives half the chromosomes of the parent nucleus; separation of the homologous chromosomes during meiosis.

Reduction of indicators (of bacteria)—Reduction of certain colored compounds resulting in loss of color.

Reflective index—The ratio of the phase velocity of light in a vacuum to that in a specific medium as measured using a refractometer.

Reflective mulch—Plastic, aluminum, or straw layer placed on the soil surface around crop plants to control airborne insect vectors.

Reflexed—(Of an edge) bent, turned, or folded back upon itself; (of nematodes) applied to ovaries and testes; (of basidiocarps) having a pileate portion with a sterile upper surface.

Refracted—Bent sharply backward from the base.

Refractive (of hyphal or cystidial contents)—Light deflecting.

Refringent spot—Shiny, light-reflecting globule.

Regeneration—Replacement of lost parts by growth.

Registered seed—The progeny of foundation or registered seed, produced and handled so as to maintain satisfactory genetic identity and purity, and approved and certified by an official certifying agency. Registered seed is normally grown for the production of certified seed.

Regular flower—One in which the corolla is made up of similarly sized and shaped petals equally spaced and radiating from the center of the flower; star-shaped flower; actinomorphic.

Reindeer lichen or moss—Mainly *Cladonia stellaris* and *C. rangiferina*; found as far north as the extreme limits of vegetation and grazed by reindeer and caribou.

Reiterated sequence—A nucleotide sequence that occurs many times in a nucleic acid.

Relative centrifugal force (RCF)—The centrifugal force induced by centrifugation relative to the force of gravity. There are two ways to calculate it: RCF (g) = $1,118 \times 10^{-8} \times r \times N^2$ or RCF (g) = $284 \times 10^{-7} \times R \times N^2$; where N = rpm; R = distance from the center of rotation in inches; and r = distance from the center of rotation in centimeters.

Relative humidity (RH)—The ratio of the quantity of water vapor pressure in the atmosphere to the quantity that would saturate it at the same temperature.

Relax—To inactivate nematodes, generally by heat prior to fixing; to kill or anaesthetize.

Remote—(Of lamellae) proximal end free and at some distance from the stipe; (of the annulus) at some distance from the apex of the stem.

Renette (of nematodes)—A ventral cell or group of cells emptying into the excretory duct(s) in Adenophorea.

Renewal (replacement) spurs—Grape canes near the trunk cut back to two buds to provide new fruiting wood in a desired location.

Reniform, renarious—Kidney-shaped; fabiform.

Rennet—The rennin-containing substance from the stomach of a calf; a preparation or extract of the rennet membrane, used to curdle milk; the lining membrane of the fourth stomach of a calf or of the stomach of certain other young animals.

Rennet curd—Coagulation of milk protein due to rennet or rennetlike enzyme and distinguished from acid curd by the absence of acid. *See* Acid curd.

Rennin—Enzyme found in certain kidneys.

Renovation—To invigorate or rejuvenate, thin plants, remove weeds, and form new plants.

Repand, repandous (of a pileus)—Having a wavy margin that is turned back or elevated.

Repeated sequence—A nucleotide sequence occurring more than once in a DNA or RNA sequence in the same or opposite orientation.

Repeating spore—Spore form produced several times in succession during a growing season in the absence of sexual reproduction; often responsible for epidemics (epiphytotics).

Repent (mycelium)—Creeping; prostrate and rooting.

Repetite—Repeatedly.

Repetition, spore germination by—Forming a new spore like the first.

Replica plating of bacteria—A technique in which the same colony or colonies can be transferred to several different plates without change in the spatial relationships between the colonies. A disk of sterile material (e.g., velveteen) is pressed on the surface of the first plate, and the adhering bacteria are "printed" on the second plate.

Replicase—The enzyme involved in the replication of viral genomic nucleic acid.

Replication (v. **replicate**)—The process by which a virus particle induces the host cell to reproduce the virus; the process by which a DNA or RNA molecule makes an exact copy of itself; to fold or bend back (replicated leaf); repetition of an experiment or procedure at the same time and place (one of several identical experiments, procedures, or samples).

Replicative form (RF)—The intracellular form of viral nucleic acid active in replication. It is usually the double-stranded form of that nucleic acid in viruses with single-stranded genomes.

Replicative intermediate (R±)—Form of nucleic acid produced during the replication of a viral genome.

Replicon—A portion of DNA able to replicate from a single origin. Since most viruses have one origin of replication, the entire genome is a replicon.

Repress—Prevent (e.g., signal-receptor interaction).

Repression—Decrease in the rate of synthesis of an enzyme specifically caused by a small molecule, usually the end product of a biosynthesis pathway, e.g., an amino acid or nucleotide.

Repressor—A protein that prevents RNA polymerase from starting RNA synthesis by binding to a specific DNA sequence upstream of the transcription initiation site.

Reproduction, sexual—Development of new plants by seeds (except in apomixis); the sexual cycle of a fungus or other microorganism.

Reproduction, vegetative (vegetative propagation)—Reproduction by other than sexually produced seed; includes grafting, cuttings, layering, etc., as well as apomixis.

Reproductocentric (of Chytridiales)—Having development of one or more reproductive structures at the center of gravity of the thallus; same as genocentric. *See* Monocentric and Polycentric.

Rescue—The reactivation of a defective virus by recombination or by complementation of the defective functions.

Resident bacteria—Pathogenic and nonpathogenic bacteria that persist and multiply on surfaces of leaves, fruit, and expanding buds of plants, without causing disease; epiphytes.

Resident phase—The stage of growth of pathogenic bacteria that colonize the foliage surface of plants.

Residual—Pertaining to or constituting a pesticide residue that persists after application in amounts sufficient to kill pests for several days to several weeks or longer.

Residue (in relation to pesticides)—Amount of chemical that remains on or in a harvested crop. It is illegal to have a pesticide residue on a crop at harvest above the set tolerance or on a crop that is not registered.

Resilient—Bending or springing back.

Resin—Sticky to brittle plant product derived from essential oils and often possessing marked odors; used in varnishes, incense, medicines, etc.

Resin-flux—A copious flow of resin in conifers due to attack by fungi, insects, etc.

Resinosis—Exudation of oleoresin (pitch) from a wound or infection on a conifer.

Resinous (of consistency or taste)—Like resin, as if impregnated with resin, or tasting like resin.

Resistance—The inherent ability of an organism (host plant) to overcome or retard, completely or to some degree, the activity (infection) of a pathogen or other damaging factor; (of a pest) able to withstand exposure to certain pesticides. Acquired resistance is a noninherited resistance response in a normally susceptible host following a predisposing treatment. *See* Continuous resistance, Discontinuous resistance, Field resistance, Horizontal resistance, Hypersensitivity, Susceptibility, Tolerance, and Immunity.

Resistant—Possessing qualities that hinder the development of a particular pathogen or the effects of other damaging factors. A plant may be slightly, moderately, or highly resistant.

Resistant sporangium—Same as meiosporangium.

Resorption—The action of again absorbing a substance previously differentiated.

Respiration—A series of enzymatic reactions (oxidation processes) in which a living organism produces energy for cellular activities, usually by oxidizing carbohydrate and releasing carbon dioxide; the oxidation of food by plants and animals to yield energy for cellular activities.

Rest period or dormancy—An endogenous physiological condition of viable seeds, buds, bulbs, or other reproductive bodies that prevents growth even in the presence of otherwise favorable environmental conditions. *See* Dormant.

Restant—Persistent.

Resting (of a spore or sclerotium)—Temporarily dormant, usually thick-walled, capable of later germination and initiating infection (e.g., chlamydospores, oospores, teliospores, sclerotia) after a resting period, frequently over winter; resistant to extremes in temperature and moisture.

Restriction endonuclease map—The cutting sites for restriction endonucleases marked on a linear or circular representation of a dsDNA molecule.

Restriction endonucleases—A group of enzymes, usually bacterial, that hydrolyze DNA at highly specific sites. In recombinant DNA experiments, restriction enzymes are used to cut dsDNA from a given organism into defined fragments of characteristic size by breaking internal bonds before it is recombined with a vector. Type I enzymes bind to the recognition site but usually cut the DNA at random sites. Type II enzymes bind at the recognition site and cut either at or close to that site. They are used extensively in recombinant DNA technology.

Restriction enzyme—Restriction endonuclease.

Restriction fragment—A polynucleotide fragment produced by cutting DNA with a restriction endonuclease.

Resupinate (of basidiocarps)—Completely effused (spread out flat and appressed to the substratum) nonpileate or reflexed parts with the hymenium facing outward; margin may be either adnate or loosely attached with upturned margins.

Resupinate-reflexed (of basidiocarps)—Attached for some distance by the back surface, the remainder extending outward like a shelf; effused-reflexed.

Retention (of surfaces)—Ability of a surface to hold a pesticide. *See* Adherence.

Reticulate—*See* Reticulum.

Reticulate-areolate—Apparently reticulate, but marked with a network of crevices.

Reticulum (adj. **reticular, reticulate,** n. **reticulation**)—Netlike or weblike structure; with netlike ridges or markings; anastomosing system of lines, veins, folds, and ridges.

Retiform—Netlike.

Retro- (prefix)—Backward.

Retroarcuate—Curved backward.

Retroculture—Reisolation of a pathogen from a host into which it had been experimentally introduced.

Retrocurved—Bent back, recurved.

Retrogressive evolution—Evolution from a structurally complex or specialized condition toward a simpler, less specialized condition.

Retrorse—Backward.

Retroserrate—Saw-toothed or sharply incised, with the teeth turned backward.

Retting—Destruction of pectin-binding material in the stems of plants, such as flax and hemp, so the fibers can be separated and used for the manufacture of thread, rope, and cloth.

rev/min—One abbreviation for revolutions per minute.

Revalidated—*See* Devalidated.

Reverse passive haemagglutination—A sensitive serological test in which red blood cells are coated with virus-specific antibody and used to test for the presence of an antigen. If virus antigen is present, the red blood cells are agglutinated. *See* Passive haemagglutination.

Reverse transcriptase—An enzyme coded by certain viruses that makes a DNA copy of a primed RNA molecule. The enzyme can also synthesize DNA from a DNA template.

Reverse transcription—Copying of the genetic information from RNA into DNA.

Reviving (of basidiocarps)—Resuming the normal fresh natural shape and functions in moist weather or when rewetted after having dried, shriveled, and become dormant in dry weather.

Revoluble—Capable of being rolled back.

Revolute—Edge rolled or turned backward, downward, or up.

Revolutions per minute (rpm)—A measure of the speed of a centrifuge. *See* Relative centrifugal force.

RF—Replicative form.

Rhabdions (of nematodes)—The plates of the cuticular lining of the stoma; compose the walls of the various divisions of the stoma in free-living Rhabditida; the major types are as follows:

1) *Cheilorhabdions*—most anterior plates of the stomatal wall; correspond to the entrance to the stoma.

2) *Protorhabdions*—the wall plates of the cylindrical stoma; may be subdivided into prorhabdions, mesorhabdions, and metarhabdions.

3) *Telorhabdions*—plates of the posterior portion of the stomal wall, serving as a connecting valve between the stoma and the esophagus.

Rhabditiform—Rod-shaped.

Rhabditoid (of nematodes)—Having characteristics of the genus *Rhabditis*. *See* also Esophagus.

Rhabditoid bursa (of nematodes)—A condition in which the caudal alae meet posteriorly and anteriorly, forming a complete oval; a wide caudal bursa.

Rhabdo virus—Small RNA animal or plant virus; characterized by a protein plus lipid protectant coat.

Rhagadiose—Deeply cracked or chinked.

Rhexolytic (n. rhexolysis, pl. rhexolyses)—Secession of conidia that involves the circumscissile splitting of the periclinal wall of the cell below the basal septum; the formation of cavities or openings or the freeing of elements by tearing or rupturing. *See* Schizolytic.

Rhinosporidiosis—Growths in the nose and other organs of humans, horses, etc. caused by *Rhinosporidium seeberi*.

Rhiz-, rhizo-, -rhiza—Combining forms meaning root.

Rhizina (pl. rhizinae)—A rootlike hair or thread; attachment organs of many foliose lichens; rhizine.

Rhizinose strand or "Rhizinenstränge" (of squamulose lichens)—A rhizinelike attachment organ that is tough and much branched. *See* Hyphal net.

Rhizobiophage—Phage isolated from *Rhizobium* species.

Rhizobium (pl. **rhizobia**)—Species of bacteria that live more or less symbiotically in the roots of leguminous plants, forming nodules, and fixing nitrogen that is used by these plants. Some species are now in the genus *Bradyrhizobium*, e.g., the soybean symbiont.

Rhizoid—Short, stiff, rootlike, radiating hyphal structure of some zygomycetes (e.g., *Rhizopus*), growing toward or into the substrate; any filamentous, branched extension of a chytrid thallus acting as a feeding organ; also, in some mosses, liverworts, etc., a hairlike appendage that penetrates the soil or other substratum, anchoring the plant and absorbing water and other substances; (of bacteria) describing growth of an irregular, branched or rootlike character, as colonies of *Bacillus mycoides*.

Rhizoidal—Of, pertaining to, made up of, or similar to a rhizoid.

Rhizomania—A disease of sugar beet (*Beta vulgaris*) caused by the beet necrotic yellow vein virus in which the rootlets proliferate.

Rhizome (adj. **rhizomatous**)—A mostly horizontal, jointed, fleshy, often elongated, usually underground stem that forms both roots and leafy shoots at its nodes; often enlarged by food storage.

Rhizomorph—Specialized type of mycelium in which several interwined strands or cords of hyphae, often dark colored, are woven into a macroscopic bundle to appear rootlike; frequently differentiated into a rind of small dark-colored cells surrounding a central core of elongated colorless cells; the dense stringlike or cordlike mass of aggregated hyphae acts as a single unit; serves as a survival organ and in the transport of food materials within the thallus.

Rhizomorphic (of basidiocarps)—With rhizomorphs.

Rhizomycelium—Filamentous, rhizoidal fungal thallus that resembles mycelium, e.g., the thallus of the Cladochytriaceae.

Rhizoplane—The surface of a root.

Rhizoplast—*See* Blepharoplast.

Rhizopodium—A branched process or pseudopodium from a plasmodium. *See* Filopodium.

Rhizosphere—The soil microenvironment immediately surrounding (within 5 mm) a living root; the microflora is frequently richer and different from that of soil away from a root.

Rhodosporous—Having light red spores.

RH—*See* relative humidity.

Rhynchosporous—Having beaked spores.

Ribonucleases (RNAases)—A group of enzymes with different activities that hydrolyze RNA. They can be used in the characterization and sequencing of RNA.

Ribonucleic acid (RNA)—Any of a number of nucleic acids containing ribose, uracil, guanine, cytosine, and adenine occurring in cell cytoplasm and the nucleus; RNA is associated with control of cellular chemical activities; the nucleic acid type of most plant viruses. See Nucleic acid and RNA.

Ribonucleoprotein—A complex composed of ribonucleic acid and protein usually linked by electrostatic bonds.

Ribonucleoside—A purine or pyrimidine base covalently bound to a ribose sugar molecule. See Nucleic acid.

Ribonucleotide—A ribonucleoside with one or more phosphate groups esterified to the 5′ position of the sugar moiety. See Nucleic acid.

Ribose—A five-carbon sugar; one of the components of RNA.

Ribosomal RNA (rRNA)—RNA molecules forming part of the ribosomal structure; they are known by their sedimentation (S) values.

Ribosome—A subcellular protoplasmic particle, made up of one or more RNA molecules and several proteins, involved in protein synthesis.

Ribosome binding site—A sequence of nucleotides in mRNA to which ribosomes will bind. These sites are three to nine bases long and precede the translation start codon by 3–12 bases. In eukaryotes, the ribosomes are thought to bind to the 5′ end of the mRNA.

Ribs (of nematodes)—In the cephalic framework, six rigid supporting structures radiate transversely, then posteriorly to fuse with the anterior edge of the basal ring and the six radial bars. Four ribs are sublateral, one dorsal, and one ventral. See Basal ring.

Rickettsiae—Pleomorphic or rod-shaped, small, fastidious bacteria (belonging to the Schizomycetes group) with a scalloped cell wall. They are intracellular parasites in the xylem and phloem of diseased plants.

Rimose (of a basidiocarp surface)—Deeply cracked; having chinks or crevices. See Rimulose.

Rimose-areolate (of the surface of a pileus)—Cracked or chinked as to mark out the surface in patches, definitely marked areas, or, almost in scales; same as tessellately-rimose.

Rimose-diffract—Widely cracked or chinked.

Rimulose—Having small cracks.

Rind—The firm outer layer of a rhizomorph, sclerotium, or other organ; cortex.

Ring test—A serological test in which the antigen solution is mixed with glycerol and placed in a tube and antibody is carefully layered on top. The antigen-antibody precipitate forms as a ring at the interface.

Ring—(Of a mushroom, agaric) same as annulus; (of liquid cultures, especially bacterial) growth at the surface and often adhering to the glass.

Ringent—Gaping; wide open.

Ringer's solution—A solution for culturing cells that contains 0.66 g NaCl, 0.015 g KCl, and 0.015 g $CaCl_2$ per 100 ml of distilled water, with the pH adjusted to 7.8 by adding $NaHCO_3$ drop by drop.

Ring-porous wood—Wood in which the pores (xylem vessels) of one part of a growth ring are distinctly different in size or numbers (or both) from those in the other part of the same ring.

Ringspot—Disease symptom characterized by yellowish or necrotic rings with green tissue inside the ring, as in some virus diseases; the rings may be irregular or indistinct due to the pattern of small veins in the leaf; the rings may spread and merge forming "oak leaf" or "watermark" patterns.

Ringworm—*See* Tinea.

Riparian, riparious—Growing on rivers or streams.

Ripening—Chemical and physical changes in a fruit that follow maturation.

RI—Replicative intermediate.

Rishitin, richitinol—Terpenoid phytoalexins from potato (*Solanum tuberosum*).

Rivose—Having sinuate channels.

Rivulose—Marked with little riverlike lines.

R-loop mapping—A technique in which single-stranded RNA is annealed to the complementary strand of partially denatured DNA. The formation of the RNA:DNA hybrid displaces the opposite DNA strand as a loop that can then be seen under an electron microscope. The DNA segment complementary to the RNA can then be mapped.

RNA—Ribonucleic acid. A nucleic acid found in the nucleus and cytoplasm involved in protein synthesis; also, the only nucleic acid (genetic material) of many plant viruses. It is made up of nucleotides in which the sugar is ribose and the bases are adenine, guanine, cytosine, and uracil.

RNA ligase—An enzyme that can join RNA molecules; it requires ATP.

RNA polymerase—An enzyme synthesizing RNA from a DNA or RNA template. *See* DNA-dependent RNA polymerase and RNA-dependent RNA polymerase.

RNA processing—The modification of RNA after transcription. *See* Posttranscriptional processing.

RNA replicae—RNA-dependent RNA polymerase.

RNA segment—A distinct piece of RNA; frequently refers to genomic segments of segmented genome viruses.

RNA transcriptase—RNA- or DNA-dependent RNA polymerase.

RNAases—*See* Ribonucleases.

RNA-dependent DNA polymerase—Reverse transcriptase.

Rock—The material forming the essential part of the earth's solid crust, including loose incoherent masses such as sand and gravel, as well as solid masses of granite, limestone, and others.

Rock hair—Pendent, brown, gray to black species of *Bryoria* that resemble human hair.

Rock tripe—Edible lichens of the genus *Umbilicaria*.

Rocket electrophoresis—A technique to measure antigens by electrophoresing antigen into an agar layer containing antibody, resulting in a rocketlike pattern of precipitation in agar.

Rocket immunoelectrophoresis—An immunological technique for determining a single constituent in a protein mixture in a number of samples. The diluted samples are applied side by side to circular wells in an agarose gel containing an antiserum to the protein of interest. Rocket-shaped precipitates that form upon electrophoresis identify the protein.

Rod (of bacteria)—A cell having a straight central axis that is longer than the diameter of the cross section of the cell.

Rod (unit of length)—5.5 yard = 16.5 ft = 198 in. = 5.02921 m = 0.003125 mi.

Rodlet—Structural unit of conidial and some hyphal walls composed of particles about 50 Å in diameter and arranged in linear series.

Roestelioid (of an aecium)—Long and tubelike having the appearance of the anamorphic genus *Roestelia*; and as in *Gymnosporangium*.

Rogue (n. **roguing**)—To remove and destroy undesired individual plants on the basis of disease infection, not true-to-type, insect infestation, or other reason, from a population. Also, a variation from the standard varietal type.

Rolling circle—A nucleic acid replication mechanism in which the template is a circular molecule found in the DNA replication of a phage and in the RNA replication of viroids.

Root—The descending axis of a plant, usually below ground, serving to anchor the plant and absorb and conduct water and mineral nutrients.

Root cap—Thimblelike mass of hard cells that fits over the apical meristem of a root and protects it from mechanical injury.

Root crown—The uppermost portion of the root system in which the major roots join together at the base of the stem.

Root devitalization—A nematode-induced cessation of root elongation caused by feeding at or near the root tip, causing stubby-root, coarse-root, and/or curly tip symptoms.

Root graft—Union of roots from two or more closely situated (within about 18 meters) trees of the same or closely related species; often an avenue of disease transmission (e.g., oak wilt and Dutch elm disease).

Root hair—Threadlike, single-celled outgrowth from a root epidermal cell through which water, nutrients, and other substances are absorbed into a plant.

Root knot—A nematode-caused disease characterized by round to irregular galls (knots) on the roots, caused by *Meloidogyne* species. Most common in sandy soils or untreated greenhouse beds, attacking over 2,000 kinds of plants.

Root nodules (of legumes)—Caused by nitrogen-fixing bacteria of the genus *Rhizobium* and *Bradyrhizobium* that live, more or less symbiotically, in the nodules of leguminous plants. Nodules on species of *Alnus, Elaeagnus, Hippophaë,* and *Myrica* are caused by members of the Plasmodiophorales.

Root pressure—Pressure developed in roots due to osmosis (active water absorption); may cause "bleeding" in stem or leaf wounds of some plants.

Root rot—Decay in roots of a living plant; sometimes caused by pathogenic fungi.

Root system—The total mass of roots of a single plant.

Root-felt (of citrus)—Disease caused by *Helicobasidium mompa*.

Rooting media—Materials such as peat moss, sand, perlite, or vermiculite in which the basal ends of cuttings are placed vertically during the development of roots.

Rootstock (understock)—Portion of the stem (trunk) and associated root system into which a bud or scion is inserted in grafting; fleshy overwintering part of a herbaceous perennial plant with buds and eyes.

Roridins—Terpinoid toxins of *Myrothecium roridum* and *M. verrucaria* that cause dendrochiotoxicosis (ill-thrift) in sheep, pigs, and humans.

Roridous—Covered with dewlike drops of liquid.

Rosaceae—Plant family to which many common plants belong, including apples and roses.

Rosaceous, rosellate, roseate, roseous—Pinkish or rose-colored.

Rosario—Arranged as beads on a string (Spanish for *rosary*).

Roseolate, roseolous—Somewhat rosy or pinkish.

Rosette—Disease symptom characterized by a short, bunchy growth habit due to subnormal elongation of internodes; cluster of flowers and/or leaves arising from a stem with shortened internodes.

Rostellate—Somewhat beaked.

Rostrate—Beaked and strongly attenuated toward the apex; bent tip of macroconidia in some Mitosporic fungi (e.g., *Microsporium canis*).

Rostriform—Beaklike.

Rostrum—Any beaklike process.

Rostrupioid (of Uredinales)—Having teliospores like *Rostrupia*.

Rosulate—In a rosette.

Rot—State of decomposition and putrefaction. The softening, discoloration, and often disintegration of plant tissue by enzymes produced by fungal or bacterial infection. Rots may be hard, soft, dry, wet, black, brown, white, etc. *See* Decay.

Rot, brown (of wood)—Decay characterized by selective degradation of cellulose and hemicelluloses, leaving a crumbly brown residue rich in undigested lignin.

Rot, white (of wood)—Decay characterized by degradation of lignin at a rate equal to or usually greater than the rate of degradation of cellulose and hemicelluloses, leaving a light-colored residue that usually contains relatively more cellulose and hemicelluloses than did the original wood.

Rotaceous—Wheel-shaped; circular and flat.

Rotation—Sequence and duration of crop cultivation in a given area.

Rotund—Rounded in outline, or orbicular, not a perfect sphere.

Rough—Describing colonies of microorganisms with an irregular, nonsmooth surface.

Rough membrane—Endoplasmic reticulum encrusted with ribosomes. mRNAs are translated by the ribosomes, and the membrane isolates and transports the proteins. *See* Smooth membrane.

Roughage—Plant materials relatively high in crude fiber and low in digestible nutrients, such as straw.

Rounded—Circular or semicircular.

Roundworms—Nematodes.

Roux flask—A glass or plastic flask for culturing cells.

rRNA—Ribosomal RNA.

R-technique (of bacteria)—Method of assessing resemblance based on the association between pairs of characters in a series of taxa. *See* Q-technique.

Rubber policeman—A glass rod with a rubber tube on one end commonly used for scraping cells off surfaces and for resuspending pellets.

Rubellous, rufous—Reddish.

Rubescent, rufescent—Becoming red or reddish.

Rubicund—Bluish red.

Rubiginose, rubiginous—Rust-colored.

Rubratoxin B—Toxic metabolite of *Penicillium purpurogeum* (syn. *P. rubrum*) that causes hepatitis in cattle and pigs.

Ruderal—Growing in waste places or rubbish; (of fungi) having a high growth rate, rapidly germinating spores, and a short life expectancy due to the exhaustion of available nutrients. *See* Sugar fungus.

Rufescent (of basidiocarps)—Reddish, particularly becoming reddish on bruising or drying.

Rugae—Wrinkles.

Rugose, rugous—Coarsely wrinkled surface; covered with coarse, netlike lines; roughened. Used as part of the names of certain virus diseases characterized by warty, roughened, or severely crinkled leaves or other plant parts.

Rugose mosaic—Severe mosaic accompanied by deformation such as leaf crinkling, curling, or roughening of the leaf surface.

Rugulose—Finely wrinkled.

Ruminants—Cud-chewing mammals (e.g., cattle, deer, goats, and sheep), characteristically having a stomach divided into four compartments.

Runcinate—Saw-toothed or sharply incised, with the teeth turned back.

Runner—A slender, horizontal stem that grows close to the soil surface. Rooting occurs at the nodes or joints in plants such as strawberry (*Fragaria*). *See* Stolon.

Runner hyphae—Thickened hyphal strands that may spread a fungus from one plant to another.

Runoff—Pesticide material carried away from the target area by the flow of surface water. Also used to describe the rate of application to a surface—"spray to runoff."

Rupestral, rupestrine—Growing among or on walls or rocks, as do lichens. *See* Saxicolous.

Ruptile—Dehiscing in an irregular manner.

Rupture, irregular (conidiomata)—Dehiscence by irregular breakdown of the upper wall, or disorganized fracture of overlying tissues.

Russet—Brownish, roughened areas on the surface (epidermis) of leaves, fruit, stems, and tubers as a result of abnormal cork formation; may result from disease, insects, or spray or other mechanical injury.

Rust—A disease caused by one of the rust fungi (Uredinales) in the Basidiomycotina; or the fungus itself; a disease giving a "rusty" appearance to a plant. The life cycle of a rust fungus may involve up to five different types of spores. Rusts may parasitize one species of plant during their life cycle (*autoecious*) or two nonrelated species (*heteroecious*). *See* Black stem rust, Blister rust, Crown rust, Leaf rust (brown rust) of barley, Red rust, Stripe (or yellow) rust of cereals and grasses. White rust is caused by members of the Albuginaceae.

S

s—Second of time.

"S" (of nematodes)—Stylet length/body diameter measured at the base of the stylet.

S—Sedimentation coefficient. *See* Svedberg units.

Saccate, sacciform, saccular—Like a bag, sac, or pouch; liquefaction in the form of an elongated sac; tubular, cylindrical.

Saccharolytic—Capable of splitting or degrading sugar compounds.

Saccule—A small sac or pouch.

Sacculiform—Like a little sac.

Saddle fungi—*Helvella* species.

Saddle-back fungus—*Polyporus squamous.*

S-adenosyl-L-homocysteine (AdoHcy, SAH)—An inhibitor of methylation, e.g., nucleic acids, and an analogue of *S*-adenosyl-L-methionine.

S-adenosyl-L-methionine (ADO, SAM)—An intracellular source of activated methyl groups including those used for RNA and DNA methylation. It stimulates transcription in some viruses and is required by class I restriction endonucleases for their initial binding.

Safener—A chemical added to a pesticide that reduces the phytotoxicity of another chemical.

Safety—The practical certainty that injury will *not* result from the use of a substance in a proposed quantity and manner.

Saffron—Yellow; crocate, croceous.

Saffron milk cap—Basidiocarp of *Lactarius deliciosus*.

Sagittae, sagittate, sagittiform—Arrowhead-shaped; triangular-elongate.

SAH—*S*-adenyosyl-L-homocysteine.

St. Anthony's Fire, Holy Fire—*See* Ergotism.

St. George's mushroom—The edible *Calocybe* (*Tricholoma*) *gambosum*.

Salient—Projecting forward.

Saline (n. salinity)—Of, relating to, or resembling salt; a natural deposit of common or other soluble salt.

Saline soil—A nonsodic soil containing sufficient soluble salts to impair plant growth.

Saline solution—A solution isotonic with body fluids, usually $0.15M$ buffered or in water.

Salmon, salmoneous, salmonicolor—Salmon-colored; pinkish with a tinge of yellow.

Saltant—A discontinuous variation of unknown origin.

Saltation (of fungi)—Mutation; dissociation. *See* Variation.

Salting out—The precipitation of proteins or nucleic acids in concentrated salt solutions. Ammonium sulfate and sodium sulfate are commonly used to precipitate proteins. *See* Soluble RNA.

Samara—Simple, dry, one- or two-seeded indehiscent fruit with the pericarp bearing winglike outgrowths on both sides, e.g., maple (*Acer*) or elm (*Ulmus*).

Sambucinin—Same as enniatin.

SAM—*S*-adenyosyl-L-methionine.

Sand—A soil particle between 0.05 and 2.0 mm in diameter.

Sanger or dideoxy method—A technique for sequencing DNA. The DNA to be sequenced is in a single-stranded form usually the result of cloning in Phage M13.

Sanguine, sanguineous—A bloody color.

Sanitation (v. sanitize)—Destruction (removal) of infected and infested plants or plant parts; elimination of disease inoculum and insect vectors; decontamination of tools, equipment, containers, work space, hands, etc.; cultural methods of disease control that reduce inoculum.

Sap hypha—A latex-bearing hypha or hyphal element; a lactifer.

Sapid, sapidous—Filled with sap, savory; having a pleasant taste.

Saponaceous—Soapy; slippery to the touch.

Sapr-, sapro- (prefix)—Rotten.

Saprobe (saprogen, saprotroph)—A general term to include organisms that are saprophytic or saprozoic.

Saprogenesis (adj. **saprogenic**)—Part of the life cycle of a disease-producing organism in which it is not directly associated with a living host. *See* Pathogenesis.

Saprophagous, saprozoic—A zoological term meaning to feed on dead or decaying matter, usually used in reference to animals, e.g., nematodes.

Saprophyte, saprophile (adj. **saprophytic, saprogenic, saprogenous, saprophilous, saprobic**)—An organism that feeds on dead organic matter commonly causing its decay; a necrophyte on dead material that is not part of a living host. *See* Parasite.

Saprot—A fungal decay of the dead sapwood of dead standing or fallen trees, stumps, and slash.

Sapwood—Young, physiologically active zone of wood contiguous to cambium; usually light-colored; outermost growth layers of xylem in woody plants.

Sarcina (pl. **sarcinae**)—A regular cubical pocket of bacterial cocci, resulting from cell division in three planes; also, the genus of cocci forming such pockets.

Sarciniform—Bundle- or packet-shaped like the dictyospore of *Stemphylium botryosum*.

Sarcodimitic—Having hyphae of two kinds (generative and sketal), in which the thick-walled, aseptate skeletal hyphae are replaced by thick-walled, long and inflating (fusiform) elements (cells).

Sarcophagous—Living on the flesh or bodies of small animals.

Sarcotrimitic—Having hyphae of three kinds (generative, skeletal, and binding or ligative), in which the generative hyphae are replaced by thick-walled inflated elements, that are similar to binding hyphae but are septate.

Sarkosyl—An anionic detergent (sodium lauroyl sarcosinate) used to disrupt virus particles or cells.

Satellite RNA—A small RNA that becomes packaged in protein shells made from coat proteins of another, unrelated helper virus, on which the satellite RNA depends for its own replication.

Satellite virus—Term first used to describe a small defective virus associated with tobacco necrosis virus (TNV) and dependent upon the TNV genome to provide for its own replication. The term is also used to describe certain nucleic acid molecules unable to multiply in a host cell without the aid of other nucleic acid molecules. *See* Carna 5-RNA.

Satratoxins—Toxins of *Stachybotrys chartarum* (syn. *S. atra*) that cause stachybotryotoxicosis in farm animals and humans.

Saturnine (of ascospores)—Having a flat edge down the middle.

Saucer—Concave liquefaction in a bacterial medium, shallower than crateriform.

Savanna—Grassland having scattered trees, either as individuals or clumps; often a transitional type between true grassland and forest.

Saxatile, saxicoline, saxicolous—Living or growing among rocks.

Sc chain—Secretory component found associated with an IgA dimer.

Scab—A roughened, hyperplastic, crustlike diseased area on the surface of a plant organ; or a disease in which such areas form. *See* Apple scab, Cherry scab, Citrus scab, Peach scab, Pear scab, Powdery scab of potato, and Scab or gummosis of cucurbits. *See also* Anthracnose and Scab of cereals.

Scab of cereals—A disease typified by a loss in weight and quality (because of mycotoxins) mainly of individual barley and wheat kernels; often with conspicuous pink colored signs. Caused by *Gibberella zeae* (anamorph *Fusarium graminearum*), *G. cyanogena* (anamorph *F. sulphureum*), *F. culmorum*, and *F.* sp.

Scab or gummosis of cucurbits (especially cucumber, *Cucumis sativus*)—Disease caused by *Cladosporium cucumerinum.*

Scabrate—Made rough or roughened.

Scabrid—Rough with delicate and irregular projections.

Scabrous—Tough; rough to the touch with short, rigid projections.

Scaffold (of nematodes)—An abnormally heavy, refractive guiding ring.

Scaffold branch—Main lateral branch off the main trunk or bole of a tree.

Scald—A necrotic condition of plant tissue (leaves, stems, flowers, and fruit), usually bleached or blanched then often brown in color, with the appearance of having been exposed to high temperature and/or excess sunlight, or standing water.

Scale (on the surface of the pileus or stipe)—A torn part of the cuticle or pellicle, that may be membranous, fibrillose, hairy, floccose, hard, erect, flat, patchlike, etc.; (of insects)—Small, inconspicuous member of the family Coccidae that is immobile throughout most of its life.

Scaliform—Ladderlike; bearing markings like the rungs of a ladder.

Scalp—To remove an excessive quantity of functioning turfgrass leaves at any one mowing; results in a shabby, brown appearance caused by exposing crowns, stolons, dead leaves, and even bare soil.

Scalpelliform—Shaped like a scalpel or lancet.

Scandent—Climbing.

Scanning electron microscope (SEM)—A technique for examining the surface features at high magnification by coating specimens with a thin layer of gold, palladium, or aluminum and studying the images (photographs) produced by a scanning electron beam.

Scape—A peduncle, rising from the ground, naked or without leaves; a leafless flower stalk.

Scaphiform, scaphoid—Boat-shaped. *See* Navicular.

Scarify—To injure the surface of seeds by chemical or mechanical means, to accelerate the passage of water and gases as an aid to seed germination.

Scariose, scarious—Papery; thin.

Scarlet (elf) cup—The fruitbody of *Sarcoscypha coccinea*.

Scheda (pl. **schedae**), **schedula** (pl. **schedulae**)—Label(s) especially of dried specimens or exsiccati.

Schiff's reagent—*See* Feulgen stain.

Schiz-, schizo- (prefix)—Division or cleavage.

Schizidium (of lichens)—A propagule formed by the upper layers of a thallus splitting off as scablike segments from the main lobes.

Schizocarp—Simple, dry, indehiscent fruit composed of two or more united carpels, that split apart at maturity, each part usually with one seed.

Schizocystidium—*See* Cystidium.

Schizogenetic, schizogenous—Formed by cracking, cleavage, or splitting. *See* Lysigenetic.

Schizogony—Process of dividing a schizont.

Schizolysis (adj. **schizolytic**)—Secession of conidia involving a splitting of the delimiting septum so one half of the cross wall becomes the base of the seceding conidium and the remaining half remains at the apex of the conidiogenous cell.

Schizont—A naked, multinucleate, vegetative thallus without a cell wall, that undergoes simple or multiple division.

Schizotype—An implied lectotype. *See* Lectotype and Holotype.

Schlieren optics—An optical system that detects gradients of concentration of a solute in a solvent using a phase plate. A differentiated curve of the gradient can be observed.

Scientific method—An approach to a problem that consists of stating the problem, establishing one or more hypotheses as solutions to the problem, testing these hypotheses by experimentation or observation, and accepting or rejecting the hypotheses.

Scintillation counter—A device in which scintillations, produced in a fluorescent material by ionizing radiation, are detected and counted by a multiplier phototube and associated circuitry.

Scintillation fluid—Usually an organic solvent containing one to three solutes that convert the kinetic energy of a nuclear particle, traversing the medium, into light photons at a maximum efficiency. The solutes (scintillations, fluors, or lumifluors) are ranked according to their level of fluorescent energy; the highest is the primary solute and the others secondary solutes. A very common scintillation fluid is composed of PPO and POPOP dissolved in toluene.

Scion—Detached bud, piece of a twig, or shoot inserted by grafting; the shoot that receives a scion is called the stock or rootstock.

Scion-stock interaction—The effect of a rootstock on a scion and vice versa, in which a scion on one kind of rootstock performs differently than it would on its own roots or on a different rootstock.

Scissile (of the flesh of a pileus)—Splitting into horizontal layers.

Sciuroid—Curved and bushy like a squirrel's tail.

Scler-, sclero- (prefix)—Hard.

Sclereid—Thick-walled, often slightly elongated or irregularly shaped sclerenchyma cell.

Sclerenchma (adj. **sclerenchymatous**)—Supporting or protective plant tissue composed of hard, short cells or elongated cells with thickened, often lignified, cell walls; usually devoid of protoplasm at maturity.

Sclerobasidium (pl. **sclerobasidia**)—Thick-walled, encysted, gemmalike probasidium of the rust fungi (Uredinales) (= teliospore) and the Auriculariales.

Sclerocarps—Modified, sclerotiumlike ascomata that lack a sexual capacity and act as sclerotia (as in *Varicosporina ramulosa*).

Sclerocortex (pl. **sclerocortices**)—A cortex of thick-walled isodiametric cells, as in *Armillaria mellea*.

Scleroid, sclerosed, sclerotic—Having a firm, hard texture; more or less lignified.

Scleroplectenchyma—Plectenchyma made up of very thick-walled conglutinate cells with narrow lumina. See Stereome.

Sclerosis (pl. **scleroses**)—A simple thickening of cell walls.

Sclerothionine—Plant growth-promoting metabolite of *Sclerotinia sclerotiorum* (syn. *S. libertiana*).

Sclerotic—Hard and sclerotialike.

Sclerotigenic—Producing sclerotia.

Sclerotium (pl. **sclerotia**)—Firm or hard, frequently rounded, and usually darkly pigmented, sterile, vegetative resting body of a fungus composed of a compact mass of usually thick-walled, interwoven, special-sized hyphal cells with or without the addition of host tissue or soil. The structure may remain dormant in soil, plant refuse, or seed for long periods and is capable of surviving under unfavorable environmental conditions. Sclerotia germinate upon return to favorable conditions to produce a stroma, fruitbody, mycelium, or conidiophores. A sclerotium usually consists of an outer, usually darkened and sclerotized rind and an inner parenchymatous medulla; (of myxomycetes), the firm resting condition of a plasmodium. See "Blackfellows" bread, Stone fungus, Tuckahoe, and Bulbil.

Sclerotized—Formed into a sclerotium; hardened dense condition of cuticular structures; in a resting condition; (of nematodes) hardened cuticular (refractive) regions.

Sclerotules—Small sclerotia.

Scobiculate—In fine grains, like sawdust.

Scobiform—Having the appearance of sawdust.

Scolecite—*See* Woronin's hypha.

Scolecospore—A long, usually septate, needle-, thread- or wormlike spore with a length/width ratio of >15:1. *See* Mitosporic fungi.

Scopate—Densely covered with bristly hairs.

Scopulate—Broom- or brushlike.

Scorch—Sudden browning and death (necrosis) of large, indefinite areas on a leaf, fruit, or stem from infection, lack or excess of some element, chemical injury, or unfavorable weather conditions. Scorch often appears suddenly as dead areas along or between the veins, margin, and tips of leaves or on fruits and stems.

Scorpioid, scorpioidal—Branching system in which laterals are curved, like the tail of a scorpion, all appearing to arise from one side of the main stem as in *Cladonia arbuscula*.

Scotospore—A dark-colored spore.

Screening test—Test to observe the response of a range of plant cultivars or types to virus infection; routine testing of organisms or chemical substances for a particular property, e.g., for antibiotic production, pesticidal effect, etc.; a test used by diagnosticians to detect or to eliminate a particular organism or group of organisms. Screening tests may be morphological, tinctorial, biochemical, or serological; may be based on sensitivity or resistance to phages or antibiotics, etc.

Scrobiculate—Roughened, furrowed, pitted; resembling sawdust; (of lichens) coarsely pitted, faveolate.

Scrotiform—Pouch-shaped.

Scrupose—Jagged, rough with very small, sharp, projecting points, e.g., fascicles of hyphae.

Sculptured (of the surface of spores, capillitium, etc.)—Having raised or incised markings such as reticulations, warts, spines, etc.

Scurfy (n. **scurf**)—Having a surface covered with thin, dry scales or branlike flakes.

Scutate—Like a round plate or shield; circular in outline.

Scutellum, scutellate (pl. **scutella**, adj. **scutellar**)—The single cotyledon of a grass embryo; the rudimentary leaflike structure at the base of the first node of the embryonic culm of a grass plant; (of fungi) small plate, shieldlike cover, or upper wall of a thyriothecium as in the Microthyriales; (of nematodes) a large phasmid, as in *Scutellonema* and other species of the Hoplolaiminae.

Scutiform—Buckle-shaped; scutate.

Scypha, scyphus (pl. **scyphae, scyphi**)—A cuplike dilation of the podetium in lichens. *See* Podetium.

Scyphoid, scyphiform—Cuplike or forming a cup.

SDI—Serological differentiation index.

SDS—Sodium dodecyl sulfate.

SDS-polyacrylamide gel electrophoresis—Electrophoresis in polyacrylamide gels of proteins denatured with the anionic detergent sodium dodecyl sulfate (SDS). Treatment with SDS usually gives proteins equal charge per unit molecular weight (MW). The proteins are thus separated according to their molecular weight.

Seceding—(Of lamellae) joined at the stipe (adnate or adnexed) then free; separating from the stipe; (of conidia) first attached to the conidiogenous cell, then separating by schizolysis or rhexolysis; secession.

Secernentea—One of two classes of the phylum Nematoda; contains the majority of plant-parasitic nematodes.

Second growth—Resumption of growth after normal growth has ceased.

Secondary cycle—Of plant disease: any cycle initiated by inoculum generated during the same season.

Secondary host—The host plant on which the asexual aphid form occurs; also used to indicate alternative hosts for pathogens. Do *not* confuse with alternate host, which involves two hosts being required for completion of the life cycle of a fungus. *See* Alternate host, Alternative host, and Primary host.

Secondary infection—Infection resulting from the spread of infectious material (inoculum) produced following a primary infection (the first infection by a pathogen after a resting period) or from subsequent infections without an intervening inactive period.

Secondary inoculum—Inoculum resulting from primary infections or from other secondary infections in the same season.

Secondary mycelium (of basidiomycetes)—The dikaryotic mycelium resulting from plasmogamy in the primary mycelium; mycelium formed from the base of a basidiocarp.

Secondary organism (pathogen)—Organism that multiplies in already diseased tissue; not the primary pathogen.

Secondary phloem—Phloem cells formed by activity of the vascular cambium found in biennials and perennials, but usually not in annuals.

Secondary root—Root developed from a root crown or node as opposed to a primary root.

Secondary septum—A cross wall that develops in a mature cell, not associated with nuclear division.

Secondary spores (of basidiomycetes)—Spores other than basidiospores.

Secondary symptom—Symptom usually of a virus infection appearing after the first or primary symptoms.

Secondary tissues—Tissues arising from a lateral or secondary meristem such as the cambium or cork cambium and increasing the diameter of a stem or root.

Secondary wall—Cell layer deposited by protoplasm upon the primary wall layer, after cell enlargement is completed.

Secondary xylem—Xylem cells formed by activity of the vascular cambium. The development of secondary xylem results in the so-called annual rings seen in most trees.

Second-year, third-year, etc., needles (of conifers)—Needles in their second, third, etc., season of life. A second-year needle is between one and two years old.

Section (land unit)—640 acres = 1 mi^2 = 258.9998 ha = 102,400 sq rod = 3,097,600 sq yard = 2.589998 km^2.

Sectoring—"Mutation" or selection in plate cultures that result in one or more sectors of the culture; the fungus or other microorganism having a changed form of growth.

Secund—Having parts directed to only one side. *See* Scorpioid.

Sedentary (sessile)—Remaining in a fixed location; stationary.

Sedentary endoparasite—Describing a parasitic nematode that enters a host, migrates to a feeding site, and then becomes nonmotile.

Sedge—A grasslike plant with triangular stems that spreads by rhizomes and overwinters as tubers.

Sedimentary rock—Rock formed from material originally deposited as a sediment, then physically or chemically altered by compression and hardening while buried in the earth's crust.

Sedimentation coefficient—Number indicating the size of a particle measured by its rate of settling in a solvent; the rate of sedimentation of a macromolecule, such as a virus, per unit centrifugal field is measured in *Svedberg units* (S). An S unit is 10^{-13} s.

Sedimentation rate—The rate at that a macromolecule, such as a virus, moves in a standard gravitational field. The rate is influenced by the mass, shape, hydration, and partial specific volume of the macromolecules as well as the temperature and viscosity of the solvent. In a standard gravitational field the sedimentation coefficient is 1 cm s^{-2}.

Seed—The mature ovule of a flowering plant containing an embryo, sometimes an endosperm, and a seed coat.

Seed (breeder)—Seed, or sometimes other vegetative propagating material, directly controlled by the originator, the sponsoring plant breeder, or institution that provides the source for the initial increase of foundation seed.

Seed (certified)—Progeny of foundation, registered, or other certified seed that is so handled as to maintain satisfactory genetic identity and/or purity and that has been approved and certified by a certifying agency.

Seed (foundation)—Seed stocks so handled as to maintain specific genetic identity and purity, such as may be designated by an agricultural experiment station. Foundation seed is the source of certified seed, either directly or through registered seed.

Seed (registered)—Progeny of foundation or other registered seed so handled as to maintain satisfactory genetic identity and purity; approved and certified by a certifying agency.

Seed coat—Testa or outer covering of a true seed.

Seed contamination—Similar to infestation.

Seed disinfectant or disinfestant—A chemical or physical agent that destroys certain disease-causing organisms carried *in* (disinfectant) or *on* (disinfestant and disinfectant) the seed. Neither is necessarily a seed protectant (which see).

Seed infection—Colonization of seed tissues by pathogens with or without conspicuous symptoms.

Seed infestation—Pathogen propagules carried on the outside of seeds or associated with seed lots.

Seed leachates—Exudates usually from germinating seeds that interact with microflora.

Seed pathology—The study of seed diseases and of plant pathogens and nonpathogens carried with seeds.

Seed potatoes—Pieces of potato (*Solanum tuberosum*) or whole tubers that are planted to produce new plants and subsequent commercial crops.

Seedbed—Soil that has been prepared for planting seeds or transplants.

Seedborne disease—Disease in which inoculum is borne in or on seed.

Seedborne inoculum—Propagules of a pathogen capable of causing disease of the seed itself or a seedling or plant derived from the seed.

Seedborne microflora—Bacteria, fungi, nematodes, mycoplasma, or, incorrectly, viruses associated with seeds.

Seedborne pathogen—An infectious agent associated with seeds and having the potential of causing a disease of a seedling or plant.

Seedling disease—A disease of young plants that may occur before or after emergence and may result in a reduced yield. *See* Damping-off.

Seedpiece—A vegetative unit of a plant used for propagation.

Seed primordia—Meristematic tissues differentiable into seed.

Seed protectant—A chemical applied to seed before planting to prevent decay and damping-off (*seedling blight*).

Seed scarification—Cutting or scratching a seed coat to facilitate the entry of water or oxygen.

Seed transmission—Passage of inoculum from an infected or infested seed to a plant.

Seed treatment—A chemical or nonchemical treatment used to control pathogen propagules associated with true seeds and seedpieces.

Seeds, true—A fertilized mature ovule consisting of an embryonic plant and a protective seed coat.

Segment (of a dictyospore)—A part of the spore cut off by an A-transseptum. *See* Septum.

Segregate (in taxonomy)—A new and separate taxon based on part of an earlier taxon.

Segregation—A genetic principle, established by Mendel, that the factors (or genes) of a pair are separated from each other in reduction division.

Seinhorst elutriator or extraction technique (Seinhorst chamber)—A method of separating nematodes from infected plant material and soil utilizing funnels, a constant fresh water spray, and suitable settling basins.

Seiospore—A dry dispersal spore generally shaken loose from a sporophore.

Selection—The process of isolating and preserving certain individuals or characters from a group of individuals or characters.

Selective pesticide—A chemical that is more toxic to some plant or animal species than to others.

Seleniform—Shaped like a full moon; round.

Self-compatible—Self-fertile; refers to a thallus that reproduces sexually by itself; ability to produce seed by self-pollination.

Self-fertile (n. **self-fertility**)—Capable of fertilization and producing viable seed after self-pollination; a condition in which sexual reproduction occurs as a result of the fusion of eggs and sperms produced by the same individual.

Self-incompatible—Self-sterile; (of fungi) refers to a thallus that cannot reproduce sexually by itself; (of higher plants) inability to produce viable seed following self-pollination. This inability is sometimes due to a pollen-borne gene that prevents pollen tube growth on a stigma with the same gene.

Self-pollination—Transfer of pollen from the stamen(s) to the stigma of the same flower, of another flower on the same plant, or other plants of the same clone.

Self-sterility—Condition in which sexual reproduction cannot be achieved by the fusion of eggs and sperms produced by the same individual.

Sellaeform—Saddle-shaped.

SEM—*See* Scanning electron microscope.

Semi- (prefix)—Half, partial. *See* Hemi-.

Semiarid—Climate in which evaporation exceeds precipitation; a transition zone between a true desert and a humid climate. The annual precipitation is usually between 250 and 500 mm (10–20 in.).

Semiconservative replication—A model of DNA-dependent DNA replication in which the synthesis uses both strands of dsDNA as templates, resulting in progeny each comprising one parent strand and one new strand. *See* Okazaki fragments.

Semiendogenous—Formed partly within a phialide.

Semiendoparasite—Describing a parasitic nematode that becomes only partially embedded in host tissue.

Semifenestra—One of a pair of openings in the vulval cone separated by the vulval bridge on cysts of the nematode genus *Heterodera*.

Semiimmersed (conidioma)—Initially immersed, finally the greater part of development appearing to be superficial, or, half superficial-half immersed.

Semimacronematous (conidiophore)—Intermediate in form between micronematous and macronematous, differing only slightly from a vegetative hypha, often ascending but seldom erect. *See* Macronematous.

Seminal receptacle (of nematodes)—*See* Spermatheca.

Seminal vesicle (of nematodes)—The enlarged tube- or pouchlike structure of the male reproductive tract that functions as the temporary sperm storage organ.

Seminicole, seminicolous—Growing on seeds.

Semiorbicular—Half-round or hemispherical.

Semioval, semiovate—Oval or ovate on one side.

Semipermeable (of membranes)—Allowing some fluids or solutes but not others to pass through. Cell membranes are semipermeable. Same as differentially permeable.

Semipersistent transmission—Transmission by a virus that circulates but does not multiply in its arthropod vector for a short time; intermediate between *nonpersistent transmission* and *persistent transmission*. The acquisition feed by the vector is short and there is no latent period, but the vector remains able to transmit the virus for hours to days.

Senesce (n. **senescence**, adj. **senescent**)—To decline or degenerate as with maturation or a physiological aging process; often hastened by environmental stress, disease or insect attack; growing old.

+ sense—Plus-sense or positive sense. In RNA viruses, the strand that functions as a messenger (mRNA).

− sense—Minus-sense or negative-sense: Nucleic acid complementary to the + strand.

Sensilla pouch—*See* Amphid.

Sensillum (pl. **sensilla**)—A group of nerve endings in the sensilla pouch of nematodes.

Sensitivity (adj. **sensitive**)—A diseased organism reacting with severe symptoms to the attack of a given pathogen.

Sensu lato (*s.l.*)—In a broad sense.

Sensu stricto (*s.str.*)—In a narrow sense.

Sepals—A division of the outer floral envelope or calyx; the modified leaflike structures, usually green, collectively forming the calyx at the base of a flower.

Separable (of cuticle, pellicle, etc.)—Not adnate.

Separating—*See* Seceding.

Separating cells—Cells that separate conidia from each other or from conidiogenous cells and are involved in rheolytic secession.

Separation—A type of propagation based on the natural breaking apart of plant segments, such as in bulbs or corms.

Separation layer—*See* Abscission layer.

Sept , septi- (prefix)—Partition.

Septal pore apparatus—*See* Dolipore.

Septal pore cap or plug—An occlusion of a septal pore. *See* Parenthesome.

Septate (n. **septation**)—With more or less regularly occurring cross walls; divided by partitions or septa; having septa.

Septiform—Having the appearance of a septum.

Septum (pl. **septa**)—A dividing cell wall or partition in a hypha or spore. Septa may be transverse, angular or oblique, longitudinal (longiseptum), an A-transeptum (forming a segment), or a B-transeptum (laid down in a segment after division by a longiseptum). *See* Primary septum and Adventitious septum.

Sepultate—Buried.

Sequence—The order of nucleotides in RNA or DNA or of amino acids in a polypeptide.

Sequencing—The determination of a sequence of a nucleic acid or protein.

Sequential epitope—An antigenic site composed of five to seven sequential peptides in their unfolded random coil form.

Seriate—Arranged in a series or connected order. *See* Uniserate.

Sericellous, sericeous—Silky or shiny, with fine, close-pressed, unidirectional hairs; or with tomentum absent or scant.

Serological differentiation index—A measure of the degree of serological cross reactivity of two antigens; the number of twofold dilution steps separating the homologous and heterologous titers.

Serology—A method using the specificity of the antigen-antibody reaction for the detection and identification of antigenic (protein) substances and the organisms that carry them; the study, detection, and identification of antigens, antibodies, and their reactions.

Serotype—(Of bacteria) an infra subspecific subdivision of a species, distinguished by serological methods, i.e., by detecting differences in the surface antigens of bacteria on addition of a specific antibody; a shortened form of serological type; (of plant viruses) a group of viruses sharing only a few of its antigens in common with another group.

Serous (of latex)—Like serum; watery; opalescent.

Serpiform—Shaped like a serpent.

Serrate—Having edges with sharp notches or teeth, like a saw blade.

Serrulate, serratulate—Minutely toothed.

Serum—The watery amber portion of the blood that remains after coagulation of red blood cells; used in serological tests for viruses. The essential difference between plasma and serum is that the latter does not contain fibrinogen. Serum is more commonly used than is plasma in immunological procedures because there is no danger of a clot forming when other materials are added to it. *See* Blood serum.

Serum neutralization—The inhibition of the infectivity of a virus by antiserum.

Serum-free medium—A culture medium for growing viruses in animal tissues or cells. The medium contains no fetal calf serum or other serum additives but does contain essential salts, amino acids, and an energy source (e.g., glucose).

Sessile—Used in reference to a leaf, leaflet, flower, floret, fruit, ascocarp, basidiocarp, etc., without a stalk, petiole, pedicel, stipe or stem; (of nematodes) permanently attached; not capable of moving about.

Seta (pl. **setae**)—Stiff, hair- or bristlelike sterile fungus appendage, a modified hypha; usually erect, deep yellow to dark brown and thick-walled; (in hymenomycetes) a sterile, hyphal, thick-walled organ like a cystidium, but dark-colored and darkening in KOH; (in mosses and liverworts) stalk that supports the capsule of the sporophyte; (of nematodes) elongated cuticular structure articulating with the cuticle; in general, tactile sensory organs usually located around the oral openings.

Setaceous—Bristlelike. *See* Setose.

Setal hypha—A thick-walled, pointed, brown hypha, found in subicular and marginal tissue of basidiocarps, mycelium in rotted wood, or cultures.

Setiform (of a conidiophore stipe or branch)—Seta- or bristle-shaped.

Setiform lamprocystidium—*See* Cystidium.

Setose—Covered with bristles or setae. *See* Setaceous, Setulose.

Setula (pl. **setulae**)—A delicate, very thin, hairlike appendage arising from the surface of a spore; (in hymenomycetes) a thick-walled, pigmented, terminal element of a tramal cystidium.

Setule (in hymenomycetes)—A thin-walled, rarely pigmented, usually lageniform cystidium on the pileus or stipe.

Setulose (of hymenial surfaces)—Covered with setae (fine bristles), especially when discernible with a 10× hand lens. *See* Setaceous.

Severity, of disease—The measure of damage done by a plant disease as distinguished from *disease incidence*, a measure of the number of individuals affected.

Sewage fungus—*Leptomitus laceus*, found in polluted water; may cause blockage of sewage filters.

Sex chromosomes—Chromosomes that determine sex.

Sex linkage—Occurrence on sex chromosomes of genes that determine characters other than sex.

Sexilocular—Having six cells or locules.

Sexual—Participating in or produced as a result of a union of nuclei; both fertilization and meiosis are usually involved; (of fungi) teleomorph or perfect state.

Sexual dimorphism—Pronounced differences in the morphological expression (form) of the two sexes of a single species.

Sexual propagation—To increase plant numbers by seed.

Sexual reproduction—Reproduction requiring the union of two compatible nuclei or gametes; reproduction involving nuclear fusion and meiosis.

Shade tolerant—Describing plants that can grow in reduced sunlight.

Shaft (of nematodes)—The distal, main body of the spicule, usually curved and often bearing a thin flange of cuticle, termed the velum.

Shaggy (of fungi)—Rough with long, compact fibrils.

Shaggy ink cap or mane—The edible and choice *Coprinus comatus*.

Shagreen (of a surface)—Minutely cobble-stoned or papillate in appearance.

Sheaflike organ (of nematodes)—The vaginal remnant in the vulval cone of cysts in the genus *Heterodera*.

Sheath—Lower tubular part of a grass leaf that clasps the culm, a membranous cover; (of nematodes) a covering of a structure; refers to a retained or extra cuticle in some species; (of fungi) a complete mucilaginous covering of conidia, conidiogenous cells, conidiophores, or paraphyses; often splitting or becoming everted to leave appendage(s) on the conidia; (of bacteria) an envelope similar to a capsule (*which see*) but surrounding a filamentous organism.

Shii-take—Principal method for the culture of the edible *Lentinula edodes* in Japan and China.

Shine-Dalgarno sequence—*See* Ribosome binding site.

Shock symptoms—The severe, often necrotic symptoms produced on the first new growth following infection with some viruses; also called acute symptoms.

Shoestring fungus—The honey agaric, *Armillaria mellea*.

Shoot—Current season's stem with its leaves. *See* Twig.

Shoot density—The relative number of shoots per unit area.

Shoot, dwarf (of gymnosperms such as pine or larch)—a lateral shoot consisting primarily of a cluster of needles at the apex of a very short stem.

Shoot, short—A dwarf shoot.

Short-day plants—Plants that initiate flowers only under short-day (= long night) conditions.

Shot-hole—Disease symptom in which small, roundish to irregular dead fragments drop out of leaves making them appear as if riddled by shot.

Shoyu (soy sauce)—Oriental sauce of soybeans (*Glycine max*) and wheat (*Triticum*) fermented by *Aspergillus*, yeasts, and bacteria.

Shrub—A perennial woody plant of relatively low stature, typically less than about 3.5 m (12 ft) tall with several stems arising from or near the ground.

Shunt—An alternate pathway, a bypass.

Sialic acid (*N*-acetylneuramic acid)—A constituent of many glycoproteins and cleaved from glycoprotein virus receptors by neuraminidase.

Siccous—Dry; juiceless; containing little or no watery juice.

Sicyoid, sicyodic (of cystidia)—Gourd-shaped; elongate-pyriform.

Sidedressing—Fertilizer or other chemical applied in a band alongside an established crop.

Siemen—A unit of electrical conductance, equivalent to a mho, that is the reciprocal of the unit of resistance, the ohm. The unit is used to determine the level of soluble salts in the soil that are damaging to plants.

Sieve cell—A phloem-conducting cell with perforations not restricted to end walls; does not occur in series or tubes.

Sieve plate—Perforated wall area between two phloem cells through which their protoplasts are connected.

Sieve tube—A series of thin-walled phloem cells forming a long cellular tube through which food materials are transported.

Sieve tube element or member—One of the cellular components of a phloem sieve tube; the sieve areas are usually on the end walls.

Sigatoka—A serious disease of banana (*Musa*) caused by *Mycosphaerella musicola*, anamorph *Pseudocercospora (Cercospora) musae*.

Sigla—A name formed from letters or other characters taken from words in compound terms (e.g., virus group names).

Sigmoid growth (or S-shaped) curve—Curve exhibited by many living organisms; it consists of a logarithmic phase, a linear phase, and a senescence phase.

Sigmoid—Curved like an open letter S.

Sign—Any indication of disease on a host plant from direct visibility of the pathogen or its parts and products (spores, mycelium, exudate, fruitbodies, or rhizomorphs). *See* Symptom.

Signal—A molecule initiating an interaction when combined with a specific receptor.

Signal peptide—Short segments, usually of 15–30 amino acids, at the terminus of a secreted or exported protein, enabling it to pass through the membrane of a cell or organelle.

Signal sequence—Signal peptide or the nucleotide sequence encoding a signal peptide.

Signal word—Word that must appear on a pesticide label to show how toxic the pesticide is. The signal words are "Danger—Poison," "Warning," and "Caution."

Silage—Forage that is chemically changed and preserved in a succulent condition by partial fermentation in the preparation of food for livestock.

"Silent" infection—An infection with no detectable symptoms; also can be termed a latent infection.

Silique—Long, dry, two-valved, pod-like fruit of Brassicaceae (cabbage or mustard family) that splits down both seams, leaving a parchmentlike center.

Silky (of fungi)—Covered with shiny, close-fitting fibrils; sericeous.

Silo—A structure for making and storage of silage.

Silt—A class of textural soil having particles between 0.05 and 0.002 mm in diameter.

Silver ear—Basidiocarp of the edible *Tremella fusiformis*.

Silvering, silverleaf—A symptom of leaves or fleshy tissues characterized by grayish or shiny lesions, often induced by exposure to certain air pollutants.

Silver-leaf disease—Foliar disease caused by *Chondrostereum (Stereum) purpureum*, primarily affecting plum, apple, and other rosaceous fruit trees.

Silviculture—The science of developing, caring for, or cultivating forests.

Simblospore—Same as zoospore or swarm spore.

Simple (of fungi)—Unbranched; lacking divisions; of one piece or series; opposed to compound.

Simple fruit—A fruit developed from a single pistil of a flower.

Simple hyphidium—An unbranched hyphidium lacking prongs and swellings.

Simple-septate (of hyphae)—With septa but lacking clamp connections.

Single diffusion—A gel diffusion test in which antigen diffuses into agar containing antiserum, resulting in a ring of antigen-antibody precipitation.

Single nucleocapsid nuclear polyhedrosis viruses—*See* SNPV.

Single radial diffusion test—Radial immunodiffusion.

Single-cycle disease—A disease whose causal agent or its progeny does not infect additional suscepts during the current growing season; monocyclic.

Single-stranded (ss)—Nucleic acid molecules (DNA and RNA) with only one linear nucleotide chain.

Sinistrorse—Twining or curling upward from right to left; anticlockwise.

Sinker—An anchorage and absorption organ of mistletoes; actually a modified root that is haustorialike and penetrates into the vascular system of the host; a number of wedgelike sinkers are formed radiating from the primary haustorium, which develops from the attachment disc.

Sinuate, sinuous—Serpentine, crooked, wavy, or flexuous; (of lamellae) notched or having a sudden curve near the stipe; emarginate. *See* Adnate.

Sinus—(Of fungi) the rounded, inward curve between two projecting lobes; (of nematodes) a curvilinear indentation; an opening or depression; a cavity.

Siphon—An aseptate hypha.

Sirenin—A sex hormone secreted by the female gamete of *Allomyces* that attracts male gametes.

Skeletal hyphae—Thick-walled, aseptate or septate, unbranched or little branched, straight to somewhat flexuous hyphae that develop from generative hyphae; false seta. *See* Cystidium.

Skeletocystidium (pl. **skeletocystidia**)—A thick-walled, sterile hymenial or imbedded element that represents the end of a tramal skeletal hypha.

Skewed—Arrangement of rows of capsomers in a left-handed (laevo, *l*) or right-handed (dextro, *d*) turn.

Skiophilous—Shade loving.

Skittle-shaped (of cystidia)—Skittle- or ninepin-shaped. Same as sicyoid or sicyodic.

Skyrin—An orange-yellow wilt toxin of *Cryphonectria (Endothia) parasitica*. *See* Luteoskyrin.

Slaframine—A toxin of *Rhizoctonia leguminicola*, the cause of slobber syndrome in livestock.

Slash—The residue of unmerchantable cull logs, tops, branches, and other material left after timber harvesting, thinning, pruning, or natural phenomena such as windstorms or snow and ice damage.

Sleep movements—Turgor movements of plants initiated by changes in light intensity.

Slime—A wet, generally sticky, substance; mucus.

Slime flux—Fermented sap exuded through a limb or trunk wound; a thick liquid from the stem or branches of trees made up of, or having a connection with, yeasts, other fungi and bacteria. *See* Wetwood.

Slime layer—A gelatinous covering of the cell wall; sometimes used synonymously with capsule.

Slime molds—Primitive organisms of the Acrasiomycota, Dictyosteliomycota, and Myxomycota; also, the superficial "diseases" caused by these organisms on low-growing plants. Found commonly on lawns, strawberry beds, seedbeds, rotting logs, tree trunks, sidewalks, organic mulches, etc. The fruiting stage is powdery.

Slime spore—Spore that becomes separated with slime from the parent cell producing it.

Slip—A herbaceous or softwood cutting.

Slug (of fungi)—The aggregated pseudoplasmodium of Dictyosteliomycota.

Slurry—A thick suspension of a finely divided material in a liquid paste. Generally used for treating seeds.

Smear—A thin layer of material, e.g., a bacterial culture, spread on a glass slide for microscopic examination. Also referred to as a film.

Smooth membrane—Endoplasmic reticulum not encrusted with ribosomes. *See* Rough membrane.

Smooth—(Of fungi) lacking any ornamentation either on the wall or as appendages; (of bacteria) colonies with an even surface.

Smut—A disease caused by a smut fungus (Ustilaginales) in the Basidiomycotina, or the fungus itself; it is characterized by masses of dark brown or black, dusty to greasy masses of teliospores that generally accumulate in black, powdery sori.

Smut ball—*See* Spore ball.

Smut spore—A dark, thick-walled resting spore of a smut fungus; often termed a teliospore; may germinate to produce a promycelium, the organ of meiosis; often in the past improperly termed chlamydospore. *See* Teliospore.

Snow blight—A smothering blight disease of conifer foliage and twigs that develops under snow.

Snow molds (of grasses)—Fungal diseases that develop under snow and as snow is melting. Pink snow mold is caused by *Microdochium (Fusarium) nivale*, gray snow mold by *Typhula* spp., Coprinus or cottony snow mold by *Coprinus psychromorbidus*, and Sclerotinia snow mold or scald by *Myriosclerotinia (Sclerotinia) borealis*.

SNPV—Abbreviation for the subtype of nuclear polydedrosis viruses in which the enveloped virions contain predominantly single nucleocapsids. *See* MNPV.

Sod—Soil (generally 2.5–7.5 cm [1–3 in.]) permeated by and held together with grass roots or grass-legume roots; plugs, squares, or strips of turfgrass with adhering soil used in vegetative planting.

Sodic soil—A soil containing more than 15% sodium that interferes with the growth of most crop plants. Also, if the total soluble salts is more than 4 mmho/cm^2, the soil is a saline sodic soil; if it is less than 4 mmho/cm^2, the soil is a nonsaline sodic soil.

Sodium dodecyl sulfate (SDS)—Also known as sodium lauryl sulfate (SLS), an anionic detergent with many uses including disruption of virus particles and cells, inhibiting nucleases during nucleic acid extractions, and denaturing proteins for gel electrophoresis.

Soft rot—Disease symptom involving primarily the parenchyma tissue of various plant organs, resulting in a slimy softening and decay; a decomposition of plant parts, e.g., fruits, roots, or stems, by extracellular enzymes produced by fungi or bacteria, resulting in the tissues becoming soft.

Soft-coriaceous (of basidiocarps)—Somewhat flexible when fresh, breaking when bent at an extreme angle; spongy and fragile when dry; hyphae loosely interwoven.

Softwood—Wood of gymnosperms; a wood lacking wood fibers.

Soil—The solid portion of the earth's crust in which plants grow. It is composed of mineral material, air, water, and both living and dead organic matter.

Soil air—The gaseous phase of the soil; the percentage of the total volume not occupied by liquid or solid.

Soil conservation—A combination of all management and land use methods that safeguard the soil against depletion or deterioration by natural or human-caused factors.

Soil fertilization—*See* Fertilization.

Soil inhabitant—A microorganism that is usually strongly competitive with other normal microflora of the soil and that often survives many years as a saprophyte in the complete absence of suitable hosts.

Soil invader—A microorganism that is poorly competitive with normal soil microflora and seldom survives over one or two years in the soil in the complete absence of suitable hosts.

Soil management—The total tillage operations, cropping practices, fertilizing, liming, and other treatments conducted on or applied to a soil for the production of plants.

Soil organic matter—The organic fraction of the soil that includes plant and animal residues at various stages of decomposition, cells and tissues of soil organisms, and substances synthesized by the soil population.

Soil pasteurization—Treating soil with heat, usually steam at 60–70° C (approximately 140–160° F), to destroy most harmful pathogens, nematodes, insects, and weed seeds without generally affecting the saprophytic soil flora. A less severe temperature treatment than soil sterilization.

Soil salinity—The amount of soluble salts in a soil, expressed as parts per million (ppm), millimho/cm, or by other convenient ratios.

Soil series—The basic unit of soil classification; a subdivision of a family, made up of soils essentially alike in all major profile characteristics.

Soil solution—The aqueous liquid phase of the soil and its solutes consisting of ions dissociated from the surfaces of soil particles and other soluble materials.

Soil sterilant, soil sterilization—A chemical that, when present in the soil, prevents the growth of plants, microorganisms, etc.; treating soil by heat (usually steam) at 100° C (212° F), with chemicals, or with gaseous fumigation to kill all living organisms in it.

Soil structure—The arrangement of primary soil particles into secondary particles, units, or peds that act as primary particles. The secondary units are characterized and classified on the basis of size, shape, and degree of distinctness.

Soil texture—The relative percentages of sand, silt, and clay in a soil.

Soil tilth—The physical condition of soil as related to its ease of tillage, fitness as a seedbed, and suitability for plant growth.

Soil transients—Parasitic organisms (e.g., most leaf-infecting fungi and bacteria) that can remain alive in soil for only short periods.

Soil type—The lowest unit in the natural system of soil classification; a subdivision of a soil series.

Soilborne—Denoting a soil source or origin of pathogens; the property of a microorganism living and surviving in the soil.

Soiling—The staining and disfiguring of paint, fruit, or other materials by pigmented fungi.

Sol—A liquid colloidal system.

Solanaceous—Plants in the family Solanaceae, which includes eggplant, pepper, potato, tobacco, tomato, and Chinese lanternplant.

Soleiform—Slipper-shaped, almost resembling an hourglass, i.e., elongate-ellipsoid.

Solitary—Arising singly at one point; not growing in the immediate neighborhood of other plants.

Solopathogenic—A pathogenic monosporidal line (e.g., of a smut fungus such as *Ustilago zeae*).

Solubility—Ability to dissolve in a liquid; measured as the maximum amount of a gas, liquid, or solid that will dissolve in a liquid. Solubility varies with temperature, hence is usually expressed at a standard temperature of 25° C.

Soluble antigen—An antigen that is virus-specific but not the virion itself; often comprises isolated structural subunits but can be nonstructural virus-coded proteins.

Soluble powder—A finely ground powder formulation of a pesticide that dissolves and forms a solution in water or other liquid.

Soluble RNA—RNA soluble in strong salt solutions, e.g., 3 M sodium acetate. Consists mainly of small species such as transfer RNA (tRNA) and ribosomal 5S RNA.

Soluble salts—Dissolved salts (anions and cations) in the soil that can become toxic to roots when exceeding certain levels.

Solute—A dissolved substance; the dispersed phase of a solution.

Solution—A homogeneous mixture; the molecules of the dissolved substance (the solute) are dispersed among the molecules of the solvent.

Solution (or soilless) culture—Aqueous nutrient solution containing essential elements for cultivating plants without soil; hydroponics.

Solvent—A substance, usually a liquid, that can dissolve other substances (solutes); the continuous phase of a solution.

Soma (pl. somata)—The assimilative body of an organism as distinguished from its reproductive organs or reproductive phase.

Somaclonal variation—Existing genetic variability in clones generated from a single mother plant, leaf, etc., by tissue culture. Variation may not be evident until after plant cells have been through aseptic culture, or the culture may force the change.

Somaclone—Plant produced by a genetic engineering technique in which single cells or protoplasts are cultured to produce an individual genetically variable from its genetically stable parent. The variation induced is described as *somaclonal*.

Somatic—Describing the vegetative phase in plants—structure or function—as distinguished from the reproductive phase; (of nematodes) refers to the body proper, as in the somatic muscles.

Somatic aberration—Mutation or abnormality in a somatic cell and its progeny.

Somatic cell—A cell other than a sex or germ cell.

Somatic hybridization—Production of hybrid cells by fusion of two protoplasts differing in genetic makeup.

Somatic musculature (of nematodes)—A single layer of more or less spindle-shaped cells attached to the hypodermis throughout their length; it functions in body movement.

Somatic tissue—Nonreproductive, vegetative tissue developed through mitosis that will not undergo meiosis.

Sooty molds—Fungi (especially families of Dothideales and their anamorphs) with dark hyphae that grow on the honeydew secreted by aphids, mealybugs, scales, and whiteflies forming a dense, superficial, sooty coating on foliage, stems, and fruit.

Soralium (pl. soralia)—Decorticate parts of a lichen thallus in which soredia are found; usually formed from medullary tissues thrusting upward through the cortical layers. Soralia are of several different types: delimited, diffuse, tuberculate, and fissural.

Sordid—Of a dirty or dingy color.

Sorede, soredium (pl. **soredia**)—One or more algal cells in a lichen wrapped in hyphal tissue, that, when set free from the thallus, are able to grow into a new vegetative thallus.

Sorocarp (of Acrasiales)—A simple, stalked undifferentiated fruiting structure.

Sorocyst (of Acrasiales)—A sorocarp without a stalk.

Sorophore (of Acrasiales)—The stalk supporting the sorocarp.

Sorus (pl. **sori**)—Compact fruiting structure, especially the erumpent spore mass in the rusts (Uredinales)and smuts (Ustilaginales); occasionally a group of fruitbodies as in Synchytriaceae; a cluster of sporangia on a fern sporophyte.

Source of inoculum—The place or object on or in which inoculum is produced.

Southern blight—*See* Crown rot.

Southern blotting—A technique in which DNA fragments, separated by gel electrophoresis, are immobilized on a membrane, e.g., nitrocellulose and nylon-based membranes. The immobilized DNA is available for hybridization with a labeled ssDNA or RNA probe.

Soy sauce—Same as shoyu.

sp. (pl. **spp.**)—Abbreviation for *species*; one kind of plant or animal life subordinate to a genus but above a race, strain, or variety. A genus name followed by *sp.* means that the particular species is undetermined. *Spp.* following the genus name means that several species are grouped together without being named individually. *See* Species.

sp. nov., sp. n., n. sp—New species.

Spadiceous—Dark brown in color.

Sparassoid—Composed of interlaced flabelliform branches to form ball-like structures.

Sparassol—The antifungal, orsellinic acid monomethyl ether from *Sparassis ramosa*.

Spathe—A large bract or pair of bracts sheathing a flower cluster—e.g., Jack-in-the-pulpit (*Arisaema triphyllum*) or palms.

Spathulate, spatulate—Shaped like a spoon, spatula or paddle; oblong with a narrowing base.

Spawn—Mycelium, especially that used for starting mushroom cultures; to put inoculum (spawn) into a mushroom bed or other substratum.

Spear—The stylet of a nematode.

Speciation—The process of differentiation into new species; process of identifying an unknown organism (e.g., nematode) to species.

Species (pl. **species**)—Any one kind of life subordinate to a genus but above a race, strain, or variety; usually the smallest unit in the classification of organisms; a group of closely related individuals of the same ancestry, resembling one another in certain inherited characteristics of structure and behavior and relative stability in nature; the individuals of a species ordinarily interbreed freely and maintain themselves and their characteristics in nature. *See* Genus and Race.

Species inquirendae—Species of doubtful status because of inadequate descriptions and lack of preserved material.

Specific—Distinct, unique.

Specific absorbance—The absorbance per unit mass of a substance usually at the wavelength of the maximum of the absorbance spectrum of that substance.

Specific activity—The activity of a radioisotope of an element, other labeled product, protein, or enzyme per unit weight present in the sample.

Specific epithet—The second element of the name of a species; the species name along with the generic name identifies the individual.

Specific infectivity—The infectivity of a virus expressed per unit weight or per cent number of particles; can be measured, for example, by plaque or pock assay or by local lesions.

Specific name—The combination formed by the generic name and the specific epithet.

Specificity (of serology)—A term defining selective reactivity between substances, e.g., of an antigen with its corresponding antibody or primed lymphocyte.

Specimen—An individual taken as a representative of a larger group, e.g., a single plant; also, a portion or sample of material. From such a specimen a culture can often be made and one or more strains (isolates) isolated.

Speck—A small spot differing in color or substance from that of adjoining tissue.

Spectrophotometry—A procedure to measure photometrically the wavelength range of radiant energy absorbed by a sample under analysis; may be by visible light, UV light, or X-rays. For viral constituents UV light is usually used in the range of 220–230 nm.

Sperm, sperm cell—A male sex cell or nucleus, a male gamete, typically motile; spermatazoa of nematodes.

Spermatheca (pl. **spermatothecae**)—The enlarged portion of the female nematode reproductive system; between the oviduct and uterus that functions as a sperm storage organ (also termed seminal receptacle); usually present in amphimictic species and absent in automictic species.

Spermatiophore—A spermatia-producing or sperm-supporting structure.

Spermatium (pl. **spermatia**)—Nonmotile, uninucleate, haploid (+ or –) "sex" cell (haploid male gamete), variously termed as pycniospore or microconidium; nonmotile, uninucleate, sporelike male structure that empties its contents into a receptive female structure during plasmogamy; spermatia are variously regarded as gametes or gametangia.

Spermatization—Placing of spermatia or microconidia on structures, such as receptive hyphae, for diploidization.

Spermatocyte (of nematodes)—A cell giving rise to a sperm cell.

Spermatogonium (pl. **spermatogonia**)—A cell in a nematode derived from the cap cell giving rise to a spermatocyte.

Spermatophyte—A seed-bearing plant.

Spermatozoa (of nematodes)—A mass of mature sperms.

Spermidine—A triamine [$H_2N(CH_2)_3NH(CH_2)NH_2$] constituent of some viruses, e.g., turnip yellow mosaic virus.

Spermodochidium—A fruitbody having spermodochia in a lysigenos cavity in the host tissue.

Spermodochium—A spermogonium without a wall.

Spermogonium, spermagone, spermagonium (pl. **spermogonia**)—Flask-shaped, walled fungus structure in which spermatia are produced, as in ascomycetes; a pycnium of a rust fungus; (of nematodes) the reported sperm-producing organ of a digonic hermaphrodite, but apparently does not exist.

Spermoplane—The surface of a seed.

Spermosphere—The microhabitat around a seed in soil. *See* Phylloplane and Rhizoplane.

Sphacelate—Withered and dark as if dead.

Sphaerocysts—Globose cells in fungal tissues, e.g., *Russula, Lactarius*, and other Basidiomycotina.

Sphaeroid, sphaeroidal (of spores)—Nearly spherical.

Sphaeroplast, spheroplast—A bacterial or plant cell from which the cell wall has been removed and may be enclosed by a modified or fragmentary cell wall; used interchangeably with protoplast but usually refers to bacterial cells. *See* Protoplast.

Sphaeropsidales—Form-order traditionally used for Mitosporic fungi in which spores are borne in pycnidia (e.g., Phomales, Phyllostictales); not accepted by many mycologists.

Sphagnophilous—Growing in *Sphagnum* or peat moss.

Sphenoid—Wedge-shaped; same as cuneate.

Spherical (of spores)—Globular; shaped like a ball; all sections through the center are circles.

Spheridium—Same as capitulum.

Spheroplast—*See* Protoplast.

Spherule—Rather large, thin-walled, sporangiumlike structure as in *Coccidioides*; *or* a multinucleate cell of a resting myxomycete plasmodium.

Spherulin—Spherule phase coccidioidin.

Sphincter (of nematodes)—A circular muscle that functions to open or close a duct or orifice.

Sphincter-Z (of nematodes)—A constriction located between the uterus and spermatheca in some species of *Xiphinema*.

Spicate, spiciform—Having the form of a spike (tail).

Spiculate—Having a small, erect point.

Spicule, spiculos—Having one or more small spines.

Spicules—(Of nematodes) paired, sclerotized structures that are the male copulatory organs; (of fungi) needlelike outgrowths.

Spiculospore—Spore formed at the tip of a pointed structure; often elongate and resembling a spike.

Spiculum—(Of nematodes) fused spicules; (of fungi, pl. **spicula**)—*See* Sterigma.

Spiculur pouch (of nematodes)—A pouch containing the spicules; often called the spicular sheath in parasitic nematodes.

Spike—1) An elongate and unbranched inflorescence that bears possible sessile flowers. Examples are some cereal or grass heads, gladioli, and snapdragons (*Antirrhinum*); 2) (in virology) projections from the surface of a virus particle usually associated with binding of the particle to the cell surface and composed of either protein or glycoprotein. Most membrane-bound particles have spikes but they are also found in some viruses not possessing membranes.

Spikelet—Spike appendage comprised of one or more reduced flowers and associated bracts arranged alternately along an unbranched axis (the rachilla); unit of inflorescence in grasses; a small spike.

Spilodium—A minute, round, blackish structure on the thallus of *Dirina massiliensis* caused by *Milospium graphideorum*. It is composed of compacted, dark-colored hyphae.

Spination (of nematodes)—Having spines on the body surface, e.g., *Criconema*.

Spindle—Larger at the middle than at the ends; (of bacteria) applied to sporangia, refers to the forms called clostridia; also describes colonies, usually subsurface forms.

Spindle organ. *See* Turbinate cell.

Spine—A narrow, sharply pointed process.

Spiniform—Spine-shaped; horny.

Spinneret (of nematodes)—A porelike opening of the caudal glands situated in the tail, located terminally or subterminally, in some nemas belonging to the Adenophorea.

Spinose, spiny, spinous—Having spines; aculeate.

Spinule—A small spine.

Spinulose—Delicately spiny; having little spines.

Spiral hyphae—Hypha ending in a spiral or helical coil, as in *Trichophyton*.

Spiral—(Of nematodes) referring to the coil-like type of amphidial opening; (of bacteria) a long, curved rod form of cell; a coiled cell.

Spirillum (pl. **spirilla**)—A curved, corkscrew, or spiral-shaped bacterium.

Spiroplasma viruses—Phagelike viruses isolated from helical mycoplasms (spiroplasmas). Isolate SV-C3 from *Spiroplasma citri* is polyhedral with a short tail. SV-CE virus, also from *S. citri*, has a polyhedral head and a long tail (morphotype B).

Spiroplasma—A genus of small, pleomorphic, one-celled, prokaryotic bacteria that lack rigid cell walls; flexuous, often helical mycoplasma present in the phloem of diseased plants. There are also spiroplasmas that live on plant surfaces.

Splash cup, splashing cup—An open, cuplike structure in certain fungi (e.g., *Cladonia, Cyathus, Trametes conchifer*) and liverworts (e.g., *Marchantia*) from which reproductive bodies are scattered by falling drops of water. *See* Bird's nest fungi.

Splendent—Glittering or shining.

Splicing—The mechanism of RNA processing in which introns are removed and the adjacent exon sequences are religated. The sequence 5' of the intron is called the splice donor, and the 3' of the intron is the splice acceptor.

Spodochrous—Having a grayish tint.

Spondaneous—Occurring without external stimulation.

Spongilliform—Sponge-formed.

Spongy (of the stipe or flesh of agarics)—Soft, tending to be watersoaked.

Spongy mesophyll—Tissue composed of loosely arranged chlorophyll-bearing parenchyma cells of diverse form found just inside the lower epidermis of the leaf.

Spongy parenchyma—The cell layer in a leaf between the palisade parenchyma and the lower epidermis; these cells are loosely packed and have thin cell walls; same as spongy mesophyll.

Spontaneous mutant—A mutant that occurs naturally.

Spora, air spora—The population ("spore flora") of airborne particles of plant or animal origin; spore content of a particular place or ecological niche.

Sporabola (pl. **sporabolae**)—Curve made by a basidiospore after its ejection from its sterigma.

Sporangiocyst (of Chytridiales)—A resting sporangium. *See* Cystosorus.

Sporangiole, sporangiolum (pl. **sporangiola**)—A small sporangium with or without a columella that contains or produces one to a few spores; characteristic of the Mucorales.

Sporangiophore—A specialized hypha (a sporophore or stalk) bearing one or more sporangia.

Sporangiosorus—A more or less compact sorus or aggregate of spherical sporangia fused together and formed from one plasmodium; also, one or more lobed sporangia formed from a single plasmodium.

Sporangiospore—A walled, nonmotile, asexual spore borne in a sporangium.

Sporangium, sporange (pl. **sporangia**)—A spore case of fungi; commonly a saclike or flasklike fungus structure whose contents are converted by cleavage into an indefinite number of endogenous asexual spores (zoospores, sporangiospores); (of bacteria) cell containing an endospore; (of ferns) the saclike structure in a sorus in which spores are produced.

Spore—A one- to many-celled, microscopic reproductive body in bacteria, fungi and some lower plants that becomes free and can develop into a new plant; may be sexually or asexually produced and of a wide variety of sizes, shapes, colors, and origins; (of bacteria) more correctly endospore, a resting phase, a resistant body within the main part of the vegetative microbial cell; (of plants) the first cell stage in the gametophyte generation.

Spore ball—Unit of dispersal comprised of a firmly aggregated group of smut spores (e.g., *Sorosporium, Tolyposporium*) or spores and sterile cells joined closely together (e.g., *Urocystis*) of varying structure.

Spore charge—The + or − electrostatic charge carried by airborne basidiospores.

Spore cyst—A cell, hollow organ, or saclike structure enclosing a mass of protoplasm that contains spores.

Spore deposit—Same as spore print.

Spore horn—*See* Cirrhus (cirrus).

Spore mother cell—A cell that, by cell divisions, produces typically four spores following meiosis.

Spore print—A deposit of basidiospores obtained from a basidiocarp (usually an agaric or polypore) allowed to fall on a sheet of paper placed below the lamellae or pores.

Spore wall—*See* Ectosporium, Endosporium, Episporium, Exosporium, and Perisporium.

Sporidesm, sporodesm—Compound spore or spore ball, of which the components are merispores. *See* Teliospore.

Sporidesmin and **sporodesmolides**—Oligopeptide toxins of *Pithomyces chartarum*, teleomorph *heptosphaerulina chartarum*, that cause facial eczema in cattle and sheep, especially in New Zealand.

Sporidiole—A small spore.

Sporidium (pl. **sporidia**)—Basidiospore of rusts, smuts, and other Basidiomycotina.

Sporocarp—A spore-producing organ or fruitbody, e.g., ascocarp and basidiocarp. Used especially for Acrasiomycota, Endogonaceae, and Myxomycota

Sporocladium—A special sporogenous branch in the Kickxellaceae.

Sporocyst—A cyst that produces asexual spores.

Sporodochium (pl. **sporodochia**)—Superficial, erumpent, cushion-shaped (pulvinate), asexual fungus fruitbody (stroma) bearing closely packed, relatively short conidiophores, pseudoparenchyma, and conidia covering its upper surface, typical of the Tuberculariaceae. *See* Pinnote sporodochium and Acervulus.

Sporogenesis—Spore development; the production of spores.

Sporogenous—Pertaining to a cell that generates spores; producing, having, or supporting spores. *See* Conidiogenous.

Sporograph—A straight-line graph obtained by plotting the ratio (E) of the length (D) to the width (d) against the length of the basidiospores of an agaric. *See* Q.

Sporont—A thallus on which spores will be produced.

Sporophore—Any structure on which spores are borne, especially a conidiophore; (of macrofungi) fruitbody of larger fungi, e.g., ascocarp, basidiocarp. *See* Hymenophore.

Sporophyll—Spore-bearing leaf.

Sporophyte—A spore-bearing plant; the diploid or asexual, spore-producing phase in the life cycle of a plant; the stage of plants, such as ferns, that produces spores; diplont, diplophase. *See* Gametophyte.

Sporoplasm—The dense spore-producing protoplasm within the epiplasm in a sporangium or ascus.

Sporostasis (adj. **sporostatic**)—Inhibition of spore germination. *See* Mycostasis.

Sporothallus—A thallus that produces spores. *See* Gametothallus.

Sporothecium—The tip of a basidium bearing basidiospores when the basidiospores are sessile.

Sporotrichosis—Lymphatic disease of humans and animals caused by *Sporothrix schenckii*.

Sporulate, sporulation—To form or produce spores.

Spot—A definite, localized, necrotic diseased area of a leaf, stem, flower, fruit, etc.

Spot anthracnose—A disease caused by species of fungi in the genus *Elsinoë*, anamorph *Sphaceloma*. *See* Anthracnose.

Spot treatment—Application of a chemical (e.g., fertilizer or pesticide) to a restricted or small area.

spp.—Species (plural).

Spray concentrate—A liquid formulation of a pesticide that is diluted with another liquid (usually water or oil) before being used.

Spray deposit—The amount of wet pesticide initially deposited per unit area of plant or other surface.

Spreader—Material added to a spray preparation to make the spray droplets spread out, increase the area covered, and improve contact between the chemical and the plant or other treated surface; also termed *film extender*.

Spreading (of bacteria)—Growth extending much beyond the point or line of inoculation, i.e., several millimeters or more.

Sprig, sprigging—A single turfgrass stem (stolon, rhizome, or tiller), usually with some attached roots, that is used in vegetative propagation.

Springwood—Early part of the yearly xylem growth ring in woody plants consisting typically of cells that are larger than those formed later in the season (summerwood).

Sprout—To germinate; to begin to germinate as a seed; to put forth buds or shoots.

Sprout cell—A cell resulting from a sprouting or budding process.

Sprout fungi—Yeasts.

Spunk—Same as touchwood or amadou. Fruitbodies of hymenomycetous fungi on trees, especially *Phellinus (Fomes) ignarius*.

Spur—A short, woody stem (branch). The principal fruiting area of many fruit trees.

Spur formation or precipitation line (syn. **partial fusion line**)—An antibody-antigen precipitation line formed when two antigenically distinct strains of a virus are placed in adjacent wells in a *gel double-diffusion test*. See Ouchterlony gel diffusion test.

Spurious—False.

sq cm—Square centimeter = cm^2 = 100 mm^2 = 0.155 $in.^2$ = 0.001076 ft^2.

sq ft—Square foot = ft^2 = 144 $in.^2$ = 0.111111 sq yard = 0.0929 m^2 = 0.003673 sq rod.

sq ha—Square hectare = ha^2 = 2.471 acres = 395.367 sq rod = 10,000 m^2 = 0.01 km^2 = 0.0039 mi^2.

sq in.—Square inch = $in.^2$ = 6.451626 cm^2 = 0.0069444 ft^2.

sq km—Square kilometer = km^2 = 0.3861 mi^2 = 247.1 acres = 100 ha^2 = 1 million m^2.

sq mile—Square mile = mi^2 = 640 acres = 1 section = 258.9998 ha^2 = 102,400 sq rods = 3,097,600 sq yards = 2.589998 km^2.

sq mm—Square millimeter = mm^2 = 0.01 cm^2 = 0.000001 m^2 = 0.00155 $in.^2$.

sq m—Square meter = m^2 = 10.76387 ft^2 = 1550 $in.^2$ = 1.195985 sq yards = 0.039537 sq rod = 1 million mm^2 = 10,000 cm^2.

sq rod—Square rod = 30.25 sq yards = 25.29295 m^2 = 272.25 ft^2 = 0.00625 acre = 0.0025293 ha.

sq yard—Square yard—9 ft^2 = 1,296 $in.^2$ = 0.83613 m^2 = 0.03306 sq rod.

Squama (pl. **squamae**)—Scale.

Squamose, squamous—Covered with or consisting of scales.

Squamule—A small, scalelike structure.

Squamulose—Having minute scales; growth form of a lichen thallus.

Square (botanical)—An unopened flower bud in cotton (*Gossypium*) with its accompanying bracts.

Squarrose—Rough with scales; (of the pileus or stipe) covered with recurved scales.

SSC—Standard saline citrate (1× SSC is 0.15 M NaCl and 0.015 M Na citrate). Various concentrations of SSC are used in Southern and Northern blotting and in hybridization experiments.

ssp.—Subspecies.

ss—Single-stranded (nucleic acid).

Stage—A phase in the life cycle; frequently equals state; for fungi, it is better reserved for circumstances in which the succession of two or more states is regular.

Staghead—Antlerlike appearance of a tree dying from the top downward; defoliated, dead, or dying major branches in the crown of a tree, usually with a wilt or root problem.

Stain—A dye used to color cells and tissues as an aid to visual inspection.

Stainsall—1-ethyl-2-(3-(1-ethylnaphtho (1,2d) thiazolin-2-ylidene)-2-methylpropenyl)-naphtho (1,2d) thiazolium bromide. It is used to stain nucleic acids in polyacrylamide cells in which RNA stains bluish purple, DNA stains blue, and proteins stain red.

Stalagmoid (of spores)—Elongate-drop-shaped or elongate-tear-shaped.

Stalagmospore—A stalagmoid spore.

Staling substances—Compounds produced by an organism that slow or stop growth of an organism in pure culture as a result of the accumulation of self-inhibiting metabolites, an adverse pH, etc. *See* Inhibitory substances.

Stalk—An indefinite name for stem, stipe, pedicel, peduncle.

Stamen (adj. **staminate**)—The pollen-producing structure of a flower consisting of an anther(s) (pollen-bearing portion) borne on a stalk or filament; a modified microsporophyll.

Staminate (male) flower—One that bears stamens (producing pollen) but no pistils.

Stand—The number of established individual plants or shoots per unit area.

Standard deviation—A commonly used measure of dispersion or spread of data values, given in the same units as the mean and thus of the original observations. It is of great value in the intuitive interpretation of the data and is often used in the estimation of confidence levels and significance tests. Statistics books contain details of its calculation and use.

Standard error—The standard deviation of the mean values calculated for each of a number of samples taken from a given population. Statistics books contain details of its calculation and use.

Staphylococccus aureus **V8 protease**—An enzyme cleaving polypeptides at the carboxyl side of asparate and glutamate residues.

Starch—A white, complex, water-insoluble polysaccharide consisting of glucose units; the principal food storage substance of plants.

Starch lesion—Local accumulation of starch in a virus-infected leaf. It is demonstrated by decolorizing the leaf and staining with iodine and can be used as an assay for virus concentration. *See* Local lesion.

Start codon—The trinucleotide in an mRNA at which ribosomes start the process of translation; sets the reading frame for the translation.

Starters—The pure cultures or mixtures of microorganisms used for starting fermenting processes.

State—(Of fungi) one phase of a pleomorphic fungus, e.g., the anamorph characterized by asexual spores and the teleomorph characterized by sexual spores; (of bacteria) "the name given to the rough, smooth, mucoid and similar variants that arise in culture . . ." (Bacteriological Code). *See* Holomorph.

Stationary phase—Interval directly following a growth phase during which the number of viable bacteria remains constant.

Statismospore—A spore not forcibly discharged. *See* Ballistospore.

Statolon—An antiviral substance that induces interferon formation, coming from *Penicillium brevicompactum* (syn. *P. stoloniferum*), whose active principle is considered to be RNA of viral origin.

Staurospore, stauroconidium—A star-shaped or branched conidium.

Staurosporous—Possessing star-shaped spores.

Steady-state infection—An infection in which both virus replication and cell multiplication occur simultaneously in cell culture. Most of the cells are infected, and virus is continuously released from them, but there is no cytopathic effect. The addition of antiviral antibody does not cure such infections.

Stele (adj. **stelar**)—The central cylinder of vascular and closely associated tissues in the stems and roots of higher plants; xylem, phloem, and (when present) pith, pericycle, and interfascicular parenchyma.

Steliogen—Structure that gives rise to the sporocarp stalk in protostelids.

Stellate, stelliform—Star-shaped; (of crystals) arranged in radiating or star-shaped patterns.

Stellate seta—Compound seta with several radiating arms; asterophysis.

Stem—The main body of a plant, usually the ascending axis, whether above or below ground in opposition to the descending axis or root. Stems, but not roots, produce nodes and buds; (in nomenclature) the part of a name to which the suffix appropriate to a category is added.

Stem pressure—Positive pressure produced under certain conditions in the stems of some plants—e.g., maple (*Acer*)—and responsible for the exudation of sap from their stems.

Stem-pitting—A symptom of some viral diseases characterized by depressions or grooves in the stem of the plant.

Stephanocyst—A typically bicellular structure found in certain basidiomycetes. The basal cell is cuplike and the terminal cell globose.

Stereome (of certain lichens)—Scleroplectenchyma, the principal thallus-supporting tissue.

Sterigma (pl. **sterigmata**)—A small, arcuate, usually pointed hyphal branch or structure supporting a spore (conidium, basidiospore) or sporangium.

Sterigmatocystin—A carcinogenic hepatotoxin, a xanthone derivative, from *Aspergillus versicolor*, a precursor of aflatoxin B1.

Sterilant—Any chemical or physical agent that destroys all living organisms in a substance, e.g., soil.

Sterile—Infertile; free from living (reproducible) microorganisms; uncontaminated; nonsporing.

Sterile conk—A hard or tough, dense mass of sterile mycelium of a wood-decaying fungus on a trunk or limb.

Sterile fungi—A group of fungi not known to produce any kind of spores.

Sterilization—The elimination of living cells or organisms (including pathogens) by means of heat, chemicals, light, etc., from soil, containers, and so forth; the killing of all forms of life; also, reduction of sexual structures or processes in an individual or race; (in media preparation) heating to the point where all living organisms are killed.

Sterilized—Free from living cells (usually microorganisms).

Sterol—Any of a group of solid polycyclic alcohols (as cholesterol and ergosterol) widely derived from plant and animal lipids.

Stichobasidium—*See* Holobasidium.

Sticker—A material added to a spray or dust to improve adherence (tenacity to a plant or other surface) rather than to increase initial deposit. *See* Adhesive.

Sticky ends—The single-stranded ends on DNA produced by many type II restriction endonucleases. They are either 5′ or 3′ extensions of the DNA molecule.

Stigma—Part of a pistil, usually the apex, that receives pollen and upon which pollen grains germinate; also, a darkened area in the upper margin near the end of the wings of certain insects.

Stigmatocyst, stigmocyst—*See* Stigmatopod.

Stigmatomycosis (of cotton [*Gossypium*] bolls, bean [*Phaseolus*] and soybean [*Glycine max*] pods, and other plants)—Damage caused by fungi, such as *Ashbya (Nematospora) gossypii* and *Nematospora coryli*, and inoculated by insects.

Stigmatopod, stigmatopodium, stigmopodium (pl. **stigmatopodia**)—A capitate, usually two-celled hyphopodium, generally dark in color, in which the end cell or *stigmatocyst* has a haustorium. A stigmatocyst in a hypha is a *node cell*. *See* Hyphopode.

Stilbaceous—Having synnemata; synnematous.

Stilbiform—Having a long stalk and a head.

Stilboid—A sterile mushroomlike structure, as in *Mycena citricolor* and other agarics, that functions as a propagule; gemma; having a stalked head like *Stilbum*. *See* Carpophoroid.

Stilbum—An erect synnema of *Stilbella* with a head of slime spores.

Stimulus—An environmental factor or change that induces a reaction in a living organism.

Stink horns—Certain smelly members of the Phallales. The vile odor is produced by the gleba at the apex of the fruitbody due to autodigestion; it attracts flies that disseminate the spores.

Stinking smut—*See* Bunt.

Stipate—Crowded.

Stipe—The stalk or "stem" of a basidiocarp (e.g., mushroom), that supports the pileus; or an ascocarp; (of conidiophores) the main stout axis or unbranched lower part; stalk of a brown alga.

Stipitate—Having a stipe or stalk (stem).

Stippling—Series of small dots or speckles in which chlorophyll is absent.

Stipule—Small leaflike structure or appendage, usually paired, found at the base of leaf petioles in many species of plants.

Stock—Portion of the stem and associated root system into which a scion is grafted; term sometimes used for an artificial breeding group and for a production planting; strain, race or group of genetically similar organisms; (of basidiomycetes) a dikaryotic mycelium. *See* Strain.

Stock cultures—Known species of microorganisms maintained in the laboratory for various tests and studies.

Stokes' radius—The radius of the spherical volume occupied by a macromolecule in solution taking into account hydration and any nonisometric features of the macromolecule.

Stolon—An elongated, slender stem ("runner") that grows horizontally just above the soil surface and is capable of developing roots and stems at the nodes and ultimately forming an independent plant; (of fungi) a horizontal hypha that sprouts where it touches the substrate and forms haustoria or rhizoids in the substrate and aerial mycelium or sporophytes above it as in the Mucorales, e.g., *Rhizopus*. *See* Runner.

Stoloniferous—Bearing or developing stolons.

Stoma or stomate (pl. **stomata** or **stomates**)—Regulated microscopic opening (pore) in the epidermis of leaves and stems, controlled by two guard cells, for passage of gases and water vapor; sometimes serves as a point of entry for pathogens.

Stomatopod, stomatopodium (pl. **stomatopodia**)—An appressorium produced from a lateral hyphal branch above or in a stoma in the penetration process. *See* Appressorium and Hyphopodium.

Stomatostyle, stomatostylet (of nematodes)—*See* Stylet.

Stone cells—Thick-walled, isodiametric sclerenchyma cells.

Stone fungus—The hard pseudosclerotium of *Polyporus tuberaster*; Pietraia fungaia. The Canadian tuckahoe is the same species. When watered, it produces an edible fruitbody.

Stool (horticultural)—Sprouts that arise from the base of the plant below ground and become roots; used for vegetative propagation.

Stool or stooling (agronomic)—Shoots that rise from below ground at the base of a plant.

Stop codon—The trinucleotide sequence at which protein synthesis is terminated due to a lack of a tRNA molecule to insert an amino acid into the polypeptide chain at that site. There are three stop codons: UAP (ochre), UAG (amber), and UGA (opal). *See* Suppressor.

Stopper—The *Neurospora* phenotype characterized by irregular cycles of stopping and renewal of growth.

Strain—Biotype; race; form; isolate; an organism or group of similar organisms that differ in minor aspects from other organisms of the same species or variety (*see* Biotype, Form, Isolate, and Physiologic race); descendants of a single isolation in pure culture; a homokaryotic mycelium; a cultivar of a bacterium; a group of plant viruses that has most of its antigens in common with another group or strain.

Stramineous—Straw-colored.

Strand—Compound hyphal structure made up of individual hyphae braided in ropelike fashion.

Strand plcctenchyma (of lichens)—Plectenchyma in strands that form the supporting tissues in a thallus.

Strand polarity—The organization of the base sequence in a single-stranded nucleic acid. Positive (+) polarity refers to single-stranded molecules that contain the same base sequence as mRNA. Negative (−) polarity molecules have a base sequence complementary to the (+)-sense strand.

Strangle fungus—*See* Choke.

Stratification—The practice of exposing imbibed seeds to cool, 2–10° C (35–50° F), or sometimes warm, temperatures for a time prior to germination in order to break dormancy; a standard practice for germinating many species of grass and woody plants.

Stratified—Arranged in layers; describing seed treated to improve germination.

Stratiform (of bacteria)—Liquefying to the walls of the tube at the top and then proceeding downward horizontally.

Stratose thallus (of lichens)—A thallus with tissues in horizontal layers.

Stratose—In distinct layers.

Straw mushroom (paddy straw or Chinese mushroom)—*Volvariella volvacea* and *V. diplasia*, widely eaten by humans in the tropics.

Streak—An elongated lesion, usually with irregular sides.

Streptococci—Cocci that divide in such a way that chains of cells are formed.

Stress (water)—A condition in which plant(s) are unable to absorb enough water to replace that lost by transpiration. This may result in wilting, cessation of growth, or death of the plant or plant parts.

Stressors—Factors that induce a manifestation of infection from a latent condition. *See* Induction.

Stria (pl. **striae**)—Narrow line, furrow, or band; (of nematodes) superficial markings of the cuticle, appearing as grooves or clefts; if present, may encircle the lips or body or extend longitudinally.

Striate (n. **striations**)—Marked with minute parallel or radiating lines, grooves, furrows, projections, or ridges.

Strict—Close; narrow and upright; very straight.

Stricture (of nematodes)—A contraction of the lumen of a passage.

Strigose—Covered with long, rigid, sharp-pointed hairs; describing coarse, stiff, appressed hyphae or hyphal strands; hispid.

Stringy rot—A type of white rot in which the decayed wood in advanced stages has a fibrous appearance.

Striolate—Finely striate.

Strip cropping—The practice of growing crops that require different types of tillage, such as rows and sod, in alternate strips along contours or across the prevailing direction of the wind.

Stripe (or yellow) rust of cereals and grasses—Disease caused by *Puccinia striiformis*.

Stripe—Elongated necrosis of tissue between vascular bundles in leaves or stems of cereals and grasses.

Stripe smut of grasses—Disease caused by *Ustilago striiformis*.

Striposomes—*See* Uncoating.

Strobiliform—Shaped like a fir or pine cone.

Strobilis—A conelike collection of sporophylls borne on a stem axis.

Stroma (pl. **stromata**, adj. **stromatic, stromatal**)—Compact mass or matrix of specialized vegetative hyphae (with or without host tissue or substratum), sometimes sclerotiumlike, in or on which fruitbodies (reproductive structures) and/or spores are usually produced; often a cushionlike mass of fungal cells or closely interwoven hyphae. *See* Ectostroma, Endostroma, Epistroma, Hypostroma, and Pseudostroma.

Stromatoid, stromoid—Stromalike.

Structural protein—A protein forming part of the structure of a virus particle.

Structural unit—Protein subunit or "protomer," the basic building block of the virus capsid.

Struma (pl. **strumae**)—Cushionlike swelling.

Stubble—Stem bases and crowns of harvested plants still rooted in soil.

Stuffed (of a stipe)—Having the inside of a different structure than the outer layer.

Stunted—Describing a plant reduced in size and vigor due to unfavorable environmental conditions; may be due to a wide range of pathogens or abiotic agents.

Stupose, stuppose—Consisting of, covered with, or bearing filaments tufted or matted together and not gelatinized.

Stupulose—Covered with fine, short hairs; finely stupose.

Style—Stalklike structure between the stigma and ovary in the pistil of most flowers through which pollen tubes grow toward the ovule.

Stylet—Relatively long, pointed, stiff, slender, hollow, feeding organ in the mouth portion of plant-parasitic nematodes and some insects (e.g., aphids) for piercing and withdrawing nutrients from plant cells. Nematodes have three types:

    a. *stomatostyle*—a spear that apparently has developed gradually through evolution by the fusion or sclerotization of the stomal wall; anterior aperture is ventrally oblique; commonly found in species of *Secernentea*.

    b. *odontostyle*—a spear evolved from a large tooth and derived prior to each molt from a single cell in the esophagus; anterior aperture is broadly oblique; found in some members of the Adenophorea.

    c. *onchiostyle*—stiff, slender, curved, grooved, toothlike stylet found in the adenophorean genus *Trichodorus*; derived through evolutionary development from an onchium and originating in the stomal wall.

Stylet aperture (of nematodes)—The anterior opening of the stylet; located obliquely dorsal in the adenophorean Dorylaimoidea and obliquely ventral in the secernentean Tylenchida.

Stylet extension, odontostyle extension, or odontophore (of nematodes)—The sclerotized structure extending from the base of the stylet to the beginning of the esophagus musculature.

Stylet knobs and flanges (of nematodes)—Basal knobs and flanges; basal protuberances at the base of the stylet (usually three, often absent); function as posterior points of attachment for the stylet muscles.

Stylet-borne—Describing a virus borne on the stylet of its insect vector; a noncirculative virus. *See* Nonpersistent transmission.

Stylospore—Historically used in various senses; currently a spore borne on a pedicel or filament.

Styptic (of taste)—Astringent.

Suaveolent—Having a sweet smell; fragrant.

Sub- (prefix)—Under; below; almost qualified; slightly; somewhat; more or less.

Subacerate—Lanceolate-subacute.

Subacute—Somewhat acute.

Subalbous—Almost white.

Subapical—Just below an apex.

Subarachnoid—Somewhat cobwebby.

Subbulbous—Somewhat bulbous.

Subcaespitose—In small tufts.

Subcarbonaceous—Almost like charcoal or cinders; slightly carbonaceous.

Subcentric—*See* Centric.

Subcoriaceous—Slightly leatherlike.

Subcortical—Beneath the cortex.

Subcrustose—Somewhat crustlike.

Subcrystalline layer (of nematodes)—A superficial, waxy, translucent layer present on some cysts in the genus *Heterodera*.

Subcuboid—Somewhat cubical.

Subculture—A culture (e.g., of bacteria or fungi) derived from another culture.

Subcutaneous—Under the epidermis or skin.

Subcutaneous injection—Injection of antigens between the skin and underlying tissues used in the preparation of antisera.

Subcuticular—Beneath a cuticle; between the cuticle and the upper epidermal wall.

Subcuticular layer, subcuticle (of nematodes)—The hypodermis.

Subcutis (of fungi)—The cutis layer below the epicutis of basidiocarps. *See* Cutis.

Subcylindrical—Almost of the same diameter throughout its length.

Subdecurrent (of lamellae or tubes)—Being attached and extending slightly down the stipe.

Subdeterminate—Limited.

Subdigitate—Having divisions, almost fingerlike, radiating from a common center.

Subdiscoid—Somewhat disc-shaped.

Subdorsal—Describing the position on the nematode body situated 30° laterally from the dorsomedian and perpendicular to the anteroposterior; dorsolateral.

Subepidermal—Under the epidermis.

Suberect—Nearly erect, but nodding at the top.

Suberin—A waxy, waterproofing, hydophobic deposit found in the walls of cork and certain other cells.

Suberize (adj. **suberized**)—To convert exposed plant surfaces into tough cork tissue; cell walls hardened by their conversion to cork (suberin).

Subfuscous—Somewhat dark.

Subfusiform (of spores)—More or less fusiform, usually with one end somewhat rounded-pointed.

Subgenomic RNA—A species of RNA less than the genomic length found in infected cells and sometimes encapsidated. When encapsidated, it is not involved in natural infection. Each species of subgenomic RNA has a different cistron at the 5' end opening it for translation.

Subglobose or subspherical—Almost spherical but slightly elongated.

Subhyaline—Nearly hyaline but somewhat colored.

Subhymenium (pl. **subhymenia**)—The generative tissue immediately below the hymenium.

Subiculum, subicle (pl. **subicula**)—A more or less dense growth of hyphae (net- or crustlike) under, on, or in which fruitbodies arise; a felted, wool-like or cottony basal stratum of hyphae.

Subirrigation—Applying water from below the soil surface, usually from a ditch, perforated hose or pipe, or by placing a potted plant on a constantly moist surface.

Sublateral—The position on the nematode body ventral to the lateral position and perpendicular to the anteroposterior axis.

Submicroscopic—Too small to be seen with a light (compound) microscope.

Suboxidation—Deficiency of oxygen that impairs normal respiration.

Subpruinose (of a surface)—Slightly powdered or pruinose.

Subsidiary—Secondary, additional.

Subsoil—Soil below the plow layer.

Subsoiling—Breaking of compact subsoils, without inverting them, using a special knifelike chisel pulled through the soil usually at depths of 30–60 cm (12–24 in.) and at spacings commonly of 60–150 cm (2–5 ft).

Subsp.—Subspecies.

Subspecies—An infraspecific population defined on the basis of more than one sharply defined character (morphologic for most organisms), and often possessing a distinct habitat, that distinguish its members from typical representatives of the species. *See* Variety and Form.

Subsquamosa—Somewhat scaly.

Substipitate—Hardly stemmed, but with a very short attachment.

Substomatal—Just below a stomatal opening.

Substrate, substratum (pl. **substrata**)—Surface or medium on or in which a microorganism is growing (attached) or living and from which it may get its nourishment; chemical substance acted upon, often by an enzyme.

Substroma—Pseudostroma in which vegetative hyphae of the fungus within the host tissue predominate.

Subtend—To extend under or be opposite to; used especially with reference to a bract or a leaf that bears a flower in its axil.

Subterminal—Situated near the end.

Subtorulose—Cylindrical with moderately swollen areas at intervals.

Subturbinate—Top-shaped with a somewhat flattened apex.

Subulate, subuliform—Rather slender and tapered to a point; awl-shaped.

Subumbonate (of a pileus)—Slightly and usually broadly raised.

Subuniversal veil—The primary universal veil or protoblem.

Subventral—The position on the nematode body situated 30° laterally from the ventromedian line and perpendicular to the anteroposterior axis; ventrolateral.

Subzonate—Marked with obscure, indefinite zones.

Succession—Orderly sequence of differing types of vegetation in a given region.

Succulent (n. or adj.)—Plant, or plant part with tender, juicy or watery tissues.

Sucker—A shoot arising from a stem or root.

Sucrase—Enzyme that brings about the hydrolysis of cane sugar (sucrose) with the formation of two hexose sugars, fructose and glucose. *See* Invertase.

Sucrose—Cane or table sugar ($C_{12}H_{22}O_{11}$); a carbohydrate formed by chemically joining a molecule of glucose ($C_6H_{12}O_6$) and a molecule of fructose ($C_6H_{12}O_5$).

Sucrose gradient, sucrose density gradient centrifugation—A technique in which particles (e.g., viruses) of different sedimentation coefficients are separated by centrifugation. A gradient of sucrose concentrations, increasing from top to bottom, is constructed in a centrifuge tube and the solution containing the particles is placed on top. On centrifugation in a swing-out rotor, the various particle species sediment as bands at different rates. *See* Isopycnic centrifugation and Rate zonal gradient.

Suctorial (of certain insects)—Adapted for sucking host fluids by means of specialized mouthparts that pierce cells.

Suffused—Spread out; diffuse; tinged.

Sufu or Chinese cheese—An oriental food made of *Actinomucor* or *Mucor* fermented soybeans (*Glycine max*).

Sugar fungus—Fungus that attacks decaying substances and is only able to utilize simple sugars, amino acids, and other relatively simple organic compounds.

Sulcate—Grooved.

Sulcule—A little furrow.

Sulfocystidium—Gloeocystidium with contents that darken in sulfovanillin.

Sulfovanillin—Solution used to detect sulfocystidia. It reacts with the contents, turning dark blue to nearly black, and is added to material after alcohol treatment (Recipe: 0.5 g vanillin, 2 ml distilled water, plus 4 ml of concentrated sulfuric acid).

Sulfur polypore, sulfur shelf-mushroom—The edible *Laetiporus (Polyporus) sulphureus*.

Sulfureous, sulphureous, sulphurine—Sulfur yellow.

Summer fallow—The tillage of uncropped land during the summer to control weeds and other pests and store moisture in the soil for the growth of a later crop.

Summer spore—A fungus spore (e.g., a conidium or urediniospore) that germinates without resting and is associated with the rapid increase and spread of the fungus during a favorable season; frequently lives for only a short time. *See* Resting spore.

Summer stage (of rusts)—The uredial stage of cereal and grass rusts.

Summer truffle—*Tuber aestivum*. *See* Truffle.

Summerwood—Part of the yearly xylem growth ring in woody plants formed late in the growing season and consisting of cells smaller than those of springwood.

Sunscald—Plant tissues "burned" or scorched from the action of intense sunlight; high temperature injury due to intense sun's rays warming the trunks and major limbs of woody plants, usually during the winter, that often results in cracking and splitting the bark. Sunscald also occurs on unshaded fruits or to house plants exposed to direct sunlight.

Super-, supra- (prefix)—Above, in either position or degree.

Supercoiled DNA—A confirmation that a dsDNA molecule can adopt when both strands of a ds molecule are covalently closed; one or both strands becomes over- or under-wound in relation to the other, with the torsional strain causing the molecules to coil into a characteristic shape.

Supercooled—Describing liquid cooled below its freezing point without formation of ice crystals.

Superficial (mycelium, conidioma, etc.)—Occurring on the surface of the substratum and easily removable; not enclosed by host or fungal tissue.

Superhelical DNA—*See* Supercoiled DNA.

Superimposed—Overlapping.

Superinfection—An attempt to infect a host with a second virus, usually a strain of the first infecting virus. It can result in 1) interference between the two viruses, the first virus preventing replication of the second (*see* Cross-protection); 2) the two viruses replicating independently (*see* Synergism); 3) recombination occurring between the two viruses; and 4) phenotypic mixing.

Superinfection exclusion—Restriction of growth (interference) of a second virus in a cell already infected with another virus. The term is usually applied to growth restrictions imposed on superinfecting phages in prokaryotic systems.

Superior (of an annulus)—Attached close to the top of the stipe.

Superior ovary—An ovary borne above the points of origin of sepals and petals from a receptacle.

Supernatant—The liquid phase of a suspension after the particulate fraction (pellet) has been removed.

Supplement (syn. **adjuvant, axillary spray material**)—A substance added to a pesticide to improve its physical or chemical properties, i.e., a spreader, sticker, safener, wetting agent, etc., but *not* the diluent. *See* Adjuvant.

Supplements (supplementary organs of nematodes)—Genital papillae, located preanal on the ventral side of the male; derived from cuticle, but may be provided with glands; function during copulation.

Suppressive soils—Soils in which certain diseases are suppressed due to the presence in the soil of microorganisms antagonistic to the pathogen or pathogens.

Suppressor—A gene that overcomes or suppresses the effects of mutations in other, unlinked genes.

Supra-—*See* Super-.

Suprabasal—Inserted just above the basal scar.

Suprahilar plage (of basidiospores, especially of *Lactarius* and *Russula*)—Area above the hilar appendage on which the eusporial ornamentation is reduced or lacking.

Supreme—Uppermost.

Surculicolous (of *Exobasidium* infections)—Circumscribed, monocarpic and systemic in annual shoots.

Surcurrent—Running up the stem or stipe, opposite of decurrent.

Surfactant—Monomolecular compound that decreases surface tension between two unlike materials (e.g., oil and water) and increases the emulsifying, dispersing, spreading, wetting, or other surface-modifying properties of a pesticide formulation. A spreader, emulsifier, dispersing or sticking agent, wetting agent, etc., used to increase coverage (wettability) of the surface being sprayed.

Suscept—Any living organism attacked or susceptible to a given disease, pathogen or toxin; an abbreviated term for "susceptible plant" or "susceptible species." *See* Host.

Susceptibility—The inability of a plant to resist the effect of a pathogen or other damaging factor.

Susceptible—Not immune; lacking resistance or having poor resistance; prone to infection. *See* Tolerance, Tolerant.

Suspension—Dispersion system in which particles of a solid are distributed in a liquid, e.g., soil particles in water; (of pesticides) a solid form (of finely divided particles) suspended in water or other liquid, solid, or gas.

Suspensor—(Of fungi) a club-shaped or conical portion of a hypha supporting a gamete, gametangium, or especially a zygospore; (of higher plants) a structure in the embryo sporophyte that attaches or forces the embryo into food storage tissue.

Sustainable agriculture—The appropriate use of crop and livestock systems, and the agricultural inputs supporting their activities, that maintain economic and social viability while preserving the high productivity and quality of the land.

Suture—Groove or seam in higher plants along which the carpels of a fruit separate at maturity; (of fungi) junction or seam of a union, e.g., plates in the Gasteromycetes; a line of opening or dehiscence; (of nematodes) a seam or impressed line formed by the union of two adjacent inarticulate margins.

Svedberg equation—A formula for estimating the molecular weight of a macromolecule.

Svedberg units—The units used to measure the rate of sedimentation of a macromolecule, subcellular component, or virus to determine its *sedimentation coefficient* (S).

SW^{20}—Sedimentation coefficient expressed in Svedberg units and adjusted to 20° C in water.

Sward—Carpet of grasses or other ground cover such as clovers.

Swarm (of bacteria)—Some motile species spread as a thin film of growth over the moist surface of a solid medium. Nonmotile bacterial species or variants do not swarm.

Swarm cell—A flagellate cell; usually applied to the motile cells of the Myxomycetes and some Chytridiales acting before or after division (as an *isogamete*).

Swarm spore, swarmer—Zoospore, as in *Pythium*.

Swath—The width of the area covered by a sprayer making a sweep or trip across a field or other treated area.

Swing-out rotor—A centrifuge rotor with buckets free to swing out horizontally when the rotor is spinning. The gravitational force is thus along the long axis of the tubes. *See* Sucrose gradient.

Sycosis (pl. **sycoses**)—A chronic fungus infection of hair follicles, especially on the face and neck; ringworm of the beard.

Sylloge—A collection or compilation.

Sym-—*See* Syn-.

Symbiont, symbiote—One member of a symbiotic relationship.

Symbiosis (pl. **symbioses**, adj. **symbiotic**)—The living together in close association of two or more dissimilar organisms (symbionts); usually applied to cases in which the relationship is mutually beneficial.

Symmetrical—Of equal morphology about a designated axis.

Symmetry (of viruses)—Structural principle of viral capsid or nucleocapsid.

Sympatric—Referring to two or more different taxa that occur in close spatial proximity; occurring in the same geographical region. *See* Allopatric.

Symphogenous (of pycnidia, etc.)—Formed from a number of hyphae. *See* Meristogenous.

Symplasm, symplast—Continuous space within the plant body that is enclosed by the plasma membrane and connected by plasmodesmata; the living part of a plant. *See* Apoplasm, Apoplast.

Sympodial—Describing continued growth or branching of a conidiophore or a conidiogenous cell arising beneath or beside the previous conidium and pushing it to one side; pertaining to proliferation of axes, in which each successive spore develops behind and to one side of the previous apex in which growth has ceased.

Sympodioconidium, sympodulospore—Spore produced on a sympodula.

Sympodula—A conidiogenous cell characterized by continued growth of a succession of apices, each of which originates below and to one side of the previous apex.

Symptom—Indication of disease by reaction of the host; the visible external or internal effects produced in or on a plant by the presence of a pathogen. *See* Sign.

Symptomatology, symptomology—The study of symptoms of disease and signs of pathogens for the purpose of diagnosis; class of symptoms typical of a particular disease.

Symptomless carrier—A plant that, although infected with a pathogen (usually a virus), produces no readily visible effects.

Symptomless virus—A virus that is present in a plant but produces no visible effects on the plant.

Syn-, sym- (prefix)—With; (in compounds, etc.) growing together; aggregation; adhesion.

syn.—Synonym(s).

Synanamorph—Term applied to any one of the two or more anamorphs having the same teleomorph, e.g., *Chalara* and *Thielaviopsis*.

Synapsis—The pairing of homologous chromosomes in meiosis.

Synascus (pl. **synasci**)—Gametangium of *Ascosphaera*.

Synchronized culture—A culture manipulated so that the division of all component cells occurs simultaneously.

Synchronospore—Spore produced simultaneously with neighboring spores.

Synchronous (conidiogenous cell)—A polyblastic cell in which all the loci develop conidia at the same time; often difficult to distinguish from sympodial development.

Syncytium (pl. **syncytia**)—A multinucleate structure resulting from continued nuclear division or cell wall breakdown and subsequent fusion of protoplasts and surrounded by a common cell wall. *See* Coenocytic, Giant cell, and Nurse cell.

Syndrome—The totality of effects produced in a plant by one disease, whether all at once or successively; pattern or sequence of disease development; a combination of symptoms and signs and/or one or more causal agents.

Synergids—Two small cells (nuclei) lying near the egg at the micropylar end of the embryo sac in an ovule.

Synergism (adj. **synergistic**)—The mutual association (living together) of two or more organisms or environmental factors acting at one time and eliciting a host response that one alone could not make; magnitude of host response to concurrent pathogens exceeding the sum of the separate responses to each pathogen. *See* Metabiosis.

Synergist (of pesticides)—Any substance that increases the efficiency or toxic effects of a pesticide; it may or may not have pesticidal properties of its own.

Syngamy—Fertilization; fusion of male and female cells (gametes) to form a zygote.

Syngonic (of nematodes)—Pertaining to hermaphroditic reproduction in which both sperm and eggs are produced by the same gonad (also termed protandry).

Synisonym—One, two, or more names with the same basionym.

Synkaryon (pl. **synkarya**)—A diploid zygote nucleus.

Synkaryotic (of a diploid nucleus)—Having 2n chromosomes.

Synnema (pl. **synnemata**, adj. **synnematous**)—Compact or fused, generally upright group of conidiophores bearing conidia at the apex only or on both apex and sides; coremium; stilbum; characteristic of the Stilbaceae. See Coremium; sometimes used for loose fascicles. There are three types:

a. *determinate*—having a terminal, nonelongated conidiogenous area, with growth stopping after sporulation has started (e.g., *Stilbella*).

b. *indeterminate*—having an elongated fertile zone involving the whole conidioma, with growth continuing after sporulation (e.g., *Doratomyces*).

c. *compound*—with determinate or indeterminate branches formed on a branched or unbranched axis (e.g., *Tilachlidiopsis*).

Synnematogenous, synnematous—Composed of several or many threads or filaments tightly bound and in some cases fused along most of their lengths; having synnemata. See Mononematous.

Synoecious—Having both sex organs on the same mycelium.

Synonym (in biology)—Another name for an organism (species or group), especially a later or illegitimate name; a rejected scientific name, other than a homonym.

Synoptic key—An artificial key in which each taxon is often given a number; all characters of the taxa are listed, with genus numbers entered for those that are positive; by eliminating successive characters, at random if desired, eventually the single number left gives the name of the taxon.

Syntype—One of a number of specimens (strains, cultures) of equal nomenclatural rank that formed all or part of an author's original material in which he/she did not designate or indicate a holotype.

Syrrotium (pl. **serrotia**)—The superficial aerophilic mycelial coil (cord) of *Merulius*.

System—A coherent body; a system of classification; taxonomy.

Systemic fungicide—Fungicide or fungistat taken up systemically by plants.

Systemic—Applied to chemicals and pathogens (or single infections) that generally spread internally throughout the plant body as opposed to remaining localized; an infection that spreads throughout a plant.

T

"T" (of nematodes)—Percent of the body occupied by each gonad of the male reproductive system, i.e., length of the gonad divided by length of the nematode times 100.

Tabacina—Tobacco-colored; pale brown.

Tabid—Dissolving; decaying.

Tablespoon (tbs.)—*See* tbs.

Taenia (pl. **teniae**) (of myxomycetes)—A band.

Taeniole—A little band.

Tail (of nematodes)—The portion of the body between the anus and the posterior terminus.

Tailed phages—A large group of unenveloped DNA viruses isolated from prokaryotes, characterized by complex virus particles consisting of a cubic head, the capsid, and a helical tail (morphotypes A, B, and C; *see* Phage). The viruses may be temperate or virulent and have been isolated from about 100 genera of bacteria, cyanobacteria, and mycoplasmas.

Tailing—The addition of stretches of nucleotides to the 3'-terminus of ds nucleic acid to facilitate cloning into a vector.

Take-all—A crown and sheath rot disease of cereals and grasses caused by the fungi *Gaeumannomyces* (*Ophiobolus*) *graminis*, *G. graminis* var. *avenae*, and *G. graminis* var. *tritici*.

Tangential (face)—The wood surface exposed when a cut is made at right angles to the rays and parallel to the long axis of the majority of cells.

Tank mix—A mixture of two or more pesticides in the spray tank at the time of application. A tank mix should be used with caution until it is clear that the ingredients are compatible.

Tannic acid agar—Malt agar extract medium with 5 g of tannic acid per liter; used to detect polyphenoloxidases.

Tannin—One of a heterogeneous group of bitter (astringent), soluble, polyphenol derivatives widely distributed in certain plant tissues, such as bark, heartwood, and others. The polyphenols condense with protein(s) to form a leatherlike substance insoluble in water.

Tapé—A fermented Indonesian food prepared by the action of *Rhizopus arrhizus* (syn. *R. oryzae*).

Taproot—The primary, elongated, deeply descending root of a plant from which the secondary or lateral roots branch.

Tar spot—A disease of plants characterized by the presence of black stromatic tissue at the surface of the host, giving the appearance of a spot of tar.

Target (of pesticides)—The plant, area, buildings, animals, or pests intended to be treated with a pesticide application.

Tartareous—Having a thick, rough, crumbling or powdery surface.

Tawny—Dull yellowish brown, about the color of a lion

Tawny grisette—The edible *Amanita fulva*. *See* Grisette.

Taxis—*See* Tropism.

Taxon (pl. **taxa**)—A taxonomic group of any rank such as family, genus, species, etc.

Taxonomy—Science or "scientific art" dealing with the systematic describing, naming, and classification of organisms based as far as possible on their natural relationships.

tbs—Tablespoonful = 3 tsp = 14.8 ml or cm^3 = 1/2 fl oz = 0.902 $in.^3$ = 0.063 cup.

TDP (thermal death point)—*See* Thermal death (or inactivation) point.

Teaspoon (tsp)—See tsp.

Technique—A method for performing an operation or test.

Teeth (of basidiocarps)—The spine- or toothlike structure covered by the hymenium.

Telamon (of nematodes)—A cuticularized thickening of the ventral cloacal wall of some males, serving as an accessory guiding structure for the spicules.

Teleblem, teloblem, teleoblema (pl. **teleblemata**)—Same as universal veil.

Teleomorph, teleomorphosis—The sexual or perfect state of a fungus, i.e., the form involved in producing meiotic spores.

Teleutosorus—An older term for telium.

Telial—Pertaining to, or possessing telia.

Teliospore, teleutospore, teleutosporodesm—Thick-walled resting or overwintering spore produced by the rusts (Uredinales) and smuts (Ustilaginales) in which karyogamy occurs; it germinates to form a promycelium (basidium) in which meiosis occurs.

Telium (pl. **telia**)—Sorus of a rust fungus producing teliospores, i.e., the terminal sorus (designated by the numeral III), of the dikaryophase; the site of karyogamy.

Telophase—The final stage in mitosis in which two sets of chromosomes are organized into daughter nuclei, and processes are initiated leading to the formation of a cell plate separating the daughter nuclei; also, the final stage in meiosis 1 and 2.

Telostom (of nematodes)—A short valve that connects the stoma to the esophagus; the basal component of the stoma.

TEM—*See* Transmission electron microscope. *See also* Ultrastructure.

TEMED—A catalyst (N,N,N'N'-tetramethylethylene diamine) that provides free radicals needed for the polymerization of monomer acrylamide and bisacrylamide to give the gel used in polyacrylamide electrophoresis.

Tempeh, tempé—Oriental food composed of *Rhizopus oligosporus* fermented soybeans (*Glycine max*).

Temperate phage—A bacteriophage that reproduces (replicates) by passing through a prophage state in a lysogenic bacterial cell; a phage that can grow in its prokaryotic host in two ways: 1) a *lytic cycle*, in which the virus replicates and the host cell is killed, and 2) a *lysogenic state*, in which the virus genome is integrated into the host on a plastid.

Temperature-sensitive mutant—A mutant not "viable" at as high a temperature as the wild-type strain.

Template—A nucleic acid molecule from which a complementary nucleic acid molecule is being synthesized.

Tenacious—Tough.

Tenacity (adherence)—The resistance of a pesticide deposit to weathering as measured by retention.

Tenacle—The circle of cilia (hairs) around the ostiole of the perithecium of certain ascomycetes that serve to collect and hold the haerangium. *See* Haerangium.

Tendril—(Of fungi) long, slender, coiling extensive mass of spores; (of higher plants) slender, coiling, leafless organ arising from a stem that serves a climbing plant {e.g., pea [*Pisum*] and grape [*Vitis*]) as a means of attachment and support; a tendril may be a modified leaf, leaflet, stipule, or stem.

Tenellous—Delicate.

Tension, soil moisture—The equivalent negative pressure of water in soil; the attraction with which water is held to soil particles.

Tentacular—Several filiform mucilaginous appendages originating from more or less the same point.

Tentaculiform—Tentacle-shaped.

Tentoxin—A cyclic tetrapeptide toxin from *Alternaria alternata* that induces chlorosis.

Tenuous—Delicate.

Tephreous, tephrous—Ash-colored; same as cinereous.

Teratogen—A chemical, exposure to which is liable to produce malformations, monstrosities, or serious deviations from the normal in fetus growth. Such a compound is described as *teratogenic.*

Teratology—Study of malformations, monstrosities, or serious deviations from the normal type in organisms.

Teratoma—A neoplasm composed of bizarre and chaotically arranged tissue that is foreign embryonically and histologically to the area in which the tumor is found.

Teratum (pl. **terata**)—An abnormal modification.

Terebrate—Having scattered perforations.

Terebrator (of lichens)—A trichogyne.

Terete—Commonly circular in cross section, but narrowing to one end, not irregular; cylindrical.

Terminal—The end of a shoot, twig, branch, or cell.

Terminal bud—A bud at the distal end of a stem.

Terminal codon—*See* Stop codon.

Terminal infection—An infection with a pathogenic microorganism or virus that occurs during the course of a disease that terminates in death of the host.

Terminal redundancy—The presence of identical nucleotide sequences at both ends of a nucleic acid molecule with linear chromosomes.

Terminal repetition—The presence of nucleotide sequences at both ends of a nucleic acid molecule that are identical (*see* Terminal redundancy) or are identical but inverted.

Terminal transferase, deoxynucleotidyl transferase—An enzyme that adds deoxynucleotide triphosphates (dNTP) to the 3′ hydroxyl groups of the DNA fragment. It is used in 3′ end labeling of DNA and in tailing DNA fragments with complementary dNPSs to facilitate cloning.

Terminus phialospore—A phialospore formed at the apex of a one-spored phialide that terminates growth of the phialide.

Ternate—In threes.

Terrace—A level, usually narrow plain bordering a river, lake, or the sea; also, a raised, more or less level strip of land, usually constructed on a contour and designed to make the land suitable for tillage and prevent erosion.

Terramycin—An antibiotic produced by *Streptomyces rimosus*.

Terreous—Earth-colored; brownish.

Terrestrial, terricole, terricolous—Growing on the ground; living on land as opposed to water.

Terverticillate (of a penicillus)—Having branching at three levels; having rami bearing metulae and phialides.

TES—A biological buffer, N-tris(hydroxymethyl)methyl-2-aminoethanesulphonic acid, with a pH range of 6.6–8.6.

Tessellate—(Of nematodes) checkered; blocklike; a type of cuticle pattern in which the longitudinal ridges are broken by annules into transverse rows of squares; (of fungi) marked with a mosaic design; checkered.

Test plant—A plant species, variety, or cultivar used for diagnosing, or evaluating the infectivity of, a pathogen. *See* Indicator plant.

Testa—The outer coat of a true seed developed from the integument(s) of an ovule.

Testaceous—Having a hard covering or shell; (of color) a light brick-red.

Testis (pl. **testes**)—That portion of the male nematode reproductive system that produces the sperm or spermatazoa.

Tetra- (prefix)—Four.

Tetracyclines—A group of broad-spectrum antibiotics, isolated from *Streptomyces* spp., that act by preventing charged transfer RNAs from binding to ribosomes.

Tetracytes—Spores resulting from meiosis.

Tetrad—A group of four cells (usually spores) produced by two divisions of a mother cell, in meiosis.

Tetradidymous—Eight-fold or with four pairs.

Tetradymous (of spores)—Having four cells.

Tetragonous—Four-angled.

Tetraploid—Having four sets of chromosomes (4n) per nucleus, twice the usual diploid number.

Tetrapolar—With incompatibility controlled by two pairs of nonlinked alleles that segregate independently; a single dikaryon gives rise to four mating types; also referred to as bifactorial. *See* Bipolar.

Tetraspore—*See* Dispore.

Tetrasporic, tetrasporous, tetraspory (of basidia)—Four-spored.

Tetratomic—Forked four times at one node.

Texospores—Ascospores covered with a layer of cells of paraphysal origin (e.g., as in *Texosporium*).

Textura (of fungi)—Hyphal tissue types (structure) in coeleomycetes and Ascomycotina. There are seven types of textura: angularis, epidermoidea, globulosa, intricata, oblita, porrecta, and prismatica.

Texture—(Of soil) the relative proportions of the various sized groups of individual soil grains in a mass of soil; refers to the proportions of sand, silt, and clay in a given amount of soil; (of plants, fungi, etc.) arrangement of the components of different tissues as compact, loose, etc.

Thalassic, thalassine—Sea-green, a bluish green.

Thall-, thalli-, thallo- (prefix)—Sprout, shoot.

Thallic (of conidiogenesis)—Describing the condition in which any enlargement of the recognizable conidial initial occurs *after* the initial has been delimited by a septum or septa. The conidium is differentiated from a *whole* cell.

Thalliform—Like a thallus in form.

Thalline exciple (margin)—Same as excipulum thallinum.

Thalloconidium—A dark brown, smooth to rugged conidium produced by a thallic conidiogenous cell with more than one wall layer and consisting of 1–2,500 cells.

Thallodic—Of, pertaining to, or belonging to a thallus.

Thallophyte—A lower plant whose vegetative (somatic) phase is devoid of stems, roots, or leaves and that usually propagates by means of spores; includes bacteria, algae, fungi, slime molds, lichens, and mosses.

Thallospore—An asexual spore without a conidiophore or conidiogenous cell or one not separated from the hypha or conidiogenous cell that produced it (an arthrospore, blastospore, or chlamydospore [and aleuriospore]); a thalloconidium (*which see*).

Thallus (pl. **thalli**)—Vegetative body of a thallophyte (nonvascular plant); a relatively simple plant body, with relatively little cellular differentiation, that lacks true stems, roots, and leaves; characteristic of algae, fungi, liverworts, etc.; (in fungi) the somatic phase.

Thamniscophagous (of endotrophic mycorrhiza)—Forming haustorial arbuscles that are finally digested by the host cells.

Thatch—A tightly intertwined layer of plant litter from accumulations of undecomposed or partially decomposed plant residues (roots, crowns, rhizomes, and stolons); in turf, situated above the soil surface and generally below the green portions.

Thec-, theca- (prefix)—Case, sac, capsule.

Thecium—The part of an apothecium containing the asci between the epithecium and hypothecium; sometimes considered equivalent to hymenium.

Thelephorous—Covered with nipplelike prominences.

Therapeutic—Having the capacity to relieve symptoms and cure established disease; curative.

Thermal death (or inactivation) point—The lowest temperature at which death of an organism occurs (usually by heating) after a selected interval of time, usually 10 min; the temperature (varying from 45 to 95° C) at which virus particles lose their infectivity or an enzyme its activity.

Thermoduric (n. **thermodrury**)—Surviving exposure to high temperatures, especially when in a dormant state (e.g., spores, sclerotia). *See* Thermophily.

Thermolabile—Destroyed by heat at temperatures below 100° C. *See* Thermophily.

Thermolysin—An enzyme that catalyzes the hydrolysis of peptide bonds on the N-terminal side of valine, leucine, isoleucine, phenylalanine, tyrosine, methionine, and sometimes alanine residues.

Thermophile (adj. **thermophilic**)—Microorganisms growing at 20–50° C or higher (optimum 40–50° C); some actinomycetes can even withstand 65–70° C.

Thermophily—Making active growth at high temperatures.

Thermostable—Relatively resistant to heat (to temperatures of 100° C).

Thermotherapy—The curing of a host (or propagative part, e.g., a seed, bulb, corm, tuber) or cell line of infection by a pathogen by heat treatment. *See* Chemotherapy.

Thermotolerant fungi—Fungi that grow at a minimum well below 20° C and a maximum about 50° C. Examples: *Aspergillus fumigatus* and *Absidia corymbifera* (syn. *A. ramosa*).

Thigmotropism—Growth curvature induced in climbing plants by the stimulus of contact.

Thin-layer chromatography (TLC)—Chromatography on thin layers of adsorbents rather than in columns; the adsorbents include alumina, silica gel, silicates, charcoal, or cellulose. *See* Chromatography.

Thinning—Removing young plants from a row to provide the remaining plants with more space to develop; also, the removal of excess numbers of fruits from a plant so the remaining fruits will become larger.

Tholus—Same as nassace.

Thorax—Insect body part between the head and abdomen that bears the legs and wings.

Thread blight—Disease with well-marked threads (hyphal strands) running over leaves and stems of mostly tropical plants, caused by species of *Corticium* and *Marasmius*; *or* a fungus causing thread blight. *See* Horse-hair blight fungus.

Thrips—Small slender insects (order Thysanoptera), usually with four long, narrow wings. Thrips have rasping-sucking mouth parts and a 10-segmented abdomen. They influence the development of disease in plants by acting as vectors of pathogens like fungi, bacteria, and the tomato spotted wilt virus.

Thrush—A throat and genital disease of humans, especially children, caused by *Candida albicans*.

Thryptogen, thryptophyte—An organism that increases the sensitivity of host tissue to an outside factor (e.g., cold).

Thymidine—A nucleoside of thymine and deoxyribose. *See* Nucleic acid.

Thymidine kinase—An enzyme catalyzing the phosphorylation of thymidine to thymidic acid and induced in cells infected with DNA viruses.

Thymine—One of the pyrimidine nitrogenous bases found in DNA. *See* Nucleic acid.

Thyriothecium (pl. **thyriotheca**)—An inverted, flattened, shield-shaped ascoma having the wall more or less radial in structure and lacking a basal plate, e.g., *Microthyrium*. *See* Catathecium.

Thyroid—Shaped like an oblong shield.

Thyrus (pl. **thyrsi**)—A type of inflorescence; also, the densely branched apices of some lichens.

Tichus—Peripheral layer of cells of perithecial walls forming a dark protective layer (e.g., in *Pleospora herbarum*).

Tiger's milk—*Polyporus sacer*, used to treat tuberculosis, colds, etc. in Malaysia.

Tigrine—Marked like a tiger.

Tile (drain)—Older pipe made of burned clay, concrete, or other material, in short lengths and commonly buried at the bottom of a trench, with open joints to collect and carry excess water from the soil. The newer polyvinyl tile is made of continuous lengths with perforations that allow water to enter the pipe for removal from the soil.

Tiller—A lateral shoot, culm, or stalk arising from a crown bud; common in grasses.

Tilth—State of soil aggregation or consistency; good tilth implies porous, friable texture. *See* Soil tilth.

Tinder fungus—*Fomes fomentarius*. *See* Amadou and False tinder fungus.

Tinea—Ringworm or other skin disease of humans and animals caused by a parasitic fungus (especially a dermatophyte). There are several types, e.g., tinea *barbae* (beard ringworm), tinea *capitis* (head ringworm), tinea *corporus* (body ringworm), tinea *cruris* (groin ringworm); tinea *favosa* = favus; tinea *imbricata* (*Trichophyton concentricum*); tinea *nigra* (cutaceous pigmented infection caused by a dermatiaceous fungus); tinea *nodosa* = piedra; tinea *pedis* ("athlete's foot," foot ringworm); tinea *ungulum* (ringworm of the nails); and tinea *versicolor* = pityriasis versicolor.

Tinophyses—Paraphysoids. *See* Hamathecium.

Tip burn, tip necrosis—Necrosis of apical or marginal tissues of plants affecting only a small percentage of the entire leaf; may be caused by internal water stress caused by wind desiccation or salt.

Tissue—A group of cells, usually of similar structure, that perform the same or related functions.

Tissue analysis (or test)—Analysis of leaf tissues for major and minor elements.

Tissue culture—An *in vitro* technique of cultivating (propagating) cells, tissues, or organs in a sterile, synthetic medium. Often used to eliminate viruses and other pathogens from vegetatively propagated plants.

Tissue types (of fungi)—*See* Textura.

Titer (in serological reactions)—A relative measure of the amount of antibody in an antiserum per unit volume of original serum. The antibody is serially diluted and antigen is added. Serum titer is indicated as the reciprocal of the highest serum dilution producing a discernible antigen-antibody reaction; also, the concentration of virus present in a preparation as measured by bioassay or a relative measure of the concentration of a specific antibody in antiserum.

Titration—The measurement of titer.

TK—Thymidine kinase.

TLCK—An inhibitor of chymotrypsin.

TM—Melting temperature.

Toadstool—A popular term for a basidiocarp, especially an inedible agaric or bolete. *See* Mushroom.

Tolerance (adj. **tolerant**)—Ability of a plant to endure an infectious or noninfectious disease, adverse conditions, or chemical injury without serious damage or yield loss; (of pesticides) the amount of chemical residue legally permitted on an agricultural product entering commercial channels and usually measured in parts per million (ppm). The tolerance in the United States is set by the federal EPA; (in serology) active state of unresponsiveness to a given immunogen; immune responses to other immunogens are normal. *See* Susceptible, Resistant, Resistance, and Vector resistance.

Tolypophagous (of endotrophic mycorrhizae)—Describing living host cells that have the ability to kill, digest, and reabsorb penetrating hyphae. *See* Ptyophagous.

Tomato leaf mold—Disease caused by *Fulvia fulva*.

Tomentose, tomentous—Downy or woolly; having a covering of short, soft, densely matted hairs or hyphae that give a texture like a woolly blanket.

Tomentum—Collective term for the layer of long, soft hairs (conglutinate hyphae) with thick walls on the upper surface of a basidiocarp that are interlaced and tangled or matted like wool.

Ton (long, U.S.) = 2,240 lb = 2,722.22 troy lb = 1.120 short tons = 1,016 metric tons = 1,016.04 kg.

Ton (metric) = 2,204.6 lb = 1.1023 short tons = 0.984 long ton = 1,000 kg.

Ton (short, U.S.) = 2,000 lb av = 2,430.56 troy lb = 0.892857 long ton = 0.907185 metric ton = 907.18486 kg = 32,000 oz.

Tonophily (adj. **tonophilic, tonophilous**)—The ability of organisms to grow under conditions of high osmotic pressure.

Tonoplast—Membrane enclosing the vacuole in the cytoplasm of a plant cell.

Top rot—Decay in the upper trunk of a living tree.

Top yeast—The yeast growing at the surface of the wort.

Topdressing—A prepared soil mix added to the soil surface, and usually incorporated into the soil by raking or irrigating; materials such as fertilizer or compost are applied to the soil surface while plants are growing.

Topotype—A specimen collected later at the exact type (or original) locality.

Topsoil—The upper layer of soil moved during cultivation and tillage.

Topworking (top-grafting)—Adding another cultivar (or cultivars) to a tree by inserting buds or grafts on its main scaffold branches.

Torfacevus—Growing in bogs.

Torn (of pores, pileus margins)—Superficially rough or jagged as if torn.

Tornate—Rounded off.

Toroid—Rounded; protuberant.

Torsive—Twisted spirally.

Tortuous—Bent or twisted in several directions.

Toruloid, torulose, torulous—Cylindrical, contorted, but having swellings at intervals like a chain of beads; moniliform; (of nematodes) knobby, having knoblike swellings.

Torulosis—See Cryptococcosis.

Totipotent—Bisexual; capable of producing a fructification; a cell capable of unregulated growth.

Touchwood—Wood rotted by fungi, especially *Polyporus squamous*; fruitbodies of *Fomes fomentarius* or *Phellinus* (*Fomes*) *ignarius* or the tinder ("amadou") made from them.

Toxic—Of, caused by, or acting as a poison; capable of causing injury to a living organism.

Toxicant—A poisonous or toxic substance.

Toxicity—The relative capacity of a substance to interfere adversely with the vital processes of an organism; quality, state, or degree of being poisonous. See LD_{50}.

Toxigenic—Toxin producing.

Toxin—Poisonous secretion produced by a living organism; a nonenzyme metabolite produced by one organism that is injurious to another. See Antibiotic.

Toxiphilous—Tolerating an impure atmosphere; favoring a polluted habitat.

Toxiphobous—Not tolerating an impure atmosphere or habitat.

Toxitolerant—Tolerant of air pollution or toxins.

Toxoid—A toxin that has been treated to destroy its toxic property without affecting its antigenic properties.

Trabecula (pl. **trabeculae**, adj. **trabeculate**)—A lamella primordium; (of *Gymnoglossum* and other gasteromycetes) plates of undifferentiated primordial tissue remaining in the developing gleba that form the branches of the dendroid columella; (of pseudoparaphyses) paraphysoids, tinophyses. See Hamathecium.

Trace elements—*See* Micronutrients.

Tracheid—Thick-walled, elongated cell of xylem tissue that is capable of water transport; has tapered or oblique end walls and lignified pitted walls adapted for conduction and support.

Tracheomycosis—A fungal disease (often a wilt) in which the pathogen is mainly confined to the xylem. *See* Hadromycosis.

Trachytectum—Same as exosporium; spore wall.

Trade name (syn. **brand name, trademark**)—Name given to a product by the manufacturer or formulator to distinguish it as their product.

Trama (pl. **tramae**)—The sterile inner tissue (hyphae) in the center of the gills or tube walls of polypore basidiocarps, a spine of species of Hydnaceae, or the dissepiment between pores of a polypore; when composed of interwoven fibers of a uniform diameter, it is called *floccose*; when the hyphae are frequently enlarged so as to give, in section, the appearance of rounded cells, it is termed *vesiculose*. *See* Context.

Tramal cystidium—A cystidium arising from the trama and extending above the hymenial fascicles. *See* Cystidium.

Transcapsidation—Encapsidation of the nucleic acid of one virus strain with the protein of another, during simultaneous infection and replication of two strains. *See also* Phenotypic mixing.

Transcript—The RNA molecule produced by transcription. The primary transcript is often processed or modified to give the mature functional RNA, e.g., mRNA, rRNA, or tRNA.

Transcription—Synthesis of RNA using one strand of DNA as the template; catalyzed by the enzyme DNA-dependent RNA polymerase; copying of a gene into RNA. *See* Reverse transcription.

Transducing phage—Defective phage containing host DNA sequences as well as, or instead of, viral genes. *See* Transduction.

Transduction—The transfer of genetic material from one bacterial cell to another by a bacteriophage particle; the phage carries part of the genome of the infected bacterium and the recipient (or transduced) cell shows characters of both bacterial parents.

Transeptate—Having all transverse cross walls.

Transfection—The successful virus infection of cells following their inoculation with viral nucleic acid.

Transfer RNA (tRNA)—The RNA that moves amino acids to the ribosome to be placed in the order prescribed by the messenger RNA (mRNA). Each tRNA has a specific trinucleotide sequence that interacts with a complementary sequence in mRNA. Also called soluble RNA (sRNA) and acceptor RNA (aRNA)

Transformation—The change in the DNA of a bacterium by absorption and incorporation of DNA fragments released from related strains of other bacteria; the change in a cell through uptake and expression of additional genetic material; the change of a normal to a malignant cell; a plant disease symptom in which a pathogen ramifies and replaces host tissue within an organ of the host, usually a floral organ; development of tissues or organs in unusual places (often called metamorphosis).

Transformation assay—A test for the frequency with which cells are rendered tumorigenic by a virus. Transformed cells grow in a manner different from normal cells, often forming plaques of heaped-up colonies.

Transformation of DNA—A genetic engineering technique in which the properties of cells may be modified by the insertion and expression of foreign DNA. See Genetic engineering.

Transgressive segregation—Genetic mechanism transferring characters (to progeny) not expressed by parents.

Translation (adj. translational)—Synthesis of proteins on ribosomes as directed by successive triplets of nucleotides moved along an mRNA.

Translocation—Distribution of a chemical from the point of absorption (plant leaves, stems, roots) to other leaves, buds, and root tips; movement of water, minerals, food, pathogens, wastes, etc., within or throughout a plant; segmental exchange between nonpaired (nonhomologous) chromosomes to form two new chromosomes.

Translucent—So clear that light rays may pass through; transmitting light without being transparent.

Transmission—The transfer or spread of virus or another pathogen from plant to plant or from one plant generation to another. Forms of transmission by physical contact, vector, or pollen are termed *horizontal* transmission. *Vertical* transmission is from mother to progeny, e.g., seed transmission of many plant viruses.

Transmission electron microscope (TEM)—An electron microscope in which an electric beam passes through a thin specimen (e.g., a preparation of virus particles negatively stained or shadowed), or a thin section of tissue mounted on a grid. The image of the specimen is magnified with electrostatic lenses and observed on a fluorescent screen or photographed.

Transmission expectancy—A transmission efficiency value calculated as the product of a given age-specific transmission rate multiplied by the age of specific survival value.

Transovarial—Through ovaries and eggs.

Transovarial transmission—Pathogen (e.g., a virus) transmission through the ovaries and eggs of the infected vector to its progeny.

Transpiration—Diffusion and loss of water vapor from aerial parts of plants, chiefly through stomata in the leaves.

Transport protein—A protein involved in the transport of small molecules around the cell.

Transposon, transposable element—A segment of chromosomal DNA that can insert itself (transpose) into the genome and integrate at different sites on the chromosomes. The ends of the transposon DNA are usually inverted repeats. Widely used to produce insertion mutations and to study gene function by examining their effect when inserted into different regions of the genome.

Transstadial—Describing a pathogen (e.g., a virus) that is retained through the molt of its insect vector.

Transverse—Crosswise.

Transverse section—Section of stem or other plant part cut at right angles to its longitudinal axis.

Transverse striae (of nematodes)—Superficial lines in the cuticle extending around the body and supporting the annules.

Tranzchel's Law—In summary, it states that the telia of microcyclic rusts adopt the habit of the parent macrocyclic species and occur on the aecial host plants of the latter.

Trapping—*See* immunosorbent electron microscopy.

Travertine—Type of calcium carbonate rock, some of which was formed by the secretion of carbonate by algae.

Tree—Woody perennial plant generally with a single main stem (trunk) that rises some distance aboveground before it branches.

Treehopper—Small, leaping, homopterous insects of the family Membracidae with piercing-sucking mouth parts. Treehoppers are known to transmit a number of plant-infecting viruses.

Trehalose—A reserve disaccharide of fungi (especially yeasts) and lichens, that is hydrolyzed by the enzyme trehalase; mycose.

Trellis rust or European pear rust—A disease on pear and junipers caused by *Gymnosporangium fuscum*.

Trembling fungi—The gelatinous Tremellales.

Tremelliform, tremelloid, tremellose—Gelatinous; like jelly or wet gelatin; like *Tremella*.

Tremorgen (adj. **tremorgenic**)—A mycotoxin inducing a neurotoxicosis or tremor in humans and higher animals, e.g., fumitremorgin, verruculotoxin.

Trenching—Physical separation of soil in a vertical plane to sever grafted roots between trees.

Tretic (of conidiogenesis)—Apparently enteroblastic protrusion of the inner wall of the conidium (tretoconidium) taking place through one or several channels in the outer wall. The channels or pores, according to which way you view them, can be seen quite easily with an ordinary compound microscope, and, although electron microscope studies indicate that conidium development is not necessarily strictly enteroblastic, their presence provides a character most useful to taxonomists. Tretoconidia are solitary or in acropetal chains.

Tri- (prefix)—Three, triple.

Triangular number—Number of identical equilateral triangles composing a face of a given icosahedron, e.g., $T = 4$, $T = 9$, $T = 16$.

Tribe—A division of a plant or animal kingdom containing a number of related genera within a family.

Trich-, trichi-, tricho- (prefix)—Hair.

Trichidium—Same as sterigma.

Trichiferous—Producing or bearing hairs.

Trichiform—Bristle-shaped.

Trichloroacetic acid (TCA, CCl_3COOH)—A chemical used at 5–10% solutions for the quantitative precipitation of nucleic acids and proteins; frequently used when the radioactive counts incorporated into macromolecules are being measured.

Trichoderm (of basidiomata)—An outer layer composed of hairlike elements that project above the surface making it tomentose, pillose, seriaceous, villose, etc.

Trichogyne—In some algae, lichens, and fungi, a projection from the female sex organ (gametophyte) that receives the male gamete (sperm) or nuclei before fertilization (karyogamy); the receptive hypha of the ascogonium, which is often long and hairlike.

Trichogynous—Of, pertaining to, or having the properties and functions of a trichogyne.

Tricholomic acid—An insecticidal amino acid derivative produced by *Tricholoma muscarium*. See Muscazone.

Trichome—A single- or many-celled outgrowth from a plant epidermal cell, usually hairlike, glandular (secretory), or spiny.

Trichomycin—An antifungal antibiotic from *Streptomyces hachijoensis* used especially against *Candida albicans*.

Trichophytin—An antigen prepared from dermatophytes especially for skin testing. Commercial trichophytin is commonly a mixture of antigens of several species of *Trichophyton* and *Microsporum*.

Trichospore—Dehiscent and deciduous, monosporous sporangium with nonmotile basal appendages characteristic of the Harpellales.

Trichothecenes—Scirpene toxins produced by *Fusarium tricinctum, F. sporotrichioides, F. poae, Trichothecium*, etc., that cause alimentary toxic aleukia in humans and farm animals.

Trichothecin—An antifungal metabolic product of *Trichothecium (Cephalothecium) roseum*; also toxic to plants and animals.

Trichotomous—Divided or forked into three parts; trifurcate.

Tricine—A biological buffer [N-tris(hydroxymethyl)methylglycine] used in the pH range of 7.0–9.2. It differs from tris in having significant binding capacity for Mg^{2+}, Ca^{2+}, Mn^{2+}, and Cu^{2+}.

Tricornate—Having three horns.

Trident—Having three teeth.

Trifid—Three-cleft.

Trifoliate leaf—Leaf with three leaflets on a petiole.

Trifoveolate—Having three hollows.

Trifurcate—Bearing three branches or forks.

Trigone, trigonous—Having three angles with plane faces between them.

Trilaminar—Consisting of three, thin, parallel layers.

Trilobed, trilobate—Three-lobed.

Trimerous—In threes.

Trimitic (of hyphal structure in basidiocarps)—With generative, skeletal, and binding hyphae.

Trinacriform—Three-pronged.

Trinomen—The three words (generic name, species name, and subspecies name) that make up the scientific name of a subspecies. In bacteriology, *variety* is often preferred to *subspecies*.

Tripartite—Three-parted; divided into threes.

Triploid—Having three sets of chromosomes (3n).

Tripoblastic (of nematodes)—Possessing three germ layers: ectoderm, mesoderm, and endoderm.

Triquetrous—Three-edged or three-cornered; having three salient angles.

Triradiate—Having three radiating arms or branches; (of nematodes) characteristic of the lumen of the esophagus.

Tris—A widely used biological buffer, Tris(hydroxymethyl)methylamine, with a pH range of 7.0–9.2.

Trisporic acid C—A hydroxy-keto acid from mated cultures of *Blakeslea trispora* that is able to induce carotenogenesis in separate strains.

Trisporous—Three-spored.

Tristichous—In three rows.

Triton 100—A nonionic detergent (iso-octyl phenoxypolyethoxy ethanol) with many uses, e.g., disruption of cells, stabilizing proteins, solubilizing aqueous samples in scintillation fluids, etc.

Triturate—To grind as with a mortar and pestle.

Trivial name—Zoological name for a specific epithet; a common name for a chemical.

tRNA—Transfer ribonucleic acid.

Troop—Group of fruitbodies, especially basidiocarps, generated usually from one mycelium.

Troph-, tropho-, -trophy—Combining forms used to denote connection to, with, or relation to, nutrition.

Trophocyst (of *Pilobolus*)—A hyphal swelling that produces a sporangiophore.

Trophogonium, trophogone (of ascomycetes)—A degenerate or nonfunctional antheridium of which the only use is supplying nourishment.

Tropism—Movement, growth, or curvature due to an external, one-sided stimulus that determines the direction of movement.

Truffle—A fleshy, edible ascoma, which generally grows underground, of *Tuber* or other Pezizales or Elaphomycetales, or a basidiocarp of Hymenogastrales; some 180 truffle-forming species of fungi are known.

Truncate—Ending abruptly as though the end was cut off squarely.

Trunk—The single main stem of a tree.

Trunk rot—Decay in the main trunk of a living tree above the stump level.

Trypacidin—An antitrypanosome antibiotic from *Aspergillus fumigatus*.

Trypsin—A proteolytic enzyme that catalyzes the hydrolysis of peptide bonds on the carboxyl side of arginine, lysine, and aminoethyl cysteine residues.

Tryptic peptide—A peptide formed from a protein by the action of trypsin. It has arginine, lysine, or aminoethyl cysteine at the C-terminus.

ts mutant—Temperature-sensitive mutant.

tsp—Teaspoonful = 5 ml = 0.17 fl oz.

Tube (of fungi)—The cylindrical hollow that bears the hymenium lining the inside of the Polyporaceae.

Tube nucleus—One of the nuclei in a pollen tube, influencing the growth and behavior of the tube.

Tuber—A short, fleshy, much enlarged, mostly underground stem, borne at the end of a rhizome, having numerous buds or "eyes" (e.g., potato [*Solanum tuberosum*], florist's cyclamen [*Cyclamen persicum*], *Aconitum*, and *Anemone*)

Tuber indexing—Propagation of a plant from a tuber or a tuber part to determine presence of a tuberborne disease.

Tubercle, tubercule—A small wartlike process.

Tuberculate, tubercular—Covered with low, rounded, wartlike processes (projections, i.e., warts or knobs) or tubercles.

Tuberculiform-frustulate—Small crowded basidiocarps, each having the approximate shape of a frustum on an inverted right pyramid.

Tuberous—Round and swollen like a tuber.

Tuberous root—A swollen, fleshy, underground root (e.g., *Dahlia* and sweetpotato [*Ipomoea batatas*]) distinguished from a tuber that is a swollen underground stem.

Tubiform—Tube-shaped.

Tubular (conidioma)—Cylindrical, not tapered to the apex.

Tubule—A small tube.

Tuckahoe (or Indian bread)—The large, generally oblong to globose, edible, subterranean sclerotium of *Wolfiporia (Macrohyporia, Poria) cocos* or *Polyporus tuberaster*. See Stone-fungus. *Wolfiforia cocos* causes a cubical root and butt rot of many old-growth conifers and hardwood trees and decay of dead trees.

Tumescent—Swelling.

Tumid—Swollen; inflated.

Tumor—Massed overgrowth of tissue or tissues that grows independently of surrounding tissues and may invade those tissues.

Tundra—Plant formation (or biome) of subarctic regions that has permanently frozen subsoil and a low vegetation of lichens, dwarf hardy herbs, and shrubs.

Tunic—*See* Exospore.

Tunica—A coat, especially a thin, white membrane surrounding the peridiole in most species of the Nidulariaceae; (of asci) mantle of bitunicate, unitunicate asci, etc.

Tunicamycin—A compound that blocks the formation of *N*-glycosidic protein-carbohydrate linkages.

Tunicate (conidia)—Covered or wrapped in a rigid layer, exterior to the cell wall, as opposed to a mucilaginous sheath.

Turbid—Cloudy; not clear.

Turbidimetry—A method of estimating bacterial growth or populations by measuring the degree of opacity (or turbidity) of the suspension.

Turbinate—Shaped like a top (an inverted cone), flattened at the apex, abruptly tapered at the base.

Turbinate organ or cell (of Cladochytriaceae)—A swelling on the vegetative thallus; spindle-organ.

Turbinulate—Like a small top.

Turgescent—Swollen.

Turgid (n. **turgidity**)—Describing cells or tissues tightly swollen, distended, plump, or rigid due to internal water pressure.

Turgor movements—Plant movements resulting from changes of water pressure in certain tissues.

Turgor, turgor pressure—Inflation of a plant cell by the fluid contents; pressure within a cell resulting from the absorption of water into the vacuole and the imbibition of water by the protoplasm. Lack of turgor pressure causes plants to wilt.

Turriform—Tower-shaped.

Twig—One-year-old stem or branch of a woody plant. *See* Shoot.

Twig scar—Scar left by the falling away of a twig.

Twiner—Plant that climbs by the twining movements of its stems about a support.

Twist—A disease of cereals and grasses caused by *Dilophospora alopecuri*.

Two-dimensional electrophoresis—A technique in which the constituents of a sample are separated by electrophoresis in one dimension on one property and in a second dimension, usually at right angles to the first, on another property; used to resolve complex mixtures of molecules.

Tylench—A common name for members of the nematode order Tylenchida or superfamily Tylenchoidea.

Tylenchoid (of nematodes)—Resembling or having the characteristics of the genus *Tylenchus*. Nematode pharynx or esophagus typically composed of a narrow procorpus, a not excessively enlarged metacorpus (the two parts of the corpus may be amalgamated into the corpus), isthmus, and basal bulb. *See also* Esophagus.

Tylicolor—Slate or dark gray.

Tylosis (pl. **tyloses**, adj. **tylose**)—A balloonlike extrusion of a parenchyma cell through a pit into the lumen of a contiguous xylem vessel or tracheid that partially or completely blocks it; the process of tylose formation. Tyloses are important in wilt diseases, e.g., Dutch elm disease, oak wilt, and tomato wilts.

Tympaniform—Drum-shaped.

Tyndallization—A process of fractional sterilization with flowing steam.

Type (in nomenclature)—The specimen or specimens on which the name of a taxonomic group (taxon) is based; an object that serves as the base for the name of a taxonomic category (e.g., a specimen is the type for a species, a species the type for a genus, a genus the type for a family, etc.).

Type species—The species designated as the type for the genus. It does not need to be the species most typical of the genus.

Type specimen or strain—This may be a slide, sometimes on a type culture, figure, or a description of the specimen (strain).

Type-specific antigen—An antigen specific to a certain type of virus. *See* Group-specific antigen.

Typing phages—Bacteriophages used in the subdivision of species, biotypes, or serotypes of bacteria into "phage types" based on their susceptibility to more or less host-range specific phages; bacteriophages used in the typing or subdivision of a bacterial taxon (usually a species but may be a serotype). *See* Phage typing.

Typonym—A name having the same type as another name that is neither its basionym nor a synisonym; a later name given to the type specimen; absolute synonym.

Typotype—The specimen used to prepare an illustration in which the latter is the type.

U

UDP—Uridine diphosphate, the 5' pyrophosphate of uridine.

-ule, -ula—A suffix forming diminutives.

Uliginose, uliginous—Rich; muddy; growing in swamps.

Ultimate—Farthest, last.

Ultra- (prefix)—Excessive or extreme.

Ultracentrifugation—Processing in a high-speed centrifuge capable of sedimenting small particles. *See* Rate zonal gradient and Isopycnic gradient.

Ultracentrifuge—A high-speed centrifuge used for determining the particle size of viruses and proteins. Observations are made under either schlieren or UV absorption optics.

Ultrafiltration—A method for the removal of all but the very smallest particles, e.g., viruses, from a fluid medium.

Ultramicroscopic—Too small to be seen with a light microscope.

Ultrasonic waves—Sound waves of high intensity (beyond the audible range) for destroying bacteria and some other microorganisms.

Ultrastructure—Submicroscopic structure of a macromolecule, cell, or tissue.

Ultraviolet radiation—Electromagnetic radiation in the wavelength range of 40–400 nm.

Ultraviolet rays—Radiations in the part of the spectrum having wavelengths from about 3900 to 200 Å.

ULV (ultra low volume)—Sprays applied at a total volume of one-half gallon or less per acre (0.4047 ha) or applied in undiluted form. *See* Low volume sprays, Full coverage sprays.

Umbel—A type of inflorescence in which flowers are borne at the end of a common stalk forming a more or less flattened or rounded cluster, e.g., carrot (*Daucus carota* subsp. *sativus*) and onion (*Allium cepa*); can be composed with subsets of umbels.

Umbellate—Having structures in umbels; umbelled.

Umbelliform—Umbel-shaped.

Umber—A dull olive-brown or dark smoky brown.

Umbilicate (especially of pilei)—With a small, rounded central depression or hollow, somewhat funnel-shaped.

Umbilicus—1) Pore in the perispore of an ascospore; 2) a central hold-fast in some foliose lichens, navel, umbo. Composed of many closely united hyphae.

Umbo—A central, raised, conical to convex mound on the upper surface of the pileus; central swelling like the boss in the center of a shield.

Umbonate (of basidiocarps)—Growth form characterized by eruption from a single, usually central, point with adherence of the effused part to the substratum and the reflexed part somewhat triangular-shaped; overall, shaped like a small, inverted, hollow cone.

Umbraculiform—Umbrella-shaped.

Umbraticole, umbraticolous, umbrosus—Growing in shady places.

Umbrinaceous, umbrinous—Umber brown in color.

Un- (prefix)—When combined with a verb expresses reversal of action or removal; when combined with an adjective, adverb, or noun means "not," giving a negative or opposite force.

Unavailable water—Water held by the soil so strongly that plant roots cannot absorb it.

Unciform—Hook-shaped.

Uncinate, uncate—Hooked; (of gill insertion) almost sinuate.

Uncoating—The removal of the outer layers of a virus particle on infection, thus releasing the viral nucleic acid. Ribosomes may be involved in uncoating the particles for some plant viruses; this process, called cotranslational disassembly, gives rise to complexes of virus particles and ribosomes, structures that have been termed "striposomes."

Unctuous—Greasy or oily to the touch.

Under cortex—The lower cortex of foliose lichens.

Underbridge (of nematodes)—A structure extending across the vulval cone of *Heterodera* cysts below and parallel to the vulval bridge.

Understock—Portion of the stem and associated root system onto which a scion is grafted. *See* Rootstock.

Undulate—Wavy, not flat or uniformly curved.

Undulating—Exhibiting a wavelike motion.

Unguicular, unguiculate—Provided with a claw.

Ungulate, unguliform—Shaped like a horse's hoof.

Uni- (prefix)—One, single.

Uniarticulate—One-jointed.

Uniaxial—Having one axis.

Unicapsid NVP—*See* SNTV.

Unicellular—Having one cell; referring to an organism the entire body of which consists of a single cell.

Uniciliate—With a single whiplike cilium.

Unicolorous—Of the same color throughout.

Uniflagellate—Having one flagellum.

Uniguttulate—With one guttula (cell inclusion).

Unilateral—Of, on, or affecting one side only.

Unilocular, uniloculate—Having a single, simple, undivided cavity (locule).

Uninucleate—Having one nucleus per cell.

Uniperithecial (of certain families of ascomycetes)—Having a single perithecium in a stroma.

Unipolar—At one end only, especially the flagellus or flagella of a bacterial cell.

Uniserial, uniseriate—Arranged in single series or row such as spores arranged in a narrow ascus, all in direct line, one above the other.

Unisexual—An organism that produces eggs or sperms but not both, or a flower that bears stamens or pistils but not both.

Unit character—A genetic principle, established by Mendel, that the various characters making up an individual are controlled in inheritance by independent determining factors.

Unitunicate—Having one definable wall or cover; an ascus in which both the inner and outer walls are more or less rigid and do not separate during spore ejection.

Universal veil—A thin, veil-like fungal membrane that entirely covers certain types of young agarics (mushrooms) and Gasteromycetes. The veil tears as the structure expands, and its remnants may be seen in the form of patches or scales on the pileus (cap) and in the form of a volva at the base; teleblem; blematogen. *See* Volva.

Unorientated—Not arranged in any particular direction.

Unstratified (of lichen thalli)—Same as homoisomerous; not layered.

Uracil—One of the pyrimidine nitrogenous bases found in RNA. *See* Nucleic acid.

Uranyl acetate—A negative stain for electron microscope samples; often used as a 1% solution that has a natural pH of 4.0.

Urceolar, urceolate—Pitcher-shaped; hollow and contracted at the mouth like an urn.

Ureaformaldehyde—A synthetic, slowly soluble nitrogen fertilizer consisting mainly of methylene urea polymers of different lengths and solubilities; formed by reacting urea and formaldehyde.

Ured-, uredi-, uredinio- (prefix)—Summer spore stage of the rusts (Uredinales).

Uredicole, uredinicolous—Growing on rust fungi.

Uredinales, Teliomycetes—The order and class of rusts, respectively; the rust fungi.

Urediniospore, uredinospore, urediospore—Binucleate, dikaryotic (n+n), asexual, one-celled, repeating or summer spore of rust fungi; borne in a uredium (uredinium). The spores may be catenulate, pedicellate, or sessile.

Uredinium, uredium, uredosorus (pl. **uredia, uredinia**)—Fruitbody (sorus) of a rust fungus that produces urediniospores formed after the aecium and before the telium in the life cycle; initiated on a dikaryotic mycelium and having spores that germinate vegetatively, i.e., not with a basidium. Uredinia have peridia, paraphyses, or neither.

Uredinoid, uredioid—Describing a rust fungus having the appearance of the anamorphic genus *Uraecium*; applies especially to aecial states having the gross appearance of uredinia but accompanied by spermogonia. Sporulation occurs after spermatization and the establishment of the dikaryophase; having spores borne singly on pedicels usually without peridium or paraphyses.

URF—Unidentified reading frame deduced from a DNA sequence but in which no protein or genetic function is known. *See* Open reading frame.

Uricle, utriculus—Bladderlike covering of certain fungi such as *Dendrogaster*.

Uridine—The nucleoside of uracil and ribose. *See* Nucleic acid.

Uridine 5′-triphosphate (UTP)—A pyrimidine nucleotide, one of the four major constituents of RNA. *See* Nucleic acid.

Urniform—Urn-shaped with a swollen base and narrow median portion expanded at the apex.

Uromorphic, uromorphous—Tail-like.

Usnic acid—A yellow, dibenzofuran derivative found in lichens that has antibiotic properties against Gram-positive bacteria and is also antifungal.

Ustic acid—A hydroxyquinol from *Aspergillus ustus*; also an antimycobacterial product from *Ustilago zeae* (syn. *U. maydis*).

Ustilagic acids—Antibacterial and antifungal metabolic products from *Ustilago zeae*.

Ustilospore, ustospore—Old names for a smut spore.

Uterus (pl. **uteri**)—The portion of the female nematode reproductive system between the oviduct and the vagina; a structurally modified portion of the oviduct in which egg development may occur.

Utriculiform, utriform—Bladder-shaped; baglike.

UV—Ultraviolet (radiation).

V

v.—Verb.

"V" (of nematodes)—Position of the vulva on the body expressed as a percentage of the total body length, i.e., distance from lips to vulva divided by the length of nematode × 100; part of deMan's formula.

V8 protease—*See Staphylococcus aureus* V8 protease.

Vaccine—A suspension of disease-producing microorganisms or viruses, modified by killing or attenuation, that will not cause disease and can stimulate the formation of antibodies upon inoculation.

Vacuole—The relatively clear, bubblelike, membrane-lined space; the cytoplasm of a plant cell that is filled with a watery solution of sugars and various other plant products and byproducts.

Vagant, vagrant (of lichens)—Unattached; erratic.

Vagiform—Without a definite figure or form.

Vagina (of nematodes)—The cuticle-lined canal that connects the female uterus with the gonopore or vulva.

Vaginate (of a stipe)—Provided with a long volva or sheath at the lower end.

Vaginiferous—Bearing a sheath.

Valence (of antibody)—The number of epitopes with which one antibody can combine. The valence of most antibodies is 2. IgM may have a valence up to 10; IgA has different valences depending on the degree of polymerization.

Valid (of taxonomic epithets)—Names published in accordance with the Code. Such names may be legitimate or illegitimate. *See* Nomenclature and Prevalid.

Valsoid—An embedded or erumpent stroma having groups of perithecia with their necks (beaks) convergent (or even parallel to the surface) and erumpent, as in *Valsa*.

Valvate—Opening by valves.

Valve—(Of nematodes) a structure that regulates the rate and/or direction of intake of materials; e.g., the esophago-intestinal valve or cardia; (of spores of *Aspergillus*) the spore half formed by the secondary thickening of the cell wall.

van Tieghem cell—A ring of glass or other material, fixed to a glass slide, over which is placed a cover slip with a "hanging drop" on the lower surface for high-power study of the microorganism under investigation.

var.—Variety.

Variability—The property or ability of an organism to change its characteristics from one generation to the other as the result of an adaptation to the environment or a mutation. *See* Genetic variability.

Variant—An organism showing some variation from the parent culture.

Variation—Differences in structural or physiological characters in an organism from those typical or common in the species to which the organism belongs.

Variecolin—An antitubercle, bacillus antibiotic from *Emericella variecolor*, anamorph *Aspergillus stellifer* (syn. *A. variecolor*).

Variegation—A general term for discoloration (presence of two or more colors) of leaves, stems, or flowers usually from genetic causes—*not* from virus infection.

Variety (botanical)—Group of closely related plants of common origin within a species that differ from other varieties in distinct morphological characters such as form, color, flower, and fruit; a subgroup within a species given a Latin name according to the rules of the International Code of Botanical Nomenclature. A taxonomic variety is known by the first validly published name applied to it so that nomenclature tends to be stable. *See* Cultivar.

Variety (cultivated)—See Cultivar.

Variolarioid—Having granular or powdery tubercules.

Vas deferens (pl. **vasa deferentia**)—The slender proximal part of the male nematode gonad that unites posteriorly with the rectum to form the cloaca (may be muscular or glandular or both).

Vas efferens (pl. **vasa efferentia**)—In nematodes, the specialized tubular part of the male seminal vesicle, connecting the growth zone of the testis and seminal vesicle.

Vascular—Pertaining to conductive (xylem and phloem) tissue in plants or a vascular parasite that grows and/or moves in those tissues.

Vascular bundle or strand—A distinct group of elongated conducting cells in a stem or leaf, consisting of primary xylem for water conduction and primary phloem for food conduction. Vascular bundles are frequently enclosed by a bundle sheath of parenchymatous tissue or fibers for mechanical support.

Vascular cambium—A meristem found in biennials and perennials that produces secondary xylem and secondary phloem cells.

Vascular element—A cell of the vascular tissue.

Vascular hypha—A hypha with dark or refractive contents.

Vascular ray—Ribbonlike aggregate of cells that extends radially in stems through the xylem and phloem.

Vascular tissue—Conducting tissue; xylem and phloem.

Vasiform—Shaped like a vessel or duct.

Vector—A living organism (e.g., insect, mite, bird, higher animal, nematode, parasitic plant, human, etc.) able to carry and transmit a pathogen (virus, bacterium, fungus, nematode) and disseminate disease; (in genetic engineering) a vector *or* cloning vehicle is a self-replicating DNA molecule, such as a plasmid or virus, used to introduce a fragment of foreign DNA into a host cell (as the receptor for DNA molecules). *See* Transmission.

Vector efficiency—A value, usually expressed as percent, indicating the proportion of vectors realized from the number tested.

Vector propensity—The probability that a vector having had the opportunity to acquire the pathogen in nature will transmit it, assuming that it probes, or otherwise goes through the act of transmitting to a noninfected plant.

Vector resistance—Resistance of a host plant to the vector of a virus. There are three basic types of vector resistance:

Antibiosis—Resistance in which the growth and multiplication of the vector on the host is inhibited.

Nonpreference—Resistance occurring because a vector prefers not to feed on a particular host. This may cause the vector to probe-feed several times shortly after landing before moving to another plant. This may reduce the transmission of a persistent virus but often increases the spread of a nonpersistent virus.

Tolerance—Ability of a host plant to withstand a vector's attack without severe damage. This type of vector resistance does not control virus spread. *See* Resistance and Tolerance.

Vegetation—The plants that cover a region; formed of species that make up the flora of the area.

Vegetative—Asexual; somatic; not forming spores or sex organs; also refers to a significant increase in size; the growth and maintenance (assimilative) stage of an organism in contrast to resting or reproductive stages.

Vegetative cycle—*See* Lytic cycle.

Vegetative reproduction—Asexual reproduction by a root, stem, leaf, or other primarily vegetative part of a plant body.

Veil—*See* Annulus, Cortina, Marginal veil, Partial or inner veil, Pellicular veil, Protoblem veil, and Universal veil.

Vein—A vascular strand of xylem and phloem in a leaf, petal, or other plant part; (of fungi) the swollen wrinkles between the lamellae of agarics or separating radially arranged tubes in boletes; (of lichens) strands of tissue on the lower surface of foliose lichens, where they may replace a lower cortex. Three types are generally recognized: *caninoid veins,* in which the strands are separated to the tips of the lobes; *polydactyloid,* in which the strands converge toward the tips of the lobes; and *molaceoid,* in which the lower surface has a few whitish interstices that faintly indicate venation.

Veinbanding—Symptom of a virus disease in which regions along the leaf veins remain darker green than the tissue between the veins.

Veinclearing—Symptom of a virus or other pathogenic disease characterized by the disappearance of green color adjacent to the veins of young leaves.

Velar—Pertaining to a veil.

Velate—Having a veil.

Velum—(Of nematodes) a delicate, cuticular membrane on the ventroanterior side of the spicule; (of fungi) the same as veil.

Velumen, velutinate, velutinous, velvety—Like velvet; thickly covered with close, short, soft, delicate hairs (hyphae). *See* Phalacrogenous.

Venate—Veined.

Venation—Arrangement of veins in a leaf blade.

Venenate—Poisonous.

Venenose—Very poisonous.

Veniform—Veinlike.

Venose, venous—Having veins.

Venter—Enlarged base of an archegonium within which an egg develops.

Ventilatorious—Fan-shaped.

Ventral—Front or lower surface, as opposed to dorsal.

Ventricose—(Of conidia) inflated; very broadly fusiform; swelling out in the middle or at one side; (of setae, cystidia) with a distinctly swollen base and abruptly narrowing apical portion.

Ventriculus (of nematodes)—A glandular region of the esophagus.

Ventricumbent—Prone; face downward.

Ventrolateral—The position on the nematode body situated 30° laterally from the ventromedian line and perpendicular to the anteroposterior axis; subventral.

Ventromedian (of nematodes)—The true middle line on the lower side of the body.

Verdant—A grass green color.

Vermicular, vermiculate, vermiform—Worm-shaped, thickened and bent in places; having the motion of a worm; marked with irregular lines like worm tracks.

Vernal—Of or belonging to spring.

Vernalization—Natural or artificial induction of early flowering by exposure of seeds to low temperatures for a certain period of time; also, the requirement for breaking bud dormancy of certain temperate woody perennials.

Vernicose—Shiny, as though varnished.

Verruca (pl. **verrucae**)—A wart or elevation.

Verrucarioid (of asci)—*See* Ascus.

Verruciform—Wart-shaped.

Verrucose, verrucous—Adorned with small, rounded, wartlike structures; appearing as a minutely roughened wall; commonly applied to a variety of roughened surfaces between echinulate and reticulate.

Verruculose—Delicately or minutely verrucose.

Verruculotoxin—A tremorgenic toxin from *Penicillium verrucuosum*.

Versatile—Describing anthers attached near the middle of the filament in such a manner that they swing freely.

Versicolor—Changing color.

Versicolored—Describing septate conidia in which one or more cells is of different pigmentation (and usually of different wall thickness) than the rest.

Versiform—Of different forms; changing form with age.

Vertex—Top of an organ.

Vertical resistance—Resistance that completely protects a host but only against specific strains of a pathogen. *See* Gene-for-gene hypothesis.

Vertical transmission or genetic transmission—The passage of the viral genome from one host generation to the next, either as a provirus or in close association with the genome of the gamete, e.g., in transovarial transmission in their insect vectors or in the seed transmission of plant viruses.

Verticillate—Having a whorl of three or more branches or sporogenous cells arising at the same level; arranged in *verticils* or whorls.

Verticillium wilt—A widespread, systemic, vascular wilt disease (hadromycosis) that attacks hundreds of different kinds of plants; caused by the fungi *Verticillium albo-atrum* and *V. dahliae*.

Veruculate—Cylindric and somewhat pointed.

Vesicle (adj. **vesicular**)—In oomycetes (e.g., *Pythium*), thin, bladderlike sac or structure produced by a zoosporangium in which zoospores are differentiated or released; subcellular membranous enclosure; bulbous head terminating the conidiophore of *Aspergillus*; (in mycorrhizae) an inflated or swollen hyphal cell (e.g., a gloeocystidium).

Vesicular bodies—Thin-walled vesicles in the subhymenium of certain hymenophytes; vesicular-arbuscular type of mycorrhiza. *See* Mycorrhiza.

Vesicular-arbuscular—Describing mycorrhizae that possess both vesicles and arbuscules. The penetrating hyphae in the cortical cells may be coiled (*pelotons*) or finely branched (*arbuscules*) haustorial branches.

Vesiculiform, vesiculate—Having the form or structure of a vesicle; sac- or bladderlike; having a swollen appearance.

Vesiculose—Full of or made from vesicles.

Vespertine—Appearing or expanding in the evening.

Vessel—A series of water-, mineral nutrient-, and sap-conducting, dead xylem tubes (cells) in the stem and root that are attached end to end.

Vessel member—A perforate, nonliving xylem cell at maturity; one segment of a vessel.

Vestibule—A chamber or passage; (of nematodes) the narrow anterior portion of the stoma, acting as the stylet guide in stylet-bearing nematodes.

Vestige (adj. **vestigial**)—Trace or visible sign left by something vanished or lost; a minute remaining amount; a small and degenerate or imperfectly developed body part or organ that remains from one more fully developed in an earlier stage of the individual, in a past generation, or in closely related forms.

Viable (n. **viability**)—State of being alive; capable of growth and development; applied usually to seeds and spores.

Vibrio—A slightly curved organism resembling a comma; a minute, threadlike bacterium.

Vicid—Slimy, sticky, mucilaginous, glutinous, lubricous, viscous. *See* Gelatinous and Viscid.

Victorin—A toxin from *Bipolaris victoriae* (syn. *Helminthosporium* and *Drechslera victoriae*) that induces symptoms of leaf blight in oats (*Avena*).

Victotoxinine—A constituent of victorin.

Villose, villous— Covered with long, soft, silky hairs (villi) that are not matted. *See* Tomentose.

Villus (pl. **villi**)—A long soft hair.

Vinaceous—With pale lavender or purplish (wine-colored) tints.

Vinescent—Turning wine-red.

Violaceous—Of a violet hue.

Violet root rot (of numerous plants)—Disease caused by *Helicobasidium brebissoni* (syn. *H. purpureum*).

Virellous—Somewhat green or greenish.

Virent—Green.

Virescence (adj. **virescent**)—State or condition of turning green; condition of a normally white or colored tissue that develops chloroplasts and turns green.

Virgate—Streaked or banded; also, long, straight and slender like a wand.

Virgin soil—Soil that has not been significantly disturbed from its original natural environment.

Virgulate, virguliform—Like a small twig, wand, or rod.

Viricide—A chemical or physical agent capable of inactivating or suppressing the replication of a virus completely and permanently.

Viridant—Becoming grass-green or verdant.

Viridin—An antifungal antibiotic from *Gliocladium virens*.

Virion—Complete virus particle consisting of nucleic acid (RNA or DNA) and a protein shell; the infectious unit of a virus.

Viroid—An infectious pathogen much smaller than a virus; an infectious, unincapsidated, free-floating, circular, single-stranded ribonucleic acid; small, low molecular weight (247–374 nucleotides) ribonucleic acid (RNA) that can infect certain plant cells, directs its own replication from host metabolites, and causes disease. Replication is in the nucleus and thought to be by rolling, circle-producing, oligomeric forms processed to unit length. The nucleic acid does not code for any proteins. Heat ranges for viroids range from narrow to wide.

Virolysis—Production of lysis or holes in the membrane of enveloped viruses by reaction with antibodies or complement.

Viroplasm—An amorphous, usually proteinaceous, cytoplasmic inclusion body without a surrounding membrane, associated with virus infection in plant cells infected with wound tumor, cauliflower mosaic, and other viruses.

Virose—Poisonous *or* having a strong and unpleasant smell.

Virosis (pl. **viroses**)—Any virus disease of plants.

Virostatic—Able to inhibit viral replication, which can be resumed when the virostatic agent is removed. DNA-containing viruses are inhibited by Actinomycin D and 5-fluorodeoxyuridase while 5-fluorouracil, thiouracil, and puromycin are virostatic to RNA viruses. *See* Base analogue.

Virucidial—A virus-inactivating agent, e.g., formalin and chlorine.

Virulence—The degree or measure of pathogenicity of a given pathogen; relative capacity to produce disease. *See* Aggressive.

Virulent—Highly pathogenic, having the capacity for causing a severe disease; able to produce all the typical symptoms of the disease in a susceptible host.

Virulent phage—A phage that produces a lytic infection in its host and cannot induce a lysigenic state. *See* Temperate phage and Lysogeny.

Viruliferous—Containing or carrying a virus; term applied particularly to virus-laden insect and nematode vectors capable of transmitting the virus.

Virus—A submicroscopic, filterable agent that causes disease and multiplies only in living cells; a submicroscopic, intracellular, obligate parasite consisting of a core of one or more infectious nucleic acid (either RNA or DNA) molecules usually surrounded by a protein coat. Viruses contain no energy-producing enzyme systems, no functional ribosomes, or other cellular organelles, which are supplied by the cell in which viruses replicate. *See* Helper virus and Latent virus.

Virus assembly—The coming together of the constituent parts of a virus particle and the formation of that particle. The assembly process can vary from the autoassembly of protein subunits around viral nucleic acid to the complicated assembly of many phages.

Virus cryptogram—A descriptive code summarizing the main properties of a virus; term is seldom used today.

Virus inactivator—A chemical that inactivates a virus, causing loss of infectivity.

Virus inhibitor—A chemical that prevents virus transmission without inactivating the virus. Inhibition may be temporary and can sometimes be avoided by dilution. Inhibitors commonly act on the recipient host rather than on the virus itself.

Virus particle—The morphological form of a virus. In some viruses it consists of a genome surrounded by a protein capsid; other virus particles have additional structures such as envelopes, tails, etc.

Virus replication—The process of forming progeny virus from input virus. It involves the expression and replication of the viral genomic nucleic acid and the assembly of progeny virus particles.

Viruslike particle—A structure that resembles a virus particle but whose pathogenicity is unproven.

Virusoid—The very small circular RNA component of some isometric RNA viruses.

Viscera—Internal organs.

Viscid, viscose—Sticky, gluey, slimy, clammy, half-liquid. *See* Vicid.

Viscidity—Stickiness; characteristic produced by a bacterial sediment in broth that rises in a coherent swirl upon shaking.

Viscosity—The internal friction of a particle in a fluid.

Vitamins—Natural organic substances synthesized by plants and necessary in minute quantities for certain respiratory and developmental processes of plants and animals; most are required by but not synthesized by animals.

Viteline—Egg yolk-yellow color; luteous.

Viticolous—Living on or in a vine (grape, *Vitis*).

Vitreous—Glasslike, transparent, hyaline.

Vittate—Having lengthwise lines, bands or ridges.

Vivipary (adj. **viviparous**)—Germination while still attached to the parent; a method of sexual reproduction (e.g., aphids) in which the young are born alive and active as opposed to eggs.

Vivotoxin—A substance reproduced internally in an infected host or by the pathogen within it, and responsible for some or all of the harmful changes produced in the course of the disease. *See* Toxin.

Void—Empty.

Volatile (n. **volatility**)—Evaporating or vaporizing readily at ordinary temperatures or exposure to air; rate of evaporation of a pesticide.

Voluble—Twining around a support.

Volunteer—Self-set plant; plant seeded by chance.

Volute—Rolled up in any way.

Volutin—A reserve substance (metachromatic, polymetaphosphate material) of fungi, especially yeasts, seen as electron-dense granules.

Volva (pl. **volvae**)—A "cup" or sheath around the base of the mature stipe or receptacle of certain agarics (mushrooms) and gasteromycetes; a remnant of the universal veil; a saclike structure that encloses the hymenial surface of *Cryptoporus (Polyporus) volvatus* at maturity.

Volvate—Having a volva.

von Magnus phenomenon—The increase in the proportion of defective virus particles produced during the repeated passage of a virus at a high multiplicity of infection.

VPg—A genome-linked virion protein.

Vulva (of nematodes)—The exterior opening (gonopore) of the female reproduction system; appears as a transverse slit or circular pore on the ventral side.

Vulval or vulvular bridge (of nematodes)—A narrow connection (persisting vulvar slit) across the fenestra of the vulval cone, forming two semifenestrae, in cysts of some *Heterodera* species.

Vulval or vulvular cone (of nematodes)—The conspicuous elevation at the posterior end of *Heterodera* cysts.

Vulviform—Like a cleft with projecting edges.

W

Wall building (of fungi)—Hyphal growth in which wall material is produced by ultrastructural secretory bodies in the cytoplasm. The three recognized types of wall building are: *apical,* where the bodies are concentrated at the hyphal tip and produce new wall by distal growth, forming a cylindrical hypha with the youngest wall material at the apex; *ring,* where the bodies are concentrated next to the cell wall at a point below the tip, in the shape of a "ring," and produce new wall by proximal growth, forming a cylindrical hypha with the youngest wall material at the base; and *diffuse,* where the bodies are present throughout the cytoplasm and produce lateral growth where the cylindrical hypha swells, altering the preexisting wall.

Walling-off—Separation of diseased from healthy tissues by barrier tissues produced by the diseased plant.

Wandering lichens—Lichens with an epigeic habit.

Warburg respirometer—An apparatus used to study enzyme reactions in which an exchange of gases takes place.

Wart disease of potato—Disease caused by *Synchytrium endobioticum*.

Wart—Any small, circumscribed excrescence or process.

Warty (of the pileus, spores, etc.)—Covered by small wartlike excrescences; verrucose.

Water mold—A primitive fungus (e.g., an oomycete) that lives in very moist soil. Some species, like *Aphanomyces*, *Pythium*, and *Phytophthora*, are able to parasitize plants; aquatic Mastigomycetes, especially Saprolegniales. *See* Aquatic fungi.

Water potential—The difference between the activity of water molecules in pure distilled water at the standard conditions of atmospheric pressure and 3° C, and the activity of water molecules in any other system; the activity of the water molecules may be greater (positive) or less (negative) than the activity of water molecules under standard conditions.

Water table—The upper surface of ground water or that level below which the soil is saturated with water.

Waterlogged—Without soil aeration due to excess water and a lack of or poor soil drainage.

Watersoaked—Describing a disease symptom of plants that appears wet, dark, and usually sunken and translucent.

Watersprout—Small, rapidly growing shoot or branch on a large stem, developed from adventitious tissues as a response to auxin shift following death of apical meristems from any cause.

Watts (W) (light)—Radiometric units expressing energy per unit of time and unit area independent of wavelength. Irradiance within the photosynthetic active region (PAR) of 400–700 nm is often referred to as W PAR.

Wavelength—The distance between two corresponding points on any two consecutive waves; for visible light it is minute and generally measured in nanometers.

Wavy (of the margin of a pileus)—Alternatively raised or sunken like waves; undulate.

Waxy (of lamellae)—Of a consistency that can partially or wholly be molded or compressed into balls.

Weathering—All physical and chemical changes produced in rocks at or near the earth's surface by atmospheric agents.

Weed—Any plant growing where it is not wanted.

Weights and measures (miscellaneous)

1 micron or micrometer (μ or μm) = 0.00039 in.

1 acre inch of water = 27,154 gal = 624.23 gal/1,000 ft.^{2}

1 g/in.^{3} = 3.78 lb/ft.^{3}
1 lb/ft^3 = 0.26 g/in.^{3}

Weights per unit area
1 oz/ft^2 = 2,722.5 lb/acre.
1 oz/yd^2 = 302.5 lb/acre.
1 oz/100 ft^2 = 27.2 lb/acre.
1 oz/1,000 ft^2 = 2.72 lb/acre.
1 lb/1,000 ft^2 = 43.56 lb/acre
1 lb/acre = 1 oz/2,733 ft^2 (0.37 oz/1,000 ft^2) = 0.0104 g/ft^2 = 1.12 kg/ha.
100 lb/acre = 2.5 lb/1,000 ft^2 = 1.04 g/ft^2.
5 gal/acre = 1 pt/1,000 ft^2 = 0.43 ml/ft^2.
100 gal/acre = 2.5 gal/1,000 ft^2 = 1 qt/100 ft^2 = 935 liter/ha.
1 qt/100 gal (approximate) = 10 ml/gal.
1 lb/gal = 120 g/liter.
1 kg/100 m^2 = 2.05 lb/1,000 ft^2 = 1 kg/ha = 89 lb/acre.
1 kg/ha = 0.0205 lb/1,000 ft^2 = 0.1 kg/100 m^2 = 0.89 lb/acre.

Western blotting—The transfer of proteins, separated on a polyacrylamide gel, to an immobilizing matrix, often nitrocellulose. Proteins are frequently transferred using electrophoresis, called *electroblotting*.

Wet bubble or white mold—A mushroom disease caused by *Mycogone perniciosa*.

Wettable powder (WP)—A powdered formulation of a pesticide containing a wetting agent that causes the powder to form a suspension in water. The designation "50WP" indicates that one half of the wettable powder preparation is active ingredient.

Wetting agent—A compound, usually a surfactant, added to a spray preparation that reduces surface (interfacial) tension and causes a liquid to contact plant surfaces more thoroughly. Effectiveness is measured by the increase in spread of a liquid over a surface area and by the "contact angle" of the liquid and surface.

Wetwood—A disease of the heartwood and older sapwood of most hardwood and some softwood trees (e.g., elms [*Ulmus*] and poplars [*Populus*]), caused by species in a number of bacterial genera; symptoms are watersoaked and discolored wood, dieback, and the presence of phytotoxic bacterial ooze. *See* Slime flux.

Wheat germ extract—A cell-free extract from wheat germ used for the translocation of eukaryotic mRNAs. *See* Rabbit reticulocyte lysate.

White agaric—*See* Female agaric.

White blister—*See* White rust.

White corpuscles—Leucocytes of the blood.

White mold or blight (of sweet pea, *Lathyrus odoratus*)—Disease caused by *Ramularia alba*; also commonly caused by *Sclerotinia sclerotiorum*, a very destructive disease of many field and garden plants.

White Piedmont truffle—*Tuber magnatum. See* Truffle.

White piedra of humans—Disease caused by *Trichosporon beigelii*.

White rust or blister—Characteristic pustular fructification or disease caused by species of *Albugo*, especially *A. candida* on crucifers.

White truffle—*Choiromyces meandriformis. See* Truffle.

White winter truffle—*Tuber hiemalbum. See* Truffle.

Whitefly—Small (about 1/12 in. long), four-winged, homopterous insect (family Aleyroidae) with sucking mouth parts and covered with a fine white powder. Whiteflies are known to transmit a number of plant-infecting viruses.

Whiteheads—A disease of cereals and grasses caused by *Gaeumannomyces graminis* and its varieties; the result of crown and sheath rot or take-all disease; dead cereal or grass heads that appear white, may be due to a variety of causes.

Whiterot—Type of wood decay resulting from enzymatic action of fungi; it degrades all components of wood and also bleaches the decayed wood in the later stages.

Whorl (adj. **whorled**)—Three or more leaves, flowers, twigs, hyphae, or other plant parts arranged in a circle and radiating from one point.

Wild strain (in virology)—A distinct virus isolate made from a natural infection.

Wild type—The naturally occurring species or taxon as opposed to morphological variants resulting from culture *in vitro* or biochemical mutants obtained from culture; prototroph; (in virology) the original parental strain of a virus that is usually laboratory-adapted and used as a basis when comparing mutants.

Wilt, wilting—Lack of rigidity and turgor or drooping of leaves, stems, and flowers from lack of water (inadequate water supply or excessive transpiration); a vascular disease caused especially by species of *Fusarium* and *Verticillium* and by bacteria (*Clavibacter, Erwinia*, and *Xanthomonas*) that interrupts the normal uptake and distribution of water by a plant.

Wilting point (permanent wilting point or PWP)—The moisture content of soil at which plants wilt and fail to recover when placed in a humid atmosphere.

Wind burn—Death and browning of tissues, most commonly on the leaves furthest from the roots; caused by atmospheric desiccation. *See* Winter desiccation.

Wind shake—Circular separation of wood along annual rings by wind action.

Windborne—Describing inoculum, microorganisms, vectors, etc., transported from place to place by wind.

Windbreak—A planting of trees or shrubs, usually perpendicular or nearly so to the principal wind direction, to protect soil, crops, homes, roads, etc., against the effects of winds.

Windrow—Hay, grain, leaves, or other plant material swept or raked into rows to dry.

Windthrow—Overthrowing and uprooting of a tree by the wind (also called windfall).

Wings—Lateral petals of a legume type of flower; (of nematodes) longitudinal elevations (longitudinal alae) in the cuticle, usually lateral but sometimes numerous and evenly spaced around the nema.

Winter annual—Plants that customarily germinate in the fall, overwinter as small plants, and complete growth, flowering, and seed production the following spring before dying.

Winter desiccation—The death or partial death of leaves or plants by drying during winter dormancy.

Winter hardiness—The ability of a plant to tolerate severe winter conditions.

Winter mushroom—*See* Enokitake.

Winter spore—A resting spore for overwintering, e.g., teliospores of numerous rust fungi.

Winter truffle—*Tuber brumale. See* Truffle.

Winterkill—Any injury to plants that occurs during the winter period.

Witches' broom—Disease symptom consisting of an abnormal, massed, brushlike development of many weak shoots or roots of mainly woody plants, arising at or close to the same point or resulting from the proliferation of buds; caused by mites, viruses, fungi (especially rusts and Taphrinales), bacteria, false mistletoes, nematodes, etc.

Wobble hypothesis—A hypothesis stating that a certain amount of variation or "wobble" is tolerated in the third position of a codon because the third base of the anticodon loop of a tRNA can bind to one more bases of the mRNA codon.

Wolf's moss—*Letharia vulpina.*

Wood—Technically, secondary, nonfunctioning xylem; popularly, the xylem of trees and shrubs.

Wood bluet—*Letharia vulpina.*

Wood's light—Ultraviolet light filtered through nickel oxide-containing glass. Used for diagnosis of certain fungi, especially agarics.

Woody—Hardy, tough, and fibrous; nonherbaceous.

Woronin bodies—The minute, rounded or elongated-oval, highly refractive bodies of certain discomycetes, found particularly in apical hyphal cells in association with septa.

Woronin's hypha (in Ascomycotina)—A coiled hypha found in some forms, in which the ascocarp later develops, and probably homologous with an archicarp; a loosely coiled hypha of large diameter at the center of a young perithecium that later gives rise to ascogenous hyphae; scolecite.

Wortmannin—An antifungal antibiotic from *Talaromyces wortmannii*, especially effective against species of *Botrytis, Cladosporium,* and *Rhizopus.*

Wort—The unfermented or fermented infusion of malt that, after fermentation, becomes beer or mash; also, a plant; herb; vegetable (usually used in combination).

Wound parasite—A pathogen that can enter a host only through wounds.

Wound—An injury to a plant in which the surface is cut, scraped, torn, or otherwise broken.

Wyerone—A phytoalexin from broad bean (*Vicia faba*).

X

X bodies—Inclusion bodies found in plant cells infected with the tobacco mosaic virus (TMV).

Xanthochroic (of hymenomycete basidiocarps)—Having a reddish or yellowish brown context, when observed in a water or acid mount, that turns dark brown on treatment with KOH.

Xanthomonad—Member of the bacterial genus *Xanthomonas*.

Xanthophyll(s)—Yellowish orange carotenoid pigments of plants, insoluble in water, occurring usually in plastids along with chlorophyll in green plants.

Xenospore—A spore dispersed from its place of origin. *See* Memnospore.

Xer-, xero- (prefix)—Dry; drought.

Xeric—Becoming dormant during dry periods and reviving again and shedding spores in moist weather.

Xerophile, xerophyte—A plant growing in soil with a scanty water supply or in soil into which water is absorbed only with difficulty.

Xerotolerant—Able to grow under dry conditions.

X-gal—A colorless compound (5-chloro-4-bromo-3-indoly-beta-D-galactoside) that, when hydrolyzed by beta-galactoside, releases a bright blue, nondiffusible dye. It is used as a color indicator for bacteria able to use lactose and in cloning vectors involving beta-galactosidase.

Xiphiiform, xiphoid—Sword-shaped.

Xylem—The complex supporting, water- and mineral-conducting tissue of vascular plants consisting of tracheids, vessel members, parenchyma cells, ray cells, and wood fibers; wood.

Xylem parenchyma—Living parenchyma cells found in the xylem.

Xylene cyanol FF—A dye used as a marker in polyacrylamide gel electrophoresis of DNA or RNA

Xylogenous, xylphilous—Preferring wood; living on wood.

Xyloma (of Dothideales)—Sclerotiumlike body producing sporogenous structures inside.

Xylostromata—Sheets of mycelium as in *Xylostroma*.

Y

Yd—Yard = 36 in. = 3 ft = 0.181818 rod = 0.9144 m.

Yeasts—Unicellular, or sometimes filamentous, commonly ascomycetous fungi that reproduce asexually by budding or produce asci not enclosed in ascocarps; a common term for fungi with this growth habit; may be ascomycetous, basidiomycetous, or imperfect.

Yellow rice—Rice discolored by *Penicillium islandicum* and rendered carcinogenic to rodents and possibly humans.

Yellow truffle—Species of *Terfezia* and *Tirmania*. *See* Truffle.

Yellowing (distinct from chlorosis)—To make or render tissue yellow that once was green.

Yellows—A plant disease characterized by yellowing and stunting of the host plant or affected parts, usually caused by xylem-limited fastidious bacteria or deficiencies and imbalances of one or more essential elements; yellows of cabbage and other Brassicaceae is caused by *Fusarium oxysporum* f. sp. *conglutinans*; peach yellows is caused by a virus.

Ypsiliform (of spores, etc.)—Shaped like the Greek letter upsilon; Y-shaped.

Z

Z-DNA—*See* DNA.

Zearalenone—A toxin produced by *Fusarium graminearum*, teleomorph *Gibberella zeae*; cause of vulvovaginitis and infertility in cattle and pigs.

Zeugite—An organ in which fertilization has occurred and the dikaryophase ends, e.g., an ascus or basidium.

Zonal centrifugation—Centrifugation in a cylindrical or bowl-shaped rotor in which the solutions are in a large central cavity instead of individual tubes or buckets. The zonal rotor is usually used for gradient centrifugation. The macromolecules (e.g., viruses) separate as concentric rings (zones) in the gradient. Zonal centrifugation is used for separating large quantities of macromolecules.

Zonate—Marked with concentric zones (bands or lines) of different colors and/or textures; targetlike; any leaf or stem development appearing in concentric rings.

Zonation (of cultures)—Regular concentric variation of texture, pigmentation, or sporulation; often associated with fluctuations in light, temperature, or other factors.

Zone lines—Narrow, dark brown or black lines visible in decayed wood, especially of hardwoods, that are generally caused by fungi; same as black line or black zone.

Zoned—Same as zonate.

Zoo- (prefix)—A combining form meaning *animal*.

Zoochoric—Dependent on animals for dissemination or dispersal.

Zoocyst—A cyst in the monads (Monadineae), a group of flagellate protozoa, that produces amoeboid or flagellate cells.

Zoogamete—A motile gamete; planogamete.

Zoogamy—Same as isogamy.

Zoogenous—Growing on animals.

Zoogloea, zoogloeal mass—Gelatinous clumps of microorganisms (usually produced by certain bacteria) in the form of a film or in suspension in liquid material; (of bacteria) colony embedded in a slimy substance.

Zoophilic (of dermatophytes, etc.)—Preferentially pathogenic for animals. *See* Anthrophilic.

Zoophilous—Disseminated or disposed by animals; zoochroic.

Zoosporangium, zoosporange (pl. **zoosporangia**)—A usually thin-walled sac (sporangium) that contains or produces zoospores endogenously.

Zoospore, zooid—An asexually produced fungus spore bearing flagella and capable of locomotion in water; a swarm spore or sporangiospore; swarmer; simblospore; planospore; planont; spores that are able to swim by movements of cilia or flagella.

Zoosporiferous—Producing zoospores.

Zyg-, zygo- (prefix)—Yoke, pair.

Zygangium—Gametangium of a zygomycete.

Zygomycota (**Zygomycotina**)—The subdivision of fungi that contains the zygomycetes.

Zygogamy—Same as isogamy (which see).

Zygomycetes—A class of fungi containing terrestrial Phycomycetes characterized by hyphoid gametes of near equal size that fuse to form a zygospore.

Zygomycosis—Mycosis caused by a member of the Zygomycetes. *See* Mucormycosis and Phycomycosis.

Zygophore (of Mucorales)—A special hyphal branch that forms copulation branches.

Zygosporangium (pl. **zygosporangia**)—A sporangium containing a zygospore; it develops following the fusion of two morphologically similar hyphoid gametangia.

Zygospore—The sexually produced, thick-walled resting spore of Zygomycetes produced by the fusion of two morphologically similar gametangia and borne on somatic hyphae; the resting spore resulting from the conjugation of isogametes.

Zygote—A diploid cell (2n) or fertilized egg resulting from the union of two haploid gametes (isogametes or heterogametes); a cell in which two nuclei of opposite sex have undergone fusion. The beginning of a new organism in sexual reproduction.

Zygotropism—A phenomenon in which two hyphae, as a result of mutual stimulation, make growth curvature and grow toward each other until they meet and fuse.

Zygozoospore—A motile zygote formed by the union of two similar cells.

Zymogenous—Describing ferment-producing organisms only active when a suitable substrate is available. *See* Autochthonous.

Zymase—An enzyme system that hydrolyzes simple sugars, breaking them down to alcohol and carbon dioxide.

Zymogram—Pattern of bands detected by electrophoretic analysis of enzyme solutions. Used in taxonomic work to characterize the esterases and heat-stable catalases; a tabulation of carbohydrate fermentations test results.

Zymologist—A student of yeasts.

Zymology—The study of yeasts.

Zymosan—An extract of yeast cell walls.